AF806012

International Series in Operations Research & Management Science

Volume 245

Series Editor

Camille C. Price
Stephen F. Austin State University, TX, USA

Associate Series Editor

Joe Zhu
Worcester Polytechnic Institute, MA, USA

Founding Series Editor

Frederick S. Hillier
Stanford University, CA, USA

More information about this series at http://www.springer.com/series/6161

Giorgio Consigli • Daniel Kuhn • Paolo Brandimarte
Editors

Optimal Financial Decision Making under Uncertainty

Springer

Editors
Giorgio Consigli
University of Bergamo
Bergamo, Italy

Paolo Brandimarte
Politecnico di Torino
Torino, Italy

Daniel Kuhn
EPFL
Lausanne, CH

ISSN 0884-8289 ISSN 2214-7934 (electronic)
International Series in Operations Research & Management Science
ISBN 978-3-319-41611-3 ISBN 978-3-319-41613-7 (eBook)
DOI 10.1007/978-3-319-41613-7

Library of Congress Control Number: 2016955942

© Springer International Publishing Switzerland 2017
This work is subject to copyright. All rights are reserved by the Publisher, whether the whole or part of the material is concerned, specifically the rights of translation, reprinting, reuse of illustrations, recitation, broadcasting, reproduction on microfilms or in any other physical way, and transmission or information storage and retrieval, electronic adaptation, computer software, or by similar or dissimilar methodology now known or hereafter developed.
The use of general descriptive names, registered names, trademarks, service marks, etc. in this publication does not imply, even in the absence of a specific statement, that such names are exempt from the relevant protective laws and regulations and therefore free for general use.
The publisher, the authors and the editors are safe to assume that the advice and information in this book are believed to be true and accurate at the date of publication. Neither the publisher nor the authors or the editors give a warranty, express or implied, with respect to the material contained herein or for any errors or omissions that may have been made.

Printed on acid-free paper

This Springer imprint is published by Springer Nature
The registered company is Springer International Publishing AG
The registered company address is: Gewerbestrasse 11, 6330 Cham, Switzerland

I would like to dedicate the book to my wife Barbara.

(Daniel Kuhn)

I would like to dedicate my work to my son Gabriele.

(Giorgio Consigli)

Preface

This volume includes chapters by several distinguished colleagues addressing different financial management and valuation problems arising in modern financial markets. When this volume was first conceived, it was motivated by an increasing heterogeneity of mathematical and methodological approaches applied to often rather similar financial optimization problems. In proposing this project to Springer, we aimed at facilitating, when appropriate, a theoretical and computational integration of those methods. To provide a relatively focused book content, the areas of strategic asset allocation, asset-liability management, and asset pricing were considered as reference topic areas. At the time we are writing this Preface, a companion special issue (SI) of *OR Spectrum* has been published (July 2015: Financial optimization: Optimization paradigms and financial planning under uncertainty, *OR SPECTRUM*, 37 (3), Springer) whose table of contents is the following:

J. Dupacova and V. Kozmik: *Structure of risk-averse multistage stochastic programs*

A.K. Konicz, D. Pisinger, K. Rasmussen, and M. Steffensen: *A combined stochastic programming and optimal control approach to personal finance and pensions*

S. Pagliarani and T. Vargiolu: *Portfolio optimization in a defaultable Lévy-driven market model*

M.H.A. Davis and S. Lleo: *Jump-diffusion asset–liability management via risk-sensitive control*

S. Desmettre, R. Korn, P. Ruckdeschel, and F.T. Seifrid: *Robust worst-case optimal investment*

M. Kopa and T. Post: *A general test for SSD portfolio efficiency*

R. Bruni, F. Cesarone, A. Scozzari, and F. Tardella: *A linear risk–return model for enhanced indexation in portfolio optimization*

A. Thiele and E. Cetinkaya: *Data-driven portfolio management with quantile constraints*

T. Driouchi, L. Trigeorgis, and Y.L. Gao: *Choquet-based European option pricing with stochastic (and fixed) strikes*

R. Cerqueti , P. Falbo, G. Guastaroba, and C. Pellizzari: *Approximating multivariate Markov chains for bootstrapping through contiguous partitions*

Jointly, this book and the SI [8] offer an extensive overview of different modeling frameworks and mathematical approaches currently adopted over a relatively wide range of financial domains. In the concluding chapter, we provide a comprehensive review of the main evidences emerging from the work carried out and, relying on the extended set of articles and chapters, assess the state of the art and suggest possible future directions of research.

Ex post we can say that the initial purpose of the volume has been achieved and the collected contributions provide a rather unique set of articles specifically in the domains of portfolio theory and asset-liability management. A general interest to long-term management problems has emerged with an explicit effort by several authors to overcome long-established modeling assumptions, whose consistency with real-world dynamics has been increasingly questioned in recent years. All chapters have gone through a rigorous refereeing process.

A rather short but effective set of characterizing elements of the included chapters may help understanding the volume's profile:

- A sort of looking forward yet back-to-basics motivation underlies several contributions: the relationship between financial risk and investment returns is considered in a very constructive and effective way also removing quite many unrealistic assumptions that in the past have led to substantial model risk and nonoptimal decision-making processes. Not only in the area of portfolio optimization, whose modern era starts with the well-known Markowitz contribution (1952), but to a certain extent also within the chapters devoted to financial engineering and life insurance models, an effort to recast the model formulation and the overall mathematical treatment emerges within more accessible and realistic frameworks, avoiding all those assumptions that in the past have certainly facilitated the mathematical development of the discipline but also to a certain extent jeopardized its practical adoption. Good examples of such generalized effort are provided in this volume by Calafiore [3, 4], Aro and Pennanen [1], Dempster et al. [10], Györfi et al. [16], and Gilli and Schumann [15] and in the SI by Dupačová and Kozmík [13], Thiele and Cetinkaya [6], Driouchi et al. [12], and Kopa and Post [19].
- As a consequence, both in this volume and in the special issue of *OR Spectrum*, data-driven approaches appear at the core of the analysis: this is explicit in Calafiore [4], Pachamanova et al. [22], and Gilli and Schuman [15] and from a different perspective in Györfi et al. [16] and within the SI in Thiele and Cetinkaya [6] and Falbo et al. [5]. Either through a nonparametric approach or by introducing alternative stochastic assumptions on market dynamics, a growing evidence of the removal of once well-established market models is confirmed. The second chapter by MacLean and Zhao [20] presents two increasingly popular financial model extensions that add to the contributions by Davis and Lleo [9] and Pagliarini and Vargiolu [23] in the special issue.

- After such an extended and stressful phase of financial instability that for the first time has affected also sovereign borrowers in developed countries, the need to extend decision horizons and address financial management problems primarily formulated as multi-period and dynamic financial planning problems is evident. This is indeed the element we have primarily considered when deciding to put the survey on multi-period risk measures by Chen et al. [7] as first chapter of the volume. Pachamanova et al. [22], Aro and Pennanen [1], Dempster et al. [10], and Mulvey et al. [21] here as well as Dupačová and Kozmík [13], Konicz et al. [18], Pagliarini and Vargiolu [23], and Davis and Lleo [9] and Desmettre et al. [11] in the special issue rely on such model formulation. In some problem types, primarily those devoted to pension funds' ALM—dynamic long-term formulations also imply a rather central role of longevity risk, one of the relevant risk sources driving nowadays market and agents' behaviors.
- As a final emerging and underlying evidence: bonds and equity are no longer sufficient, and risk management is complex. A prolonged period of high volatility and regime switching and historically low interest rates and vanishing safe market sectors have had profound effects on agents' risk preferences and institutional investors, strategies. As mentioned above, longevity risk is affecting pension funds, management options and their liabilities, but increasingly the source of financial risk is linked to prolonged negative economic cycles. The contributions by Pachamanova et al. [22], Mulvey et al. [21], Aro and Pennanen [1], and Giandomenico and Pinar [14] in this volume as well as those of Driouchi et al. [12] and Konicz et al. [18] in the SI are motivated by the evidence of a growing role in investors' portfolios of assets other than equity and bonds: specifically alternative investments, inflation-linked securities, and life insurance products, with relevant implications on financial institutions hedging strategies and the overall market liquidity.

The chapters are included in the volume following a sequence, which aims at conveying a research and thematic path that we wish to clarify.

The chapter by Chen et al. [7] provides a thorough methodological survey of risk measures theory within dynamic investment theory, setting the stage for the following specifications and extensions. The derivation of a unified set of necessary and sufficient conditions to have multi-period, time-consistent, coherent risk measures represents a central, key contribution of the chapter. The concept of time-consistent investment policies is attracting increasing interest within the formulation and solution of complex optimal portfolio management problems: in recent years, however, different formulations have been attached to the concept. This starting chapter and the first article in the SI, by Dupačová and Kozmík [13], clarify the modeling and methodological implications of such consistency, adopting a discrete time formulation as reference model framework. The chapter immediately following, by MacLean and Zhao [20], has a different aim, but addresses directly the issue of how risk has to be modeled at least in liquid equity markets. MacLean and Zhao [20], in a chapter with strong statistical emphasis, summarize the results of a long-dated research effort, clarifying the implications of regime switching models leading to mixtures of probability distributions, whose statistical characterization is addressed in a rigorous way, against discontinuous return models also leading typically to probability mixtures. The effort to read the data and characterize the markets' stochastic dynamics is associated with what we can refer to as a first step to a model-based optimization approach in which model risk is still an issue. The third chapter by Calafiore [4] moves from such evidence to propose an approach to portfolio allocation which is model-free and relies on data analysis and a clever probabilistic characterization of market returns to present a sequence of results on optimal open- and closed-loop policies in dynamic markets. The last part of Calafiore's chapter is devoted to yet data-driven robust portfolio allocation models and the so-called scenario approach to portfolio optimization. The message is that quite a lot can be done and may lead to superior results, without the need to introduce in the investment problem a stochastic return model. The adoption of portfolio policies will return later in the volume as central element of Mulvey et al. [21].

Robust optimization is the approach now for some years adopted by Pachamanova et al. [22] to address portfolio optimization problems, extended here, however, to a dynamic ALM model specifically for a pension fund. Here, we have an interesting chapter combining a characterization of the underlying market uncertainty based either on historical data or a factor model, an enterprise-wide asset-liability management problem by a pension fund, and a computational study comparing robust and stochastic optimization approaches. The numerical evidence is extended, and this chapter responds to the benchmarking philosophy put forward in the original volume proposal to the publisher. Here, we have a company-wide asset-liability management, and the complex risk sources faced by the decision maker, an institutional investor, are captured in one instance through a factor model. Aro and Pennanen [1], in the following chapter, formulate and solve also an ALM problem, in this case focusing directly on the pension fund liability and the need for the pension fund manager to hedge such liability in an optimal way. Here, the hedging strategy is in discrete time while the return process has continuous state space. The optimization problem aims at determining the minimum required initial wealth needed to hedge the pension liabilities in a market in which perfect hedging and portfolio replication is typically not possible. This chapter addresses thus a fundamental pricing issue for life insurance contracts by postulating an incomplete market and focusing on the extended set of risk sources faced by the pension manager. Aro and Pennanen [1] show, as an aside, that with no need of specific heroic assumptions on the underlying probability space, the associated (imperfect) hedging problem can be formulated and solved relying on convex analysis. Longevity risk is at the center of the analysis.

Asset pricing motivates the contribution by Giandomenico and Pinar [14], where the authors rely on a discrete approach, based on a non-recombining tree, to formulate and analyze the valuation problem for an American option carrying multiple exercise dates: it is an extension of the classical pricing problem with only one possible stopping time, in which the authors generalize the seminal paper by King [17] on contingent claim analysis. As in Aro and Pennanen [1], we have also here a pricing problem formulated as minimal cost problem to hedge a given liability: not only the optimization problem is interesting on its own but also such formulation emphasizes the importance of the hedging problem faced by the derivative writer; this is to be regarded as a necessary condition for the market to develop. Aro and Pennanen [1] in this respect clarify that indeed in modern markets, which are sometimes characterized by poor liquidity, perfect hedging tends to be a pure theoretical construction: partial hedging becomes then the reference market condition and market incompleteness the associated stochastic concept, if one wants to link the analysis to typical mathematical finance concepts. Incompleteness refers to the lack of sufficient financial contracts to hedge every risk source embedded in the contract. It is also true that often life insurers are either unwilling to undertake complex hedging strategies (and rely on high management and underwriting fees) or adopt indirect hedging positions based on correlation analysis. From a methodological viewpoint, a distinctive positive element of Giandomenico and Pinar's [14] article (in addition to the proposed methodology which motivates the contribution) is represented by the initial rigorous description of the probability space and the definition of a scenario tree process for the underlying uncertainty. It is the same underlying statistical formulation typically underlying multistage stochastic programs as those considered by Chen et al. [7], Mulvey et al. [21], and Dempster et al. [10], who, respectively, the latter two, in their chapters address first the implications of optimal portfolio policies in illiquid markets and the second a fundamental methodological implication when dealing with scenario trees with limited branching and the market is incomplete. Among all chapters included in this volume, the one by Mulvey et al. [21], even if with rather complex underlying methodological implications, is the contribution where an advanced knowledge of financial economics and now a days financial markets and agents' policies is mostly needed. Interestingly, the authors present evidence coming from university endowments to clarify a widespread evidence, which is the growing role in modern portfolios of illiquid positions and their implications. The key, innovative motivation of the contribution lies in the construction of a replicating market index able to generate a benchmark for hedging problems as well as portfolio selection problems overcoming at the same time the illiquidity issue arising in markets such as private equity, commodity, and, we add, real estate markets. Mulvey et al. [21] provide a convincing motivation and an effective overview of current challenges faced by institutional investors, after constructing an index with given desirable risk-reward properties that link their study to multistage stochastic programming to suggest a possible approach to formulate and solve a strategic asset allocation problem in the presence of this new asset class of illiquid instruments.

When it comes to formulate a dynamic stochastic programming problem, related to the issue of market incompleteness, Dempster et al. [10] address a key methodological issue in financial optimization when, to ease and allow the solution of large-scale problems, an approximation of typically continuous probability measure by means of discretely sampled scenario trees is needed. The approximation will both lead in general to an approximation bias and result into relatively unstable first-stage decisions. The analysis on implementable decisions is limited to first-stage, root-node decisions and the authors here introduce a method to limit and generate consistent and stable first-stage decisions in the presence of small and coarse scenario trees, as often the case, resulting anyway on large-scale and computationally very challenging programs. Interestingly, Dempster et al. [10] tackle the small sample approximation problem from the viewpoint of a robust optimization approach where the sample is reinterpreted as a problem of incomplete data. This contribution adds to previous studies in which rather than considering the decision space, focus on criteria to best approximate the probability space by introducing appropriate metrics and information measures and devising appropriate scenario reduction and generation methods (see Bertocchi et al. [2] in the same collection).

The following two chapters by Györfi et al. [16] and Gilli and Schumann [15] provide good examples of what we have initially referred to as a valuable effort to move forward in finance theory by recalling some fundamental rules and results that cannot be ignored when addressing maybe theoretical issues but strictly related to applied finance. The chapter by Györfi et al. [16] focuses on a well-known portfolio optimization approach, the growth optimal strategy, based on assumption of log-optimal portfolios to present in a very effective and readable way the set of results that can be called upon to motivate over long-term horizons such decision paradigm and under which market conditions such strategy does indeed satisfies also risk constraints, from which the title given by the authors, the growth optimal investment strategy, is secure too. The chapter presents an extended and useful set of probabilistic and statistical results to show that, applying and deriving a set of inequalities from large deviation theory, it is possible to study the rate of convergence of a log-optimal portfolio return to a target return and that the time in which such target can be achieved under the worst possible market circumstances is still bounded. Before our concluding chapter, the chapter by Gilli and Schumann [15] also motivates the need to address portfolio, only asset, investment problems by considering jointly the issue of model accuracy and realism and the one on the problem solvability. The latter in particular is obviously needed to facilitate the practical adoption of a methodology, but it is also meant not to induce to achieve the problem solution a modification of otherwise realistic model assumptions. Remaining in the area of one-period optimization problems, the authors analyze in a very accessible way the key elements of heuristic solution techniques that can be fruitfully adopted to yield optimal portfolios when other solution approaches are not viable, and the need not to modify the original model assumptions is taken into consideration. The chapter provides an excellent wrap-up of the many issues addressed in the volume, from the formulation of a mathematical description of a portfolio selection problem, consistent with agents' behavioral properties, to the treatment of the stochastic model elements and finally its solution. A case study with an interesting application of a heuristic is presented, relying on the so-called threshold accepting approach, and the authors provide extended evidence of the potential of heuristic methods under several portfolio optimization problem specifications. As other chapters in the volume, independently from the specific application [15], rightly we would add, emphasize the need not to terminate a portfolio selection problem with its solution but to back-test and validate its solution with an appropriate statistical and scenario analysis: only such analysis can validate the theoretical framework and lead to a method practical application.

The concluding chapter aims at consolidating the state of the art through a unified model formulation clarifying the key elements of the chapters included here below and in the special issue and the associated theoretical and applied contributions. An overall assessment of the state of the art on the different financial topics is provided.

A sincere thanks goes to the authors, the publisher, and the colleagues at Springer, as well as to Prof. Camille Price, scientifically responsible of this Springer series devoted to Operations Research and Management Science, who followed and stimulated this work.

Bergamo, Italy — Giorgio Consigli
Lausanne, Switzerland — Daniel Kuhn
Torino, Italy — Paolo Brandimarte
September 2016

References

1. H. Aro, T. Pennanen, Liability driven investment in longevity risk management, in *Optimal Financial Decision Making under Uncertainty*. International Series in Operations Research and Management Science (Springer, Cham, 2016)
2. M. Bertocchi, G. Consigli, M.A.H. Dempster (eds.), *Handbook on Stochastic Optimization Methods in Finance and Energy. New Financial Products and Energy Market Strategies*. Fred Hillier International Series in Operations Research and Management Science (Springer, US, 2011)
3. R. Bruni, F. Cesarone, A. Scozzari, F. Tardella, A linear risk-return model for enhanced indexation in portfolio optimization. OR Spectr. **37**(3), 735–759 (2015)
4. G.C. Calafiore, Scenario optimization methods in portfolio analysis and design, in *Optimal Financial Decision Making under Uncertainty*. International Series in Operations Research and Management Science (Springer, Cham, 2016)
5. R. Cerqueti, P. Falbo, G. Guastaroba, C. Pelizzari, Approximating multivariate Markov chains for bootstrapping through contiguous partitions. OR Spectr. **37**(3), 803–841 (2015)
6. E. Çetinkaya, A. Thiele, Data-driven portfolio management with quantile constraints. OR Spectr. **37**(3), 761–786 (2015)
7. Z. Chen, G. Consigli, J. Liu, G. Li, T. Fu, Q. Hu, Multi-period risk measures and dynamic risk control, in *Optimal Financial Decision Making under Uncertainty*. International Series in Operations Research and Management Science (Springer, Cham, 2016)
8. G. Consigli, P. Brandimarte, D. Kuhn, Editorial: financial optimization: optimization paradigms and financial planning under uncertainty. OR Spectr. **37**(3), 553–557 (2015)
9. M.H.A. Davis, S. Lleo, Jump-diffusion asset-liability management via risk-sensitive control. OR Spectr. **37**(3), 655–675 (2015)
10. M.A.H. Dempster, E.A. Medova, Y.S. Yong, Stabilizing implementable decisions in dynamic stochastic programming, in *Optimal Financial Decision Making under Uncertainty*. International Series in Operations Research and Management Science (Springer, Cham, 2016)
11. S. Desmettre, R. Korn, P. Ruckdeschel, F.T. Seifried, Robust worst-case optimal investment. OR Spectr. **37**(3), 677–701 (2015)
12. T. Driouchi, L. Trigeorgis, Y. Gao, Choquet-based European option pricing with stochastic (and fixed) strikes. OR Spectr. **37**(3), 787–802 (2015)

13. J. Dupačová, V. Kozmík, Structure of risk-averse multistage stochastic programs. OR Spectr. **37**(3), 559–582 (2015)
14. M. Giandomenico, M. Pinar, Pricing American options: binomial and trinomial non-recombinant trees and two exercise rights, in *Optimal Financial Decision Making under Uncertainty*. International Series in Operations Research and Management Science (Springer, Cham, 2016)
15. M. Gilli, E. Schumann, Constructing portfolios with heuristics, in *Optimal Financial Decision Making under Uncertainty*. International Series in Operations Research and Management Science (Springer, Cham, 2016)
16. L. Györfi, G. Ottucsak, H. Walk, The growth optimal strategy is secure too, in *Optimal Financial Decision Making under Uncertainty*. International Series in Operations Research and Management Science (Springer, Cham, 2016)
17. A.J. King, Duality and martingales: a stochastic programming perspective on contingent claims. Math. Programm. Ser. B **91**, 543–562 (2002)
18. A.K. Konicz, D. Pisinger, K.M. Rasmussen, A combined stochastic programming and optimal control approach to personal finance and pensions. OR Spectr. **37**(3), 583–616 (2015)
19. M. Kopa, T. Post, A general test for SSD portfolio efficiency. OR Spectr. **37**(3), 703–734 (2015)
20. L.C. MacLean, Y. Zhao, Asset price dynamics: shocks and regimes, in *Optimal Financial Decision Making under Uncertainty*. International Series in Operations Research and Management Science (Springer, Cham, 2016)
21. J.M. Mulvey, W.C. Kim, C. Lin, Optimizing a portfolio of liquid and illiquid assets, in *Optimal Financial Decision Making under Uncertainty*. International Series in Operations Research and Management Science (Springer, Cham, 2016)
22. D. Pachamanova, N. Gulpinar, E. Canakoglu, Robust data-driven approaches to pension fund asset-liability management under uncertainty, in *Optimal Financial Decision Making under Uncertainty*. International Series in Operations Research and Management Science (Springer, Cham, 2016)
23. S. Pagliarani, T. Vargiolu, Portfolio optimization in a defaultable Lévy-driven market model. OR Spectr. **37**(3), 617–654 (2015)

Contents

Chapter 1
Multi-Period Risk Measures and Optimal Investment Policies

Zhiping Chen, Giorgio Consigli, Jia Liu, Gang Li, Tianwen Fu, and Qianhui Hu

Abstract This chapter provides an in-depth overview of an extended set of multi-period risk measures, their mathematical and economic properties, primarily from the perspective of dynamic risk control and portfolio optimization. The analysis is structured in four parts: the first part reviews characterizing properties of multi-period risk measures, it examines their financial foundations, and clarifies cross-relationships. The second part is devoted to three classes of multi-period risk measures, namely: terminal, additive and recursive. Their financial and mathematical properties are considered, leading to the proposal of a unifying representation. Key to the discussion is the treatment of dynamic risk measures taking their relationship with evolving information flows and time evolution into account: after convexity and coherence, time consistency emerges as a key property required by risk measures to effectively control risk exposure within dynamic programs. In the third part, we consider the application of multi-period measures to optimal investment policy selection, clarifying how portfolio selection models adapt to different risk measurement paradigms. In the fourth part we summarize and point out desirable developments and future research directions. Throughout the chapter, attention is paid to the state-of-the-art and methodological and modeling implications.

Keywords Multi-period risk measures • Time-consistency • Dynamic risk control • Recoursive risk measures • Portfolio optimization • Information processes • Bellman's principle

---

Z. Chen (✉) • J. Liu • G. Li • T. Fu • Q. Hu
Department of Computing Science, School of Mathematics and Statistics, Xi'an Jiaotong University, Xian Ning West Road 28, 710049 Xi'an, Shaanxi, People's Republic of China
e-mail: zchen@mail.xjtu.edu.cn; liu.jia@stu.xjtu.edu.cn; lg.2009@stu.xjtu.edu.cn; twfu@stu.xjtu.edu.cn; huqianhui123@stu.xjtu.edu.cn

G. Consigli
Department of Management, Economics and Quantitative Methods, University of Bergamo, Via dei Caniana 2, 24127 Bergamo, Italy
e-mail: giorgio.consigli@unibg.it

© Springer International Publishing Switzerland 2017

G. Consigli et al. (eds.), *Optimal Financial Decision Making under Uncertainty*, International Series in Operations Research & Management Science 245,
DOI 10.1007/978-3-319-41613-7_1

1.1 Introduction

Medium to long-term investment management problems are often formulated as dynamic portfolio selection problems, in which investment decisions are allowed to change over time. Such choice becomes standard in presence of time- and state-dependent constraints. Due to market frictions such as trading costs and regulatory as well as tax constraints, furthermore, dynamic decision problems are increasingly formulated as discrete, multi-stage, rather than continuous time, control problems. Key to the achievement of an effective risk control are the properties of the adopted risk measure over multiple periods. Compared to a static situation, as we see in what follows, it is not a trivial task to establish suitable multi-period risk measures which satisfy reasonable and practically relevant properties.

This chapter reviews the state of the art on multi-period risk measures and their inclusion in optimal portfolio selection problems. We see that, due to their complexity, dynamic risk measures have been considered only occasionally in multi-stage problems to date. A relevant stream of research has gone indeed into the construction and theoretical properties of multi-period risk measures. Therefore, we will devote the first part of this chapter to analyze alternative formulations and mathematical properties of a qualified set of existing risk measures. Basic properties are considered first in Sect. 1.2. While in Sect. 1.3 current multi-period risk functions are derived, introducing the canonical distinction between *terminal wealth*, *additive* and *recursive* risk measures. In what follows, we will only consider dynamic risk measures in discrete time. For the sake of convenience, however, we will use the terms *multi-period* and *dynamic* (risk measure) interchangeably to highlight that despite the discrete-time approximation, underlying time is actually evolving continuously. The application of *multi-period* risk measures to *multi-stage* portfolio selection models is discussed in Sect. 1.4. Finally, we provide a brief summary and point out open research problems in Sect. 1.5.

1.2 Dynamic Risk Control

A dynamic risk measure can be studied with reference to events defined in a canonical probability space $\omega \in (\Omega, \mathcal{F}, P)$, where Ω is a sample space, $\mathcal{F}$ is a σ-algebra, and P assigns to any event B in $\mathcal{F}$ a probability measure $P(B)$. The investment horizon is $\mathcal{T} := \{0, 1, 2, \cdots, T\}$ with T consecutive periods. For each $t \in \mathcal{T}$ the σ-algebra $\mathcal{F}_t \subset \mathcal{F}$ denotes the set of all events $\omega \in \Omega$ corresponding to information available at time t. $\mathcal{F}_0 = \{\emptyset, \Omega\}$ and $\mathcal{F}_0 \subseteq \mathcal{F}_1 \subseteq \cdots \mathcal{F}_T$. In a dynamic setting we require any time-t (random) loss $X_t \in \mathcal{L}_t$ to be adapted to $\mathcal{F}_t$, i.e. be $\mathcal{F}_t$-measurable, where $\mathcal{L}_t = \mathcal{L}_p(\Omega, \mathcal{F}_t, P)$, $p \in [1, +\infty]$ denotes the class of p-th integrable random functions on Ω.

From the notion of random loss X_t we move to the definition of a loss process over the periods from t to T by introducing $X_{t,T} = (X_t, \cdots, X_T) \in \mathcal{L}_{t,T}$, where

$\mathcal{L}_{t,T} = \mathcal{L}_t \times \cdots \times \mathcal{L}_T$ is a corresponding product space. Following [68] a conditional risk function can be defined in period t as $\rho_{t,T}(\cdot) : \mathcal{L}_{t,T} \rightarrow \mathcal{L}_t$, thus providing a risk estimate at time t of a loss process with realizations in $t, t+1, \ldots, T$. As time evolves from current time $t = 0$ to T the risk function $\{\rho_{t,T}\}_{t=0}^{T}$ will provide an estimate of the risk associated with the residual time to horizon: this is a multi-period risk measure. As a special case, $\rho_t(\cdot) : \mathcal{L}_{t+1} \rightarrow \mathcal{L}_t$ represents the corresponding single-period risk mapping.

In the existing literature, multi-period risk measures have been discussed with respect to cash-flow, or return or wealth processes, where losses can be identified in monetary or percentage terms, see for example [2, 38, 60, 64, 68]. Without any loss of generality we will only refer to loss processes in what follows.

Relying on the definition of $\rho_{t,T}$ we can introduce a corresponding multi-period risk control problem. Consider in particular an investment problem defined over a discrete planning horizon $\mathcal{T}$ on which we define a stochastic state process $\{\xi_t(\omega)\}_{t=0}^{T}$, a control process $\{u_t\}_{t=0}^{T}$ and a random loss process $\{X_t\}$ whose evolution will depend on $\{u_t\}$ and $\{\xi_t\}$. In dynamic models we typically consider for each $t \in \mathcal{T}$ a sequence of controls and random events, the former always facing an uncertain outcome and the latter always determining a new risk exposure. The control problem from a financial viewpoints aims for given sequence of actions and random events to minimize the cost associated with the loss process. In the general case at any point in time the dependence of X_t on X_{t-1}, u_{t-1} and ξ_t may not be linear and it will depend on a functional $g_t : \mathcal{L}_{t-1} \times \mathcal{U}_{t-1} \times \mathcal{L}_t \rightarrow \mathcal{L}_t$. We require g_t to be convex for each t. Then a generic formulation of an optimal risk control problem [34, 60, 73] is:

$$\min_{u_t} \rho_{0,T}(X_{0,T}) \tag{1.1}$$

$$s.t.\ X_t = g_t(X_{t-1}, u_{t-1}, \xi_t),\ t = 1, \cdots, T, \tag{1.2}$$

$$u_t \in \mathcal{U}_t,\ t = 0, \cdots, T-1, \tag{1.3}$$

$$u_t \lhd \mathcal{F}_t,\ t = 0, \cdots, T-1. \tag{1.4}$$

Here, the objective function $\rho_{0,T}$ in (1.1) describes the overall multi-period risk exposure, for $t = 0, 1, \ldots, T$ induced by a sequence of loss evaluations and controls (and subsequent resolution of uncertainty as described in (1.3)). The state equation provides the core set of dynamic relationships captured by this formulation. Different specifications of this feed-back equation will be considered depending on the problem at hands: the dynamic wealth equation in [53], the cash balance condition in [76] or the linear recourse constraint in [34]. The control space $\mathcal{U}_t$ may vary with time and it will depend on lower and upper bounds as well as turnover constraints which will determine the problem feasibility region. As mentioned we generally require this space to be convex and smooth. The non-anticipativity condition is described in (1.5) by $u_t \lhd \mathcal{F}_t$: this condition is equivalent to the condition of $\mathcal{F}_t$-measurability of u_t for each $t \in \mathcal{T}$. In multi-stage stochastic programs this condition makes sure that the dynamic decision process relies only on information available at that time, excluding partial or total foresight.

Leaving aside the constraints (1.3)–(1.5) widely adopted in dynamic programs (as multi-stage stochastic programs), the model instance (1.1)–(1.5) will change for different selections of the dynamic risk measure and its properties. It is useful to distinguish between properties, such as the measure's *coherence*, which reflect axioms regarded as necessary conditions for an appropriate risk assessment methodology and those, like *time consistency* who are relevant specifically in relationship with the optimal control problem. Among the former, the axiomatic theory of risk measures has been developed extensively with respect to static risk measures and we provide a set of definitions and results aimed at the measures' extension into a dynamic framework. Among the latter we consider in particular the issues of convex and time consistent risk measures and/or associated optimal policies. Here next we analyze a set of properties and their impact on the optimal control problem (1.1) formulation, by following a building-up approach to clarify a set of mathematical and financial implications.

1.2.1 Key Properties of Dynamic Measures

A rich stream of research was motivated in recent years by the effort to extend into a dynamic framework, results previously established for static risk measures. We present first a possible refinement of risk measures' properties lading to their adoption within dynamic models. Of specific interest the conditions leading to the measures' *dynamic coherence*.

1.2.1.1 Extension of Risk Measures' Axioms

The following definition allows an extension of otherwise well established properties into a dynamic context. A dynamic risk measure $\{\rho_{t,T}\}_{t=0}^{T}$ is *dynamically coherent* if for each $t = 0, 1, \cdots, T$ the following properties hold:

- *Monotonicity*: given two loss processes $X_{t,T}$ and $Y_{t,T}$: $X_{t,T} \leq Y_{t,T} \Rightarrow \rho_{t,T}(X_{t,T}) \leq \rho_{t,T}(Y_{t,T})$.
- *Translation invariance*: if, for any $X_{t,T}$ and $m_t \in \mathcal{L}_t$, $\rho_{t,T}(X_t + m_t, X_{t+1}, \cdots, X_T) = m_t + \rho_{t,T}(X_t, X_{t+1}, \cdots, X_T)$.
- *Sub-additivity*: if, for any $X_{t,T}$, $Y_{t,T}$, we have $\rho_{t,T}(X_{t,T} + Y_{t,T}) \leq \rho_{t,T}(X_{t,T}) + \rho_{t,T}(Y_{t,T})$, $t = 0, 1, \cdots, T$.
- *Positive homogeneity*: if, for any $X_{t,T}$ and $\lambda > 0$, $\rho_{t,T}(\lambda X_{t,T}) = \lambda \rho_{t,T}(X_{t,T})$, $t = 0, 1, \cdots, T$.

As in the static case sub-additivity and positive homogeneity imply the measure's convexity but all four properties are needed for coherence. The latter implies the former but not the opposite. Already in a static, one-period model the variance, lacking such condition, is known to be convex but not coherent. Here above we define monotonicity with respect to a loss process, which is weaker than that defined on terminal wealth or related to cumulated losses (like that in [68]). Indeed monotonicity with respect to the cumulated sum of period-wise losses implies the monotonicity with respect to the loss process. However the converse is in general not true [40].

The translation invariance property implies that adding a certain amount to the current loss will increase the risk by the same amount. There are several weaker versions of *translation invariance* [21, 40, 49, 68, 80]. For instance $\{\rho_{t,T}\}_{t=0}^{T}$ is said to satisfy the *translation invariance* property if $\rho_{t,T}(X_t, X_{t+1}, \cdots, X_T) = X_t + \rho_{t,T}(0, X_{t+1}, \cdots, X_T)$ for all $X_{t,T} \in \mathcal{L}_{t,T}$, which is used to derive a time consistent dynamic risk measure in [68]. A particularly weak condition posed to satisfy translation invariance is given in [49]: $\rho_{t,T}(X_t, 0, \cdots, 0) = X_t$ holds for all $X_t \in \mathcal{L}_t$, $t = 0, 1, \cdots, T$. It reflects a simple intuition: with no future losses, the resulting dynamic risk is equal to the current loss.

The concept of translation invariance allows us to interpret the risk measure as a capital requirement to make a position acceptable: by adding (respectively, subtracting) a deterministic loss to (from) an initial position and interpreting such loss as a liability (asset) to be invested in the reference instrument will simply increase (decrease) the investment risk by that deterministic amount. Strictly related from a mathematical viewpoint to the monotonicity and translation-invariance of a dynamic risk measure is the so-called *local property* condition:

Local property: A dynamic risk measure $\{\rho_{t,T}\}_{t=0}^{T}$ is said to satisfy the *local property* if $1_A \rho_{t,T}(X_{t,T}) = 1_A \rho_{t,T}(1_A X_{t,T})$ for any random loss process $X_{t,T} \in \mathcal{L}_{t,T}$ and $A \in \mathcal{F}_t$.

For any t and any given subset A of the current σ-algebra $\mathcal{F}_t$, this property states that it is sufficient to evaluate the risk measure within such set rather than filtering the events first and derive their risk estimates afterwards. In a dynamic setting, since t varies, this condition also implies that under the local property condition, it is sufficient to work with an $\mathcal{F}_t$-adapted risk measure since that risk exposure should only depend on future information. As a basic property of dynamic risk measures, the local property was discussed in [19, 46]. In [48], the local property was introduced under the name of regularity, and in [70, 72], it was introduced as time consistency, which is nevertheless different from the usual definition of time consistency adopted in the literature on dynamic risk measures.

As it is well-known, sub-additivity describes a basic rule of investment theory, according to which portfolio diversification may only have a positive impact on the risk exposure. In a dynamic setting sub-additivity implies that if two loss processes are pooled together, the resulting risk exposure should not increase. For any $t = 0, 1, \cdots, T$:

$$\rho_{t,T}(X_{t,T}) \leq \rho_{t,T}(X_t, 0, \cdots, 0) + \rho_{t,T}(0, X_{t+1}, 0, \cdots, 0) + \cdots + \rho_{t,T}(0, \cdots, 0, X_T).$$

In a dynamic model sub-additivity implies that period-wise investment diversification across stages won't increase the risk exposure.

Under the assumption of sufficient liquidity, we can assume that risk increases in proportion to the increase of the investment. Positive homogeneity describes this property. Under the framework of acceptable risk measures (such as coherent risk measures or convex risk measures), a risk measure is positive homogeneous if and only if its corresponding acceptance set is a cone [1].

We can extend the concept of risk measure convexity and portfolio diversification to a dynamic setting. By induction, we expect diversification to lead to a dynamic risk reduction [32, 37, 60, 69].

Convexity: A dynamic risk measure $\{\rho_{t,T}\}_{t=0}^{T}$ is *convex* if, for any $X_{t,T}$, $Y_{t,T}$, $\rho_{t,T}(\lambda X_{t,T} + (1-\lambda)Y_{t,T}) \leq \lambda\rho_{t,T}(X_{t,T}) + (1-\lambda)\rho_{t,T}(Y_{t,T})$, $\lambda \in [0,1]$, $t = 0, 1, \cdots, T$.

The convexity of a risk measure implies that its minimization over a convex set is a convex programming problem. For a dynamic risk measure, convexity further ensures that the period-wise diversification of investments among stages can reduce or at least not increase the risk exposure. Convexity can be deduced from sub-additivity and positive homogeneity. Under the framework of acceptable risk measures, a risk measure satisfies convexity if and only if its corresponding acceptance set is convex [1].

1.2.1.2 Dynamic Risk and Information Processes

When studying the evolution of the risk exposure of a given financial portfolio in multi-stage models, it may be appropriate to analyze risk measures by explicitly introducing their dependence on evolving information processes. These as mentioned above in our framework are captured by *filtrations*, i.e. increasing sequences of σ-algebra. Consider two information processes, e.g. two filtrations: $\{\mathcal{F}_0, \mathcal{F}_1, \cdots, \mathcal{F}_T\}$ and $\{\mathcal{F}'_0, \mathcal{F}'_1, \cdots, \mathcal{F}'_T\}$. Let us denote by $\rho_{t,T}(X_{t,T}\,|\mathcal{F}_t, \cdots, \mathcal{F}_T)$ the conditional risk measure under the first information process at time t, and by $\rho_{t,T}(X_{t,T}|\mathcal{F}_t{}', \cdots \mathcal{F}_T{}')$, the same measure under the second information process at time t. The concept of information monotonicity [60] can be described as follows:

Information monotonicity: If, for any two filtrations $\{\mathcal{F}_0, \mathcal{F}_1, \cdots, \mathcal{F}_T\}$ and $\{\mathcal{F}'_0, \mathcal{F}'_1, \cdots, \mathcal{F}'_T\}$ satisfying $\mathcal{F}_s \subseteq \mathcal{F}'_s$, $s = t, \cdots, T$, we have

$$\rho_{t,T}(X_{t,T}\,\left|\mathcal{F}_t{}', \cdots \mathcal{F}_T{}'\right.) \leq \rho_{t,T}(X_{t,T}\,|\mathcal{F}_t, \cdots, \mathcal{F}_T),\ t = 0, 1, \cdots, T.$$

Then $\{\rho_{t,T}\}_{t=0}^{T}$ is called *information monotone*.

Information monotonicity is used to distinguish risk measures under different filtrations. This property expresses the idea that for a given portfolio, increasing available information will never lead to increasing risk exposures but typically to more effective risk control . Unlike in the canonical monotonicity axiom presented in previous section, here the loss process is the same but we assume that the associated filtrations may differ. The two axioms hold jointly if the filtrations are assumed to be those *generated* by the processes and one loss process dominates the other and/or one filtration is richer than the other even if the two loss processes are equivalent.

Focusing on conditional distributions, the distinction between *distribution*- and *law*-invariant risk measures is also relevant. The former was introduced in [80] with respect to terminal payments. The latter in [51], where the risk estimate must depend only on the unconditional law at initial time.

Distribution- versus law-invariant risk measures: A dynamic risk measure $\{\rho_{t,T}\}_{t=0}^{T}$ is:

- *distribution-invariant*: if, for any $X_T, Y_T \in \mathcal{L}_T$ with the same conditional law at any time t, $0 \leq t \leq T$,

$$\rho_{t,T}(X_T) \leq 0 \Leftrightarrow \rho_{t,T}(Y_T) \leq 0, \; t = 0, 1, \cdots, T. \tag{1.5}$$

- *law-invariant*: if, for any $X_T, Y_T \in \mathcal{L}_T$,

$$\rho_{t,T}(X_T) = \rho_{t,T}(Y_T), \; 0 \leq t \leq T \tag{1.6}$$

Unlike the distribution-invariance property which focuses on the two positions' conditional laws, law-invariance considers the exact risk estimates and it is regarded as a stronger condition. According to [80], any distribution-invariant dynamic risk measure can be represented by a sequence of static risk measures. Since it states that the value of a dynamic risk measure should only depend on the conditional distribution of the random variable, law invariance is useful for the computation of dynamic risk measures based on historical data.

1.2.2 Time Consistency

The research on time consistency of dynamic risk measures has attracted much attention in recent years (e.g., [2, 21, 41, 65, 68, 77, 80]). Generally, time consistency can be examined from two perspectives: in relation to multi-period risk measures or in relation to optimal investment policies. The first notion originates directly from the evolving relationship, as time goes by, between the risk generated by a random process and it's filtration. The second concept, instead, is motivated by the inherent relationship between an optimal control and the resolution of uncertainty: when solving a finite-horizon stochastic control problem we seek the definition of an optimal policy over a given time horizon. As time goes by such strategy will be revised according to the collected new information and so on until the beginning of the last period. The concept of time consistency in both cases has to do with the *consistency* over time of the first risk evaluation and/or the first optimal control solution (the one taken under full uncertainty). Subsequent knowledge on the evolution of an underlying stochastic process is not expected to affect neither the risk assessment nor the optimal control.

More rigorously, dynamic time consistency guarantees that investors' preferences implied by a dynamic risk measure remain consistent over time. Some researchers (see [2, 66]) discussed dynamic time consistency in the framework of multi-period coherent risk measures, while the same problem for multi-period convex risk measures is examined in [32, 37]. Similar definitions have also been proposed by Wang [78], Epstein and Zin [35], and Kovacevic and Pflug [49].

Characterizing the relationship of risks across time periods becomes thus an important step when modeling dynamic investment problems. Several notions of time consistency have been proposed in the literature. We review the main results in the following two sub-sections. For given, desirable, risk measures' statistical and mathematical properties the focus is now on the interaction between risk measure evolution in time and portfolio revisions as induced by a given control strategy.

1.2.2.1 Time Consistency of Multi-Period Risk Measures

The concept of dynamic time consistency, as originally investigated by Wang [77], was based on a rather simple insight: given two investment positions A and B, if A is riskier than B under a specific risk measure *at* some future time, then A is riskier than B under the same measure from today's perspective. Such simple idea calls however for the introduction of the notion of risk assessment forward in time under a given probability setting, conditionally on the information available at that time. From an economic viewpoint, the concept intends to rule out the possibility that as we assess a risky event, the risk assessment may change as we move from the future backwards in time. Mathematically:

Dynamic time consistency with respect to terminal payoffs: If, for any $0 \leq \tau < \theta \leq T$ and $X_T, Y_T \in \mathcal{L}_T$, the condition on the *forward risk measures* $\rho_{\theta,T}(X_T) \leq \rho_{\theta,T}(Y_T)$ implies that $\rho_{\tau,T}(X_T) \leq \rho_{\tau,T}(Y_T)$, then the multi-period risk measure $\{\rho_{t,T}\}_{t=0}^T$ is *dynamically time consistent*.

In case of equality [39] the resulting definition means that, if two payoffs carry the same riskiness in every state of nature tomorrow, then the same conclusion can be drawn about their riskiness today. These two definitions are equivalent if a dynamic risk measure satisfies monotonicity and translation invariance [39]. The more general form of time consistency of a risk measure and its dynamic behavior has to do however with the risk evaluation of an underlying random process whose dynamic, as captured by an evolving filtration, is expected not to affect previous estimates. By considering a loss process over an entire investment horizon, Ruszczyński [68] proposed a more general notion of dynamic consistency: if $X_{\theta,T}$ will be at least as good as $Y_{\theta,T}$ from the perspective of some future time θ, and $X_{0,\theta}$ and $Y_{0,\theta}$ are identical between today and θ ($0 < \theta$), then $X_{0,T}$ should not be worse than $Y_{0,T}$ from today's perspective. Extending such concept to any $0 \leq \tau < \theta$, in our notation, we have:

Dynamic time consistent risk measures: Let $0 \le \tau < \theta \le T$ and $X_{\tau,T}, Y_{\tau,T} \in \mathcal{L}_{\tau,T}$ for all τ. If for any τ the conditions

$$X_k = Y_k, k = \tau, \cdots, \theta - 1,$$
$$\rho_{\theta,T}(X_{\theta,T}) \le \rho_{\theta,T}(Y_{\theta,T})$$

imply

$$\rho_{\tau,T}(X_{\tau,T}) \le \rho_{\tau,T}(Y_{\tau,T}),$$

then $\{\rho_{t,T}\}_{t=0}^{T}$ is called *dynamic time consistent.*

Actually, it is proved in [68] that, if a dynamic risk measure satisfies dynamic time consistency, monotonicity and translation invariance, then with respect to either the terminal wealth or the loss process, the resulting multi-period risk measure can be equivalently expressed in the following recursive form:

$$\rho_{\tau,T}(X_{\tau,T}|\mathcal{F}_\tau) = \rho_{\tau,t}\left[X_{\tau,t-1}, \rho_{t,T}(X_{t,T}|\mathcal{F}_t)|\mathcal{F}_\tau\right]$$

for $0 \le \tau < t \le T - 1, \rho_{T,T}(X_T) = X_T$.

In some papers, the above recursive relationship is also used to define time consistency; see, for example, [21, 46].

The research on the time consistency of an optimal strategy can be introduced by focusing on the role played by information processes: for dynamic risk measures the flow of information is captured by the process filtration. In a dynamic risk-control problem such as problem (1.1) a (minimal) information process, whose evolution can not be neglected, is associated with the state variable ξ_t. In general, as shown here below, the *dynamic programming* recursion provides a link between the two concepts of time consistency. From a financial viewpoint the first type of time consistency appears relevant in risk management applications where the evolution over time of a market or credit risk exposure is key to the decision maker. The second type is instead concerned with the effectiveness of a risk *hedging* strategy aimed at controlling dynamically the evolution of such exposure.

1.2.2.2 Time Consistency of Optimal Investment Policies

In a multi-period problem, an investor will determine his investment strategy, or the pre-committed optimal policy [3] over the entire investment horizon, so to optimize a given financial objective. Such policy may be in the form of an optimal decision rule or scenario-tree based optimal contingent plan or in other forms. After solving the *first* problem at $t_0 = 0$, resulting into an optimal strategy over $\mathcal{T} := \{t_0, t_1, \ldots, t_n\}$, when standing at t_1, such pre-commitment may no longer be optimal with respect to the corresponding sub-problem. The time consistency of the optimal policy as determined in t_0 would imply its optimality at later stages

$[t_1, t_n]$. Adopting the notation of problem (1.1), such condition would imply an $\mathcal{F}_t$-measurable $u_t \in \mathcal{U}_t$ leading to a minimum $\rho_{0,T}(X_{0,T})$ for $t = 0, \cdots, T-1$ that would preserve optimality when evaluated at $t = 1$ over the remaining stages: notice that Bellman's optimality principle would lead to such result and indeed this is the approach suggested by several authors [5, 26, 79] to enforce the dynamic consistency of the optimal strategies. In practice, the optimal policy of a multi-period portfolio selection problem may fail to satisfy time consistency. For example, Cuoco et al. [28] proved that an optimal investment policy with multi-period VaR in the objective function is not time consistent. From a methodological viewpoint, as suggested by Boda and Filar [9], an optimal investment policy time consistency needs to be evaluated backwards in time, as from Bellman's principle, and forward in time focusing on the subproblems solutions consistency as information reveals. Under the coherent risk measure framework, it is proved in [2] that, if a multi-period risk measure is time consistent, the resulting optimal investment policy must satisfy the Bellman's optimality principle.

The discussion on risk measures' and optimal control' time consistency may be specialized to discrete probability spaces, as in the case of a random *tree process*, canonical in multi-stage stochastic optimization problems. In [70], the time consistency of an optimal policy is defined with respect to scenarios: at every state, optimality of our decisions should not depend on scenarios which we already know cannot happen in the future. In a probability space generated by a tree data process, the time consistency of a contingency plan must thus be associated with the conditional behavior of the optimal strategy. Focusing on the first stage only for instance: for a given optimal decision tree, as determined at the root node at $t = t_0$ and *conditionally* on one of the children nodes actually realizing at $t = t_1$, the resulting sub-tree strategy, as anticipated at $t = t_0$ should preserve its optimality.

1.2.3 Discussion

The relationship between the two concepts of time consistency, of multi-period risk measures and of optimal risk-control, or investment policies, deserves a comment. In most cases, the time consistency of optimal controls $u_t \in \mathcal{U}_t$ relies on the time consistency of the multi-period risk measure $\rho_{t,T}(X_{t,T})$: as such they jointly lead to a dynamically consistent optimization problem. In the axiomatic theory of dynamic measures, much emphasis is put on the measures dynamic coherence outside an optimal stochastic control framework. Lack of dynamic coherence due to non-convexity, as in the case of Value-at-Risk, is relevant since it will lead as in the static case, to a non convex dynamic program. A dynamically consistent risk *ranking*, furthermore is key to the control problem and can be associated with a risk measures dynamic monotonicity. Chen et al. [15] show that, to ensure that an optimal sub-policy derived from the dynamic programming principle is also optimal over the entire investment horizon, the corresponding conditional risk mapping should satisfy the monotonicity property. Furthermore, Xin et al. [82] examined the relationship between time consistency of robust multi-period risk measures and that of optimal investment policies: if there is at least one optimal policy satisfying the Bellman's principle, the problem is called *weakly* time consistent; if every optimal policy is time consistent, the problem is called *strongly* time consistent.

We have considered above the basic properties of dynamic risk measures as resulting from an extended literature survey. Among those properties, time consistency is the most frequently mentioned in recent years. Its relevance is evident from the perspective of filtration-adapted measures. By introducing different acceptance sets, time consistency can be analyzed under increasing restriction rules. Some of the analyzed properties are dynamic extensions of corresponding static measures' properties: monotonicity, convexity, translation invariance, sub-additivity, and positive homogeneity. While inheriting the meaning of the corresponding static properties, dynamic extensions contain new implications with respect to information processes over the decision horizon. The local and the law-invariance properties as well as information monotonicity are defined in relationship with filtrations evolution, a core concept in (dynamic) information theory. In the following sections, relying on the above theoretical setting, we consider alternative risk measures and specifications of the optimal control problem (1.1) presented in the literature and propose a comprehensive approach to optimal risk control.

1.3 Multi-Period Risk Measures

Existing multi-period risk measures can be classified in three groups: *terminal wealth*, *additive* and *recursive* risk measures. A terminal wealth risk measure can be viewed as a conditional risk mapping with respect to a terminal wealth or equivalently to a total loss cumulated over a finite time horizon. Additive risk measures arise when measuring the risk of losses separately in different periods and then aggregate them. Finally recursive formulations originate from an assessment of the dynamic risk exposure over time and over a given horizon based on a sequence of recursive risk estimates. We see that indeed, the following dynamic measures generic specifications are possible and will ease the analysis that follows:

$$\text{Terminal} \qquad \rho_{t,T}(X_{t,T}) = \mu_t\big(d_t(X_{t,T})|\mathcal{F}_t\big) \tag{1.7}$$

$$\text{Additive} \qquad \rho_{t,T}(X_{t,T}) = \textstyle\sum_{s=t}^{T} \beta^{t,s}\big(\mu_s(d_s(X_{s,T})|\mathcal{F}_t)\big) \tag{1.8}$$

$$\text{Recursive} \; \rho_{t,T}(X_{t,T}) = \mu_t(X_t, \rho_{t+1,T}(X_{t+1,T})) \; t = 0, \cdots, T-1, \tag{1.9}$$

In (1.7) and (1.8), $d_t : \mathcal{L}_{t,T} \rightarrow \mathcal{L}_T$ is a monotonic function called a *cash - flow aggregator* while $\mu_t : \mathcal{L}_T \rightarrow \mathcal{L}_t$ in (1.7) is a risk mapping assumed to satisfy monotonicity and translation invariance and in (1.8) is a conditional risk mapping satisfying the same properties. Cash-flow aggregators are in general simple functions common in financial mathematics to evaluate a stream of cash flows at a given (future) point in time. We may assume that they provide a point estimate of time-varying losses at T. The risk measure μ_t will then generate the risk estimate as from today. A terminal wealth risk measure can be viewed as a direct extension of the static risk measure to the multi-period case. It is easy to understand for investors but can hardly capture dynamic variations of investment risk. The formulation (1.8) requires the additivity of μ_t but allows a more flexible risk estimation process as s varies: to derive a risk estimate, say at time t, all period-wise losses prior and including t are added together and then discounted through the factors $\beta_{t,s}$ for $s = t, t+1, \cdots, T$. The additive formulation may accommodate the previous one, as final term in the summation, but it is more general. The key difference between the two relates to the fact that in the case of terminal risk measures we first aggregate period losses forward in time and then apply the risk measure while under the mapping additivity property we aggregate the risks directly. Most of terminal wealth risk measures are not dynamic time consistent [9, 20, 26, 41].

A direct way to construct additive multi-period risk measures is therefore to add all stage-wise risk measures together, say

$$\rho_{t,T}(X_{t,T}) = \sum_{s=t}^{T} \beta^{t,s} \mu(X_s|\mathcal{F}_t),$$

where $\mu : R \to R$ can be chosen as any terminal wealth risk measure, $\beta^{t,s}$ is a discount factor.

The reason for calling additive this class of measures comes from the linear additivity of terminal measures when $\mu(\cdot)$ is the expectation operator: $E\left(\sum_{s=t}^{T} \beta^{t,s} X_s \middle| \mathcal{F}_t\right) = \sum_{s=t}^{T} \beta^{t,s} E(X_s|\mathcal{F}_t)$. Notice that in most risk control problems we require sub-additivity of the multi-period measure to represent investors' risk aversion. In which case we have the interesting relationship $\mu\left(\sum_{s=t}^{T} \beta^{t,s} X_s \middle| \mathcal{F}_t\right) \leq \sum_{s=t}^{T} \beta^{t,s} \mu(X_s|\mathcal{F}_t)$.

The stage-wise separability of additive risk measures make them suitable for dynamic programming formulations as in [12]. From (1.9), with terminal condition $\rho_{T,T}(X_{T,T}) = X_T$ as t varies, a backward recursion step is defining the risk assessment procedure. Recursive measures are more general than additive ones in the sense that they include most of additive risk measures as special cases. Meanwhile, most of the additive measures in the literature can be reformulated in a recursive form, but not the converse.

Compared to additive risk measures, recursive risk measures are more general. Several types of risk mappings μ indeed allow a recursive separable formulation:

$$\rho_{t,T}(X_{t,T}) = \sum_{s=t+1}^{T} E\big[\mu_s(X_s|\mathcal{F}_{s-1})|\mathcal{F}_t\big]. \tag{1.10}$$

When given this form, we can link easily the risk control problem to Bellman's optimality principle, confident of it's time consistency as further investigated below. Consider in particular

- For $\mu_s(X_s, \rho_{s+1,T}(X_{s+1,T})) = \mu_s(X_s|\mathcal{F}_t) + \beta^{s,s+1}\rho_{s+1,T}(X_{s+1,T})$, $s = t, \cdots, T-1$, where again $\beta^{s,s+1}$ is a discount factor, then the recursive risk measure becomes the classical separable functional $\rho_{t,T}(X_{t,T}) = \sum_{s=t}^{T}(\beta^{t,s}\mu_s(X_s|\mathcal{F}_t))$, proposed in [60];
- For $\mu_t(X_t, \rho_{t+1,T}(X_{t+1,T})) = E[\rho_{t+1,T}(X_{t+1,T})|\mathcal{F}_t], \mu_s(X_s, \rho_{s+1,T}(X_{s+1,T})) = \mu_s(X_s|\mathcal{F}_{s-1}) + E[\rho_{s+1,T}(X_{s+1,T})|\mathcal{F}_s],$, $s = t+1, \cdots, T-1$, $\mu_T(X_T) = \mu_T(X_T|\mathcal{F}_{T-1})$, then the recursive risk measure falls in the class of separable functions (1.10);
- For given specification of a Kreps-Porteus style risk measure $\phi_t(x, y) = \mu_t(x) + y$, we have:

$$\rho_{t,T}(X_{t,T}) = \phi_t\big(X_t, E\big[\rho_{t+1,T}(X_{t+1,T})|\mathcal{F}_t\big]\big),$$

  then it also reduces to a separable expected conditional function of the type (1.10).

We will not examine rigorously the mathematical properties of the above three classes of multi-period risk measures, but just analyze a qualified set of them from the perspective of associated risk control problems. Interested readers can refer to papers like [2, 31] and [37].

1.3.1 *Statistical Estimates of Dynamic Risk Measures*

We summarize next how canonical statistical measures, such as variance, value-at-risk (VaR) and conditional VaR (CVaR), or entropic measures fall in the outlined framework. They may be accommodated within the above three classes and we are interested to their resulting properties can be inferred, mostly in terms of coherence and time consistency.

1.3.1.1 Variance

The terminal variance $var(Z_T|\mathcal{F}_t)$, $Z_T := \sum_{t=1}^{T} X_t$ is defined as

$$var(Z_T|\mathcal{F}_t) = E\big[(Z_T - E(Z_T|\mathcal{F}_t))^2|\mathcal{F}_t\big] = E[Z_T^2|\mathcal{F}_t] - \big(E[Z_T|\mathcal{F}_t]\big)^2.$$

From the properties of the static variance operator, we know that the terminal variance does not satisfy translation invariance, positive homogeneity, monotonicity, and hence convexity, but it is sub-additive. It is not additive The terminal variance time inconsistency is directly derived from the conditional variance property [29]:

$$var(Z_T|\mathcal{F}_t) = E[var[Z_T|\mathcal{F}_\tau]|\mathcal{F}_t] + var[E[Z_T|\mathcal{F}_\tau]|\mathcal{F}_t].$$

The conditional variance in period t is given by the conditional expectation of the conditional variance in period τ and an adjustment term. As the adjustment term depends on the strategy via its behavior not only on $(t, \tau]$, but also on $(\tau, T]$, which restrict the dynamic risk measure to give the same risk preference at both period t and period τ. This leads to the time inconsistency of the terminal variance. Nevertheless the adoption of financial returns' variance as risk measure both in static and dynamic *mean-variance* (MV) trade-off models is widespread and indeed also within dynamic programming formulations. This is primarily due to the attractiveness of the variance as statistical measure to capture financial risk and the natural formulation of the MV dynamic problem as a quadratic programming problem.

1.3.1.2 Value-at-Risk

The terminal VaR can also be defined with respect to the cumulative loss: $VaR_\alpha(Z_T|\mathcal{F}_t)$

$$VaR_\alpha(Z_T|\mathcal{F}_t) = \inf_{z\in\mathcal{L}_t} \left\{ z \,\middle|\, P(Z_T \geq z|\mathcal{F}_t) \leq \alpha \right\},$$

where α is the confidence level and Z_T is the sum of losses from t to T. From the properties of static VaR, as already mentioned, it can be deduced that the terminal VaR satisfies monotonicity, positive homogeneity, and translation invariance. After being introduced as risk management standard for tail risk estimation in several regulatory frameworks (Basel I, II, III for banking intermediaries as well as Solvency I and II for insurance companies) this risk measure has become the core estimation task within the financial industry. Its adoption in risk control applications however has been jeopardized by the violation of the sub-additivity property and, under sufficiently general market assumptions, its' counter-intuitive lack of diversification incentive. It is not a time consistent risk measure either. The introduction of the Conditional VaR, however, has led to a relevant stream of research and applications after the seminal contribution of Rackafellar and Uryasev [64] since unlike VaR, this risk measure is coherent and its control implies the control of the Value at Risk as well. It is defined as the average loss beyond the VaR with a given confidence interval.

1.3.1.3 Conditional Value-at-Risk

The terminal CVaR can also be defined with respect to cumulative losses at the horizon: $CVaR_{\alpha}(Z_T|\mathcal{F}_t)$. CVaR is equal to the optimum value of a linear programming problem:

$$CVaR_{\alpha}(Z) = \inf_{z \in R}\{z + \frac{1}{1-\alpha}E(Z-z)^{+}\}$$

Due to its good mathematical properties, the extension of CVaR to a multi-period setting has received increasing interest in recent years [14, 36, 76]. CVaR is widely adopted for tail risk minimization and it's dynamic properties are determined by the assumptions on the time behavior of the confidence interval: when pre-determined at fixed levels (say 99 %) VaR dynamic inconsistency propagates to CVaR. When instead a random evolution is assumed for α and for all t, $\alpha_t \lhd \mathcal{F}_t$ then the CVaR can be decomposed [59] leading to a time consistent formulation:

$$CVaR_{\alpha_t}(Z_T|\mathcal{F}_t) = \text{ess inf}_{W_\tau} E[W_\tau \cdot CVaR_{\alpha_t \cdot W_\tau}(Z_T|\mathcal{F}_\tau)|\mathcal{F}_t],$$

where the infimum is among all densities $W_\tau \lhd \mathcal{F}_\tau$ with $0 \leq W_\tau, \alpha_t \cdot W_\tau \leq 1$ and $E(W_\tau|\mathcal{F}_t) = 1$.

The CVaR measure can be given an additive formulation:

$$\rho_{t,T}(X_{t,T}) = \sum_{s=t}^{T} \beta^{t,s} CVaR_{\alpha}(X_s|\mathcal{F}_t).$$

Accordingly, the assessment of the tail risk generated by the process $X_{t,T}$ can be attained by summing together the sequence of $\mathcal{F}_t$-measurable conditional VaR's as t varies.

A related but different additive representation of dynamic CVaR focuses on its average behaviour over the risk measurement horizon:

$$\rho_{t,T}(X_{t,T}) = E\left[\sum_{s=t+1}^{T} CVaR_{\alpha_s}(X_s|\mathcal{F}_{s-1})\Big|\mathcal{F}_t\right]. \tag{1.11}$$

After Pflug [60], this type of additive risk measures is called the *separable multi-period acceptability functional.* It is proved in [49] that under this formulation the dynamic CVaR is time consistent and law invariant, and can thus be used to make time consistent investment decisions. We can exploit this separability to yield a CVaR recursive formulation, consistent with the generic form in (1.9). We have:

$$\begin{aligned}\rho_{t,T}(X_{t,T}) =& CVaR_{\alpha_t}\{X_{t+1} + \cdots \\ &+ CVaR_{\alpha_{T-2}}[X_{T-1} + CVaR_{\alpha_{T-1}}(X_T|\mathcal{F}_{T-1})|\mathcal{F}_{T-2}]\cdots|\mathcal{F}_t\}.\end{aligned} \tag{1.12}$$

Notice that the nested structure here above is consistent with a recursive risk evaluation and associated controls. Two interesting bounds on dynamic CVaR have been proposed separately by Pflug and Römisch [60] and Xin and Shapiro [83]. In the first case, still working backwards and exploiting CVaR sequential additivity we have:

$$CVaR_\alpha\left(X_{t+1}+\cdots CVaR_\alpha\left(X_{T-1}+CVaR_\alpha\left(X_T|\mathcal{F}_{T-1}\right)|\mathcal{F}_{T-2}\right)\cdots|\mathcal{F}_t\right)$$
$$\leq \alpha^{-(T-t-1)}E\left[\sum_{s=t+1}^{T} CVaR_\alpha(X_s|\mathcal{F}_{s-1})\Big|\mathcal{F}_t\right].$$

The bound is relatively strict over a limited set of time-periods. Yet in a recursive framework Chen and Liu [14] generalize the recursive CVaR measure to allow conditionality with respect to financial markets regimes under a regime switching framework:

$$\rho_{t-1,T}(Z_T) = E_{M_{t-1}}\left(CVaR_{\alpha_t,M_t}(\rho_{t,T}(Z_T))\right),$$

where M_t denotes a market regime in period t, $E_{M_{t-1}}$ is expectation conditional on being in regime M_{t-1} at $t-1$ and $CVaR_{\alpha_t,M_t}$ is a one-period conditional CVaR function under regime M_t with confidence level α_t. This formulation agrees with (1.12) when conditionality is ruled by transitions across regimes. In case of a single regime in the market we go back to the original formulation.

1.3.2 Coherent and Time Consistent Risk Measures

When a time-inconsistent risk measure is adopted in a multi-stage portfolio selection problem, this will lead in practice to a time-inconsistent strategy. That would imply that for given stochastic assumptions and risk process evolution, a strategy defined as optimal today may no longer be optimal when standing at some time point in the future. Several papers [4, 6, 20, 28, 47, 52, 79] focus on time consistent alternatives or modifications of the terminal variance, VaR and CVaR. More results on this issue are discussed in Sect. 1.4.1, with applications to multi-stage portfolio selection problems.

Several approaches have been studied to derive conditions for dynamic time consistency. Of particular interest are the approaches introducing explicitly some restrictions on the risk measure' probability space, those associated with *distortion* measures and those exploiting the risk measures' conditional separability.

In the first group [63] derives a terminal wealth risk measure by introducing a *sup*-criterion within a set of probability measures:

$$\rho_{t,T}(X_{t,T}) = \max_{P\in\mathcal{P}} E_P\left[\sum_{s=t}^{T}\beta^{t,s}X_s|\mathcal{F}_t\right].$$

Here $\beta^{t,s}$ is a discount factor from s to t, $t \leq s \leq T$. If the set of probability measures $\mathcal{P}$ is closed, convex, and consistent, then the terminal wealth risk measure $\rho_{t,T}(\cdot)$ can be shown to be coherent and dynamically time consistent, and vice versa. A set of probability measures P is consistent if the set is closed under arbitrary pasting of conditional probabilities and marginal distributions from this set. The consistency property for a set of probability measures is also called stability under pasting by Artzner et al. [2]. Delbaen [30, 31] characterizes the consistency in terms of martingale theory.

Distortion measures defined in [16] are generated by mappings $\phi : \mathcal{L}_t \to \mathcal{L}_t$ which combined with $\mu : \mathcal{L}_T \to \mathcal{L}_T$ and the expectation E will determine a risk measure with the required properties: $\rho_{t,T}(Z_T) = \phi\left[E\left(\mu(Z_T)\,|\mathcal{F}_t\right]\right.$, where $Z_T = \sum_{s=t}^{T} X_s$.

An example of terminal distortion measure $\rho_{t,T}(Z_T) = \phi\left[E\left(\mu(Z_T)\right)\right]$ where $\mu : \mathcal{L}_T \to \mathcal{L}_T$ and $\phi : \mathcal{L}_T \to \mathcal{L}_t$ are monotonic functions, is represented by the terminal entropic risk measure. In [18, 21], the following risk measure based on the entropic function is proposed:

$$\rho_{t,T}(Z_T) = \frac{1}{\gamma}\log E\big[e^{\gamma Z_T}|\mathcal{F}_t\big],$$

where $\gamma > 0$ is a constant, typically *individual*-specific risk aversion coefficient. The entropic risk measure $\rho_{t,T}(\cdot)$ is time consistent. Interest in this type of risk measure comes from the possibility to link through the exponential utility the characterization of an individual risk aversion to the definition of the risk measure itself. The entropic measure is convex but not coherent. In this case we have then a dynamically time consistent convex risk measure. By selecting the distortion functions μ and ϕ as convex functions, we can obtain different forms of terminal distortion risk measures which are convex and law invariant. In fact, as proved in [51], the only dynamic risk measure which is convex, law invariant, time consistent and relevant is the entropic risk measure.

Similar to the above terminal distortion risk measures, Chen et al. [16] define an *additive distortion* risk measure as follows:

$$\rho_{t,T}(X_{t,T}) = \sum_{s=t}^{T} \beta^{t,s}\phi\big(E(\mu(X_s)|\mathcal{F}_t)\big). \tag{1.13}$$

Where ϕ and μ are mappings satisfying sub-additivity. The additive distortion risk measure provides an upper bound to the corresponding terminal distortion risk measure: $\phi\left\{E\left[\mu\left(\sum_{s=t}^{T}\beta^{t,s}X_s\right)\middle|\mathcal{F}_t\right]\right\} \leq \sum_{s=t}^{T}\beta^{t,s}\phi\big(E(\mu(X_s)|\mathcal{F}_t)\big)$.

Risk measures additivity may lead to the definition of recursive risk measures suitable for dynamic programming formulations of the associated stochastic control problem, significantly increasing the likelihood of time consistent decision policies. As discussed above, the CVaR is conditionally separable. Another *separable expected conditional* (sec) function which is additive, was proposed in [60]:

$$\rho_{t,T}(X_{t,T}) = \sum_{s=t+1}^{T} E\big[\mu_s\,(X_s|\mathcal{F}_{s-1})\,|\mathcal{F}_t\big].$$

Here μ_s, $t+1 \leq s \leq T$, is a single-period acceptability function, sometimes also referred to as conditional risk mapping from s to $s-1$.

Recursive risk measures have recently attracted a lot of attentions due to their time consistency [68] and intuitive economic interpretation [67]. Their complex structure has to a certain extent limited their adoption in real-world market contexts: curse of dimensionality has also limited the possibility to achieve a realistic cardinality of the investment universe. Even if, due to recent market troubles and industry requirements, relative to one-period static ones, the adoption of dynamic optimization approaches is growing, by now there are only few cases which actually apply recursive risk measures to large-scale multi-stage portfolio selection problems [14, 41, 50, 56].

Risk measures' single and multi-period separability as well as conditional separability lead to dynamic programming formulations of the associated risk control problems [33, 75] which implies the time consistency of the resulting optimal strategy. We present next an example of combined recursive risk measure formulation and associated optimization problems.

Example: Carpentier et al. [12] considered costs of both intermediate periods and the final period, and employed the following dynamic investment policy selection problem:

$$\begin{aligned} &\min_{u_t \in \mathcal{U}_t} E\left[\sum_{t=0}^{T-1} f_t(X_t, u_t, W_{t+1}) + f_T(X_T)\right], \\ &\text{s.t. } X_t = g_t(X_{t-1}, u_{t-1}, \xi_t),\ t = 1, \cdots, T, \end{aligned}$$

where the cost f_t at stage t is a function of the control variable (portfolio) $u_t \in \mathcal{U}_t$, the state variable X_t and the exogenous noise variable ξ_{t+1}, the cost at the final stage f_T is a function of the final stage X_T, and the dynamic relationship between states of two adjacent stages is described by the function g_t. Using the dynamic programming technique, the optimal cost V_t at stage t under the state $X_t = x$ can be found backwardly under the Markovian setting. Concretely, the dynamic equations are as follows:

$$\begin{aligned} V_T(x) &= f_T(x), \\ V_t(x) &= \min_{u_t \in \mathcal{U}_t} E\big[f_t(x, u_t, \xi_{t+1}) + V_{t+1}(f_t(x, u_t, \xi_{t+1}))\big],\ t = 0, 1, \cdots, T-1. \end{aligned}$$

Such a recursive form corresponds to the time consistency of an optimal investment policy [2].

In the final section we analyze several optimization models and summarize, in view of the above discussion, their characterizing elements. Up to this point we have been mainly interested in studying risk measures' theoretical and applied properties and their suitability for the formulation of risk control problems. In Sect. 1.4, we consider instead established stochastic optimization problem formulations and link them to risk measures' properties.

1.4 Dynamic Risk Control and Risk Measures Selection

We consider next the pro's and con's of commonly adopted optimization approaches from the perspective of the measures' coherence and time consistency, thus mainly formulated as stochastic dynamic programming models such as from Eqs. (1.1)–(1.4).

From a financial viewpoint, the two concepts of time consistency are related the first to the very characterization of a random loss process within a given probability space and the second to the properties of the resulting optimal strategy. These are indeed associated with two distinctive areas of financial management: the first with risk management applications—whose relevance over the years has increased due to revisions of regulatory frameworks (e.g Basel I, II, III, Solvency I and II and similar)—and the second with optimal risk control or risk minimisation approaches. In the problem specification (1.1)–(1.4), we have provided a general mathematical framework capturing both areas of interest. The definition of $\rho_{0,T}(X_{0,T})$ is next adapted to different risk-reward optimization paradigms as emerging from an extended literature survey. Accordingly the definition of the process $X_{0,T}$ will vary. The measurability condition $u_t \lhd \mathcal{F}_t$ for $t = 0, 1, \cdots, T-1$ and the decision space characterizations $u_t \in \mathcal{U}_t$ still hold and qualify the type of control problem. Notice that contrary to canonical dynamic stochastic programming formulations, whose outcome is an optimal control in the form of a contingency plan [7] or optimal decision tree, when adopting a recursive dynamic programming formulation u_t takes the form of an optimal policy rule or stochastic control.

We will first review some classic dynamic models, such as mean-variance models, mean-terminal CVaR and mean-additive CVaR models. Due to their time inconsistency, their solution will generate time inconsistent optimal policies, which lead to sub-optimality in intermediate periods. In order to overcome the time inconsistency problem, we introduce two revision methods: one based on the definition of a weak form of time consistency and revising directly the (time inconsistent) investment policy; the other based on new time consistent risk measures to guarantee the time consistency of the optimal investment policy selection model. A set of investment models based on time consistent, and furthermore coherent multi-period risk measures is then introduced. Finally, we discuss some practical issues about problem solutions.

1.4.1 Mean-Variance Models

The seminal work of Markowitz [55] on the single period MV model is the foundation of modern portfolio theory. Merton [57], Perold [58] and Yoshimoto [84] further considered the dynamic MV model and its efficient solution with or without short-selling constraints. In a dynamic setting, the terminal variance has been frequently adopted as risk measure in the associated risk control problem. The decision maker seeks a minimization of the portfolio dispersion around the mean at a given time horizon through a dynamic revision of his investment policy. The resulting optimization problem takes the form of a stochastic quadratic programming problem, whose solution within a dynamic programming formulation, dates back, to the relatively recent contribution by Li and Ng [53]. Including a self-financing constraint, Li and Ng [53] considered the following multi-period problem:

$$\begin{aligned} \max_{u_t} \;& E[W_T] - \lambda Var[W_T] \\ \text{s.t.} \;& W_{t+1} = r_t^0 W_t + R_t' u_t, \; t = 0, 1, \cdots, T-1. \end{aligned}$$

Here the state equation (1.2) is specified as a self-financing constraint with respect to the wealth process $\{W_t\}$, the interest rate process $\{r_t^0\}$ for the riskless asset, the excess return process $\{R_t\}$ of the risky assets, and the portfolio process $\{u_t\}$. The objective function includes two criteria with a trade-off parameter $\lambda \in [0, +\infty)$ between the mean and the variance of the terminal wealth W_T.

In presence of a self-financing constraint, cash inflows and outflows are ruled out over the given horizon and the model is suitable to solve asset pricing and associated hedging problems. In this context it is of interest that the above multi-period MV problem can be embedded into a separable parametric auxiliary problem:

$$\max_{u_t} \; E[-\omega W_T^2 + \lambda W_T] \tag{1.14}$$

$$\text{s.t.} \; W_{t+1} = r_t^0 W_t + R_t' u_t, \; t = 0, 1, \cdots, T-1, \tag{1.15}$$

here λ is the trade-off parameter introduced above and ω is an auxiliary parameter typically needed to rescale the value of the squared wealth process.

Since the expectation operator is separable, the auxiliary problem can be solved by the dynamic programming technique. It is proved in [53] that given the solution u^* of the auxiliary problem, a necessary condition for u^* to be optimal with respect to the original problem is $\lambda = 1+2\omega E[W_T]|_{u^*}$. Therefore, we can obtain the optimal policy of the original problem by utilizing u^*: in a MV setting such policy will do the job of controlling the quadratic error and typically carry a sufficient diversification and dynamic efficiency, in the sense of not being dominated according to portfolio theory, but it will not be time consistent due to the time-inconsistency of the variance.

Recently, adding the no-shorting constraint, thus in more practical terms, the multi-period MV problem has been considered in [27]. Zhu et al. [85] proposed a discrete-time MV model with bankruptcy constraints and derived analytically an optimal investment policy. Besides the dynamic programming technique, the multi-stage stochastic programming method has also been used to solve multi-period allocation problems under the MV framework [24, 43, 62], mostly using stochastic quadratic programming algorithms.

From the perspective of dynamic portfolio theory [25], a key difference in the problem formulation refers to the assumptions on the return generating process. Serially and mutually independent return processes are in general consistent with optimal myopic policies. Financial returns' serial independence is sometimes assumed in multi-period portfolio selection problems and leads to a dynamic programming formulation. In general however, the serial dependence and a multivariate model of financial returns are required in operational context leading to the adoption of multi-period models in general and multi-period MV in particular, when assuming a continuous Gaussian noise process. The recursive dynamic programming formulation requires the underlying return process to be Markovian.

Consider in particular from [10] a dynamic optimization problem where I denotes the set of the states of the market. It is assumed that the market state forms a Markov chain and for each state $i \in I$, the return of a risk-less asset is denoted by $r^0(i)$, while $R_t(i) = (R_t^1(i), \cdots, R_t^n(i))'$ represents a random excess return vector of n risky assets, and $u_t(i) = (u_t^1(i), \cdots, u_t^n(i))'$ denotes the portfolio vector. With these notations, the multi-period MV problem with the dynamics of the market state is formulated as follows.

$$\begin{aligned} \max_{u_t(i)} \quad & E[W_T | i_0 = i^0, W_0 = w^0] - \lambda Var[W_T | i_0 = i^0, W_0 = w^0] \\ \text{s.t.} \quad & W_{t+1}(i) = r^0(i) W_t + R_t(i)' u_t(i), \; t = 0, 1, \cdots, T-1. \end{aligned}$$

Here i^0 and z^0 are the initial state and the initial wealth, respectively. By using a similar approach as that for the multi-period MV problem in [53], an explicit optimal investment policy of the multi-period MV problem is derived in [13]. Wei and Ye [81] further considered a multi-period MV problem in a Markovian market by incorporating a bankruptcy constraint, and obtained the optimal investment policy by using the similar method as above.

Among Markovian decision models, in this context, it is worth recalling the analysis based on market regimes: these models, in [11] for instance, provide relevant generalizations of the above classical MV framework recently formulated and solved as stochastic dynamic programs. Except for considering the Markovian stochastic market, the multi-period optimal investment policy selection models are also examined with the robust counterpart, for example, the multi-period MV model is extended in [42] to the worst-case analysis with multiple return and risk scenarios, which can generate a robust strategy to avoid excessive losses.

Despite the inclusion of no-shorting constraints, bankruptcy constraints, as well as more accurate market models have been considered in a multi-period MV set-up, even in the case of robust representations, these models still suffer from the time-inconsistency of the terminal variance. Moreover, under the normality assumption, even if the risk control is generalized to account for bankruptcy events, the resulting optimal control will still be Gaussian and control of tail risk is attained within the limits of such assumption. From a financial standpoint, the explicit introduction of a tail risk measure such as the conditional value-at-risk overcomes that drawback.

1.4.2 Time Inconsistent Mean-CVaR Models

Due to its tractability, several papers discuss multi-stage portfolio selection problems under the terminal CVaR measure. A generic problem formulation, still following the reward-risk trade-off approach, can be described as follows:

$$\begin{aligned} \max_{u_t \in \mathcal{U}_t} \ & E[W_T] - \lambda CVaR_\alpha(W_T) \\ \text{s.t.} \ & W_{t+1} = r_t^0 W_t + R_t' u_t, \ t = 0, 1, \cdots, T-1. \end{aligned}$$

By linearizing the CVaR, this model can be transformed into:

$$\begin{aligned} \max_{u_t \in \mathcal{U}_t, z \in \mathcal{R}} \ & E[W_T] - \lambda \left(z + \tfrac{1}{1-\alpha} E(W_T - z)^+ \right) \\ \text{s.t.} \ & W_{t+1} = r_t^0 W_t + R_t' u_t, \ t = 0, 1, \cdots, T-1. \end{aligned}$$

In practice, $\mathcal{U}_t$ is a decision space accommodating several types of constraints such as lower and upper bounds on specific asset classes, turnover constraints and others. We assume that the resulting decision space is convex. The stream of contributions employing terminal CVaR as risk measure within dynamic optimization problems is extended since the original contribution by Rockafellar and Uryasev [64]. We consider here only the subset resulting into recursive dynamic programming formulations. In a stochastic programming framework, Topaloglou et al. [76] considered a two-stage international portfolio management model by minimizing the terminal CVaR. When extending to a multi-stage model, the resulting stochastic programming formulation suffers the curse of dimensionality for increasing stages and scenario paths. To overcome such drawback, a different CVaR model is proposed in [67], where the investment risks of two adjacent periods are defined dynamically, hence, the corresponding multi-stage control problem could be solved relying on a dynamic programming recursion.

Because of its good properties, there are many papers discussing multi-period portfolio selection problems under different variants of the additive CVaR measure. When adopted to control multi-period investment risk, the corresponding optimal investment problem can be formulated as:

$$\begin{aligned} \max_{u_t \in \mathcal{U}_t} \ & E[W_T] - \textstyle\sum_{t=1}^{T} \lambda_t CVaR_\alpha(W_t) \\ \text{s.t.} \ & W_{t+1} = r_t^0 W_t + R_t' u_t, \ t = 0, 1, \cdots, T-1, \end{aligned}$$

here $CVaR_\alpha(W_t)$, $1 \le t \le T$, in the objective function can also be linearized [34, 64].

As for the solution method, Fábián [36] designed a decomposition algorithm based on the penalty function to solve a multi-stage portfolio selection problem with additive CVaR measure. By using the dual representation of the additive CVaR measure, the *stochastic dual dynamic programming* (SDDP) solution method in [61] can be applied to solve the problem. Although the dual form of the additive CVaR measure is more tractable, for large and realistic problems the efficiency of the solution method must still be assessed. Besides, current numerical results on multi-period risk control problems under additive risk measures [45, 61, 76] are still limited to problems of unrealistic dimension. Efficient decomposition algorithms are still needed in this context.

Additive risk measures can control period-wise risks over the investment horizon. However, additive risk measures are simple extensions of terminal risk measures. The resulting optimal investment policy still suffers the time inconsistency problem. Since all the models in the two sub-sections leads to a time inconsistent policy, it becomes interesting to investigate the impact of time consistency and consider how to ensure it.

1.4.3 Time Inconsistency and Time Consistent Revisions

When a time *inconsistent* risk measure is adopted in a dynamic problem, this will lead in practice to a time inconsistent strategy: the focus is on the relationship between future decisions regarded as optimal today and decisions that will be derived as optimal when formulating and solving the problem in the future. Maintaining the same probabilistic assumptions on the underlying stochastic processes.

In principle when time goes by it is by no mean sure that, say 1 month from now, the risk manager will face the same market conditions as those expected today at the end of the month. But assuming that this will actually be the case and indeed relying on the market condition that will materialize in 1 month time, then the pre-committed strategy optimality should still hold. From this perspective, *ex-post*, it is possible to evaluate the cost of inconsistency by analyzing the period-wise divergence between the value functions and the optimal policies as determined *ex-ante* along specific market scenarios and *ex-post* once those scenarios have realized. In either cases the problems are solved under uncertainty on the future market evolution [41, 67].

In detail, Rudloff et al. [67] quantified the impact of time inconsistency by computing a related sub-optimality gap defined by the difference in the objective function value as evaluated using planned (thus ex-ante) and implemented (ex-post) policies, respectively. By adopting a CVaR model, they perform sensitivity analysis on the sub-optimality gap for different planning horizon and risk-aversion levels: they conclude that the cost of inconsistency depends on the assumed market scenario with increasing marginal costs as we move from extreme to average scenarios. Several papers [15, 17, 50], furthermore, suggest a comparison between the market performances of a time inconsistent against a time consistent policy, reporting that, *ceteris paribus*, consistently over time the latter tends to overperform the former under the same probabilistic assumptions.

To overcome the time inconsistency problem, several authors [4, 6, 20, 28, 47, 52, 79] focus on time consistent alternatives or modifications of the above models. Generally speaking two perspectives can be adopted: either considering directly the optimal policies and enforcing conditions for time-consistency or introducing modifications of the risk measures and then deriving time-consistent policies. Time consistent revisions of optimal investment policies will be discussed in what follows. Dynamic control models based on time consistent risk measures will be discussed in the next subsection.

In presence of a time inconsistent risk measure such as the variance, a relatively acceptable and commonly used approach to enforce a time consistency principle amounts to relaxing the conditions for (strong) time consistency. Weaker formulations are thus introduced. Cui et al. [26], for instance, propose a weak time consistency based on Bellman's optimality principle where the trade-off parameter between risk and return can change over time still within a mean-variance framework. They show that a pre-committed optimal policy satisfies the introduced weak time consistency in any intermediate period as long as the investor's wealth path falls in a compact state space. If, instead, there is a positive probability that the wealth will exceed a threshold, the weak time consistency will no longer hold and the consistency may be recovered by allowing withdrawal of monetary surpluses from the wealth. Such revision on optimal policy is rather intuitive: when the time consistency holds, we directly use the pre-committed optimal policy; when it doesn't, we revise the pre-committed optimal policy by withdrawing the money from the market.

Another interesting way to derive, from an otherwise inconsistent measure, a time consistent optimal policy is to allow the risk measures' parameters to change dynamically over time. For instance, the nested decomposition method for terminal CVaR recently proposed by Pflug and Pichler [59] considers a time-varying random of confidence level. Such nested decomposition method can help the investor to adopt time-consistent strategies under the terminal CVaR criterion. Meanwhile, it may bring mathematical tractability for complex multi-stage risk-averse problems.

1.4.4 Time Consistent Models

The time inconsistency issue can be better dealt with by modifying directly the risk measurè or, even better, adopting new time-consistent and coherent dynamic risk measures. Based on the concept of separable expected conditional mapping, a time consistent multi-period MV model was introduced in [15]:

$$\begin{aligned} \min_{u_t \in \mathcal{U}_t} \quad & \sum_{t=1}^{T} E[\rho_{t-1}^{MV}(W_t)] \\ \text{s.t.} \quad & W_{t+1} = r_t^0 W_t + R_t' u_t, \ t = 0, 1, \cdots, T-1. \end{aligned}$$

where $\rho_{t-1}^{MV}(W_t) = Var(W_t|\mathcal{F}_{t-1}) - \lambda_t E[W_t|\mathcal{F}_{t-1}]$, and λ_t is a period-wise risk-averse parameter which can reflect an investor's changing risk attitude. An optimal solution to the above problem can be found by adopting the dynamic programming technique. Numerical experiments in [15] support the intuition that an optimal policy obtained under the above time consistent MV model is superior to the one presented by Li and Ng [53] when neglecting the time-consistency issue. Out-of-sample results are collected analyzing average returns, returns variance and mean/variance ratios generated by the resulting optimal policies and the evidence is consistent. Although the time consistency is satisfied under such a time-varying combination of mean and variance, the coherence is still missing due to the variance lack of sub-additivity.

To yield an optimal control of extreme losses, it is worthwhile to consider the case of CVaR- or other downside risk measures-based approaches. Limited by their formulations, additive risk measures could not fully reflect the dynamic dependence of random information across periods, so a recursive risk measure appears necessary.

Using CVaR as reference risk measure, Kozmík and Morton [50] consider a risk-averse multi-period investment problem with the following nested formulation:

$$\min_{u_1} c_1^\top u_1 + \rho_2^{CVaR}\left[\min_{u_2} c_2^\top u_2 + \cdots + \rho_T^{CVaR}\left[\min_{u_T} c_T^\top u_T\right]\right],$$

where the first and the t-th period ($2 \leq t \leq T$) minimizations are constrained by $A_1 u_1 = b_1, \ u_1 \geq 0$ and $A_t u_t = b_t - B_t u_{t-1}, \ u_t \geq 0, \ t = 2, \cdots, T$, respectively. Here, ρ_t is a weighted sum of the conditional expectation and CVaR associated with a random loss Z:

$$\rho_t^{CVaR}(Z) = (1 - \lambda_t)E[Z|\mathcal{F}_{t-1}] + \lambda_t CVaR_{\alpha_t}[Z|\mathcal{F}_{t-1}], \ 2 \leq t \leq T,$$

with a weighting factor λ_t ($2 \leq t \leq T$). Under the assumption that the underlying random process is serially independent, the SDDP algorithm is used in [50] to solve the problem.

Under very risky market conditions $\lambda_t = 1, \forall t$, the resulting optimal control will minimize jointly the CVaR and the VaR under rather general assumptions on the underlying market dynamics. Alternatively for $\lambda_t \to 0$, more risky strategies will be attained. In either cases, the resulting strategy will be time consistent and as time progresses, will persist *along* an optimal path as determined at the initial time.

As a further example of risk control under complex market dynamics, we consider the recursive CVaR minimization model presented in [14] where a regime switching technique, a time series model and a multi-factor model are simultaneously utilized to capture asymmetric and leptokurtic market movements observed in the market. The resulting multi-period portfolio selection problem can be described as follows:

$$\begin{aligned}
&\max_{u_t \in \mathcal{U}_t} E(W_T) \\
&\text{s.t. } E_{M_0}[CVaR_{\alpha_1,M_1}(\cdots E_{M_{T-1}}[CVaR_{\alpha_T,M_T}(W_T)]\cdots)] \le \delta, \\
&\quad\; W_{t+1} = r_t^0 W_t + R_t^{'} u_t,\ t = 0,1,\cdots,T-1,
\end{aligned}$$

here M_t denotes the market regime in period t, which follows a Markovian process, u_t is an M_t-adapted portfolio and $CVaR_{\alpha_t,M_t}$ is the one-period conditional CVaR function under regime M_t with the confidence level being α_t. Under the proposed two-level information structure, the above multi-stage portfolio selection problem can be formulated as a second order cone programming problem, which can be efficiently solved in polynomial time. Numerical experiments show the proposed multi-period risk function is good at balancing risks among stages and the performance of the optimal investment policy is stable as the regime switches.

From the above nested CVaR examples, we can find that the recursive form of a dynamic risk measure and the coherence of the adopted static measure guarantee the time consistency of the resulting dynamic risk measure. Inspired by this, we can deduce a generic form of an optimal control model based on a recursive risk measure as follows:

$$\min_{u_t \in \mathcal{U}_t} \left\{\rho_1[\cdots \rho_{T-1}[\rho_T[W_T|\mathcal{F}_{t-1}]|\mathcal{F}_{t-2}]\cdots]\right\} \tag{1.16}$$

$$\text{s.t. } W_{t+1} = r_t^0 W_t + R_t^{'} u_t,\ t = 0,1,\cdots,T-1. \tag{1.17}$$

In (1.16) a nested sequence of conditional risk measures $\rho_1 \otimes \rho_2 \otimes \ldots . \otimes \rho_T$ ensures the suitability of a recursive approach. A sequence of coherent ρ_t $(1 \le t \le T-1)$ will lead to a recursive dynamic programming formulation and *hence* to a time consistent risk control policy. Matmoura and Penev [56], for instance, considered such a model by using a higher moment coherent risk measure to control tail losses:

$$\rho_t(X|\mathcal{F}_{t-1}) = \inf_{Y \in \mathcal{L}_{t-1}} \left\{ Y + \frac{1}{\alpha} E\left[(-X-Y)_+^p \,|\mathcal{F}_{t-1}\right]^{1/p} \right\},\ 1 \le t \le T-1,$$

where $\alpha \in (0,1)$ and $p \ge 1$. The authors introduced a recursive algorithm to solve the above nested optimization problem, and provide an optimal investment policy which is more risk-averse under increasing ps than that obtained with CVaR.

Thanks to its dynamic time consistency, a multi-period optimal investment policy determined under a recursive risk measure usually has superior and robust performance, as available empirical results in [14, 56, 56] show. However, the implementability and computational performance of this class of problems varies greatly depending on the adopted information structure, the assumptions on the underlying wealth/loss process, and the suitability of the associated risk measure. From both an implementation and methodological viewpoint, the adoption of simulation and bootstrapping methods to validate risk measurement approaches and refine a stochastic program in the generic form (1.1)–(1.4) is increasing [14, 34]. Indeed as time progresses $t = 0, 1, \cdots, T$ yet assuming only $\mathcal{F}_t$-measurable controls u_t and for given state-space characterization and optimal control u_t, (1.1)–(1.4) describes a stochastic system that can be fully simulated and the resulting properties of the optimal risk measure trajectory and control considered from the perspective of their time consistency to facilitate their practical adoption.

1.4.5 Practical Solution Methods for Optimal Dynamic Risk Control

We see from the above review that the dynamic programming technique provides a suitable approach to solve financial planning problems carrying a recursive risk structure and typically with a feasible region generated by a Markovian market process and a not too extended decision universe. Dynamic programming is used for instance, combined with an embedding technique, to solve a multi-period MV problem first in [53] and then in [4, 10, 26, 27]. Besides the MV problem, an optimal investment policy model with bankruptcy constraints was solved analytically through the dynamic programming technique in [54, 85] by applying Chebyshev inequality. Extremely large even if sparse and linear constraint operators may easily jeopardize the use of dynamic programming. In such cases, dynamic stochastic programming appears the only viable alternative.

Other than stochastic dynamic programming backward recursion, there are mainly two solution methods for multi-period risk control problems under more complex risk measures: the SDDP algorithm and those algorithms based on the scenario tree technique.

The adoption of the SDDP method requires in general that the underlying statistical processes are stage-wise independent and can only solve multi-period problems under additive or recursive risk measures with specific structures. By using the dual representation of the additive CVaR, for instance, SDDP is applied in [61] to solve a multi-period portfolio selection problem under the additive CVaR measure.

By assuming that the underlying random process is stage-wise independent, the SDDP algorithm is also used in [50] to solve a multi-period stochastic program with risk control under the recursive CVaR measure. The theoretical foundation for the application of SDDP algorithm is the time consistency of the dynamic risk measure, which guarantees the satisfaction of Bellman's optimality principle. However, Philpott and de Matos showed in [61] that the Bellman equations for the multi-period portfolio selection problem under the additive CVaR measure can not be solved explicitly, but could be solved through linear programming. Correspondingly, for multi-period problems under other coherent risk measures, the

Bellman equations are associated with convex programming problems. Hence, when dealing with decomposable recursive formulations of an optimal control problem such as (1.16), in presence of coherent risk measures, an approximation by cutting planes can be performed and it is expected to lead to convergence but sometimes at a relevant computational cost to compute the value function bounds. Several researchers have investigated the application of the SDDP algorithm in multi-period stochastic programs [45, 50, 61].

In the meanwhile, solution approaches based on discrete scenario tree representations have also been widely used to solve multi-period problems based on dynamic risk measures [7, 22, 24, 43, 62, 76]. The stream of contributions in this area is wide. We are here interested only on the subset of optimization approaches applied to multi-period problems with coherent or time consistent risk measures.

For example, Topaloglou et al. [76] considered a two-stage international portfolio management model by minimizing the terminal CVaR. Due to the coherency and piece-wise linearity of CVaR, the model can be reformulated as a linear programming problem relying on a discrete scenario tree process and hence can be efficiently solved in polynomial time. However, due to the time inconsistency of terminal CVaR, the time consistency of the optimal strategy can not be guaranteed here. If time consistent risk measures, the recursive CVaR in [14] for example, are adopted, the resulting portfolio selection problems relying on the discrete scenario tree would preserve the time-consistency of the optimal strategy. When CVaR is replaced by other nonlinear coherent risk measures, the portfolio selection problems relying on the scenario tree can be formulated as large-scale convex programming problems. To solve these programming problems, we can utilize the Markovian block-diagonal structure of the scenario tree based model and adopt some scenario decomposition methods like those in [23, 44].

There is still work to do with the scenario tree technique when applied to risk-averse portfolio selection problems. The search for coherence and time-consistency should not jeopardize a computationally efficient solution of the problem, particularly in consideration of the required approximation steps induced to facilitate numerical tractability. Additional evidence is in this respect needed, also taking into account solution approaches based on decision rules o approximate dynamic programming. The use of robust optimization techniques as in Shen and Zhang [74] has also been shown to provide a good approach to cope with scenario tree instability and possible inconsistency of the underlying risk measure. Furthermore, the robust optimization technique can be adopted to describe the ambiguity of the distribution, which leads to the distributionally robust optimization. In this respect, an adjustable robust approach is proposed in [8] to solve the optimal investment policy selection model under a recursive risk measure, which allows one to solve the multi-period robust problem in a computationally efficient way by adapting the dynamic programming technique [71]. See on this issue the Editors' survey chapter in this volume.

In a word, the dynamic programming technique is helpful for finding the analytical optimal solution of a multi-period risk control problem. However, it relies on rather strict conditions to satisfy the principle of optimality. In contrary, the scenario tree technique can be flexibly adopted to solve complex multi-period investment policy selection problems. However, the computation complexity caused by the curse of dimensionality is a main obstacle to apply the scenario tree technique to an investment problem with many assets and/or many periods. SDDP, a newly proposed algorithm as a compromise between the dynamic programming technique and the scenario tree technique, shows its efficiency in solving many multi-period stochastic optimization problems. However, the research on SDDP is still far from complete.

1.5 Conclusions and Future Research

Due to its theoretical and practical importance, the construction of a multi-period risk measure and its application to multi-stage portfolio selection problems have attracted much attention in the last decade, and has become a hot topic in operations research and financial management. Similar to traditional utility functions, a multi-period risk measure describes the influence of an uncertain loss process on investors' risk attitudes. The construction of multi-period risk measures is however rather demanding from both theoretical and practical viewpoints. In this chapter we have conducted a methodological survey of the state-of-the-art on dynamic risk control optimization problems, by focusing on a qualified, maybe not exhaustive, set of multi-period risk functions, whose theoretical and applied properties have been considered. The concept of risk measures and optimal control time-consistency has emerged in recent studies as the characterizing feature to be considered within dynamic risk control problems as those studied here. We have started in Sect, 1.2, linking the analysis to the axiomatic theory of static risk measures and their extension into a dynamic framework.

In Sect. 1.3 existing multi-period risk measures have been classified into three categories: terminal wealth risk measures, additive risk measures, and recursive risk measures. Among them, the terminal wealth risk measure is the closest to the traditional static risk measure and it has been investigated and applied extensively in risk control and more general financial management problems. However, the optimal strategy obtained by controlling the terminal wealth risk has been shown not to be time consistent in general. Such evidence has led to increasing research efforts focusing on additive and recursive risk measures and their adoption within dynamic risk control problems. We have seen that additive risk measures, once the risk function additivity is satisfied, are typically easy to compute and handle in risk control problems by using the dynamic programming technique. Nevertheless, additivity restricts possible formulations of a multi-period risk measure by fixing the relationship between risk functions of adjacent periods and in doing so may not be suitable to reflect evolving risk attitudes in full generality. Recursive risk measures have thus been considered specifically as a natural way to handle time consistent measures in sufficiently general risk control problems. Unfortunately, the calculation and application of recursive measures within dynamic optimization problems has been shown also to lead to complex numerical issues. To the best of our knowledge the introduction of coherent and time-consistent risk measures in dynamic control problems represents a current area of research with open issues, as it is the production of a sufficiently robust numerical evidence in support of such theoretical developments.

The following open issues are worth remarking:

- As we know, time consistency is an important property for multi-period risk measures. Nevertheless, some authors [65, 80] argued that dynamic time consistency is a very strong assumption. Then, how to introduce a proper form of weak time consistency so that it can not only characterize suitably the relationship of risks across periods, but ensure a resulting optimal investment policy satisfying the basic consistency requirement? Also, given a maybe weak form of time consistent measure in the problem, what are the implications of adopting a discrete rather than a continuous probability space characterization?
- Under a coherent risk measure framework, strong time consistency of a multi-period risk measure leads to the time consistency of the optimal investment strategy. However, more generally, is weak-time consistency sufficient to such purpose? Theoretical and computational validation on dynamic control problems is needed maybe with a focus on the *cost of time-inconsistency* relying on associated statistical measures.
- When dealing with dynamic risk measures, the strong operational link between risk measurement and risk management applications cannot be overlooked: this is key to the practical adoption of time-consistent risk measures (instead of maybe time-inconsistent measures currently adopted as the Value-at-Risk). The effectiveness of both the risk assessment step, relevant also for regulatory purposes, and the risk hedging step, relevant for the internal risk manager, must still be established and appears necessary in financial applications. From a methodological viewpoint this is expected to facilitate simulation- and optimization-based joint methods.
- The impact of discrete representations and approximation of dynamic risk measures whose properties have been established assuming continuous probability spaces as the interplay between solutions of stochastic dynamic programming problems in recursive form and associated tree representations o robust representations are also open issues.

More generally the establishment of specific properties of dynamic risk measures, maybe resulting in the proposal of new risk measures, along with their introduction in multistage stochastic control problems appears strongly motivated in practice by continuously evolving regulatory frameworks and persistent instability conditions in financial markets.

Acknowledgements The authors are grateful to two anonymous reviewers for their detailed and stimulating comments, which have helped us to improve the chapter significantly in both content and style. This research was supported by the National Natural Science Foundation of China (Grant Numbers 70971109, 71371152 and 11571270).

References

1. P. Artzner, F. Delbaen, J.M. Eber, D. Heath, Coherent measures of risk. Math. Financ. **9**(3), 203–228 (1999)
2. P. Artzner, F. Delbaen, J.M. Eber, D. Heath, H. Ku, Coherent multi-period risk adjusted values and Bellman's principle. Ann. Oper. Res. **152**, 5–22 (2007)
3. S. Basak, G. Chabakauri, Dynamic mean-variance asset allocation. Rev. Financ. Stud. **23**(8), 2970–3016 (2010)
4. S. Basak, A. Shapiro, Value-at-risk based risk management: optimal policies and asset prices. Rev. Financ. Stud. **14**(2), 371–405 (2001)
5. B. Bäuerle, A. Mundt, Dynamic mean-risk optimization in a binomial model. Math. Meth. Oper. Res. **70**, 219–239 (2009)
6. A. Berkelaar, P. Cumperayot, R.R.P. Kouwenberg, The effect of VaR-based risk management on asset prices and volatility smile. Eur. Financ. Manag. **8**, 65–78 (2002)
7. M. Bertocchi, G. Consigli, M.A.H. Dempster, *Handbook on Stochastic Optimization Methods in Finance and Energy: New Financial Products and Energy Market Strategies*. Fred Hillier International Series in Operations Research and Management Science (Springer, New York, 2011)
8. A. Ben-Tal, L. El Ghaoui, A. Nemirovski, *Robust Optimization* (Princeton University Press, Princeton, 2009)
9. K. Boda, J.A. Filar, Time consistent dynamic risk measures. Math. Meth. Oper. Res. **63**, 169–186 (2006)
10. U. Çakmak, S. Özekici, Portfolio optimization in stochastic markets. Math. Meth. Oper. Res. **63**, 151–168 (2006)
11. E. Çanakoğlu, S. Özekici, Portfolio selection in stochastic markets with exponential utility functions. Ann. Oper. Res. **166**, 281–297 (2009)
12. P. Carpentier, J.P. Chancelier, G. Cohen, M.D. Lara, P. Girardeau, Dynamic consistency for stochastic optimal control problems. Ann. Oper. Res. **200**(1), 247–263 (2012)
13. U. Çelikyurt, S. Özekici, Multiperiod portfolio optimization models in stochastic markets using the mean-variance approach. Eur. J. Oper. Res. **179**, 186–202 (2007)
14. Z.P. Chen, J. Liu, Time consistent risk measure under two-level information structure and its application in dynamic portfolio selection. Optimization Online (2015). http://www.optimization-online.org/DB_FILE/2015/05/4906.pdf
15. Z.P. Chen, G. Li, J.E. Guo, Optimal investment policy in the time consistent mean-variance formulation. Insur. Math. Econ. **52**, 145–156 (2013)
16. Z.P. Chen, J. Liu, G. Li, Z. Yan, Composite time consistent multi-period risk measure and its application in optimal portfolio selection. Working paper, Xi'an Jiaotong University (2015)
17. Z.P. Chen, J. Liu, G. Li, Time consistent policy of multi-period mean-variance problem in stochastic markets. J. Ind. Manag. Optim. **12**(1), 229–249 (2016)
18. P. Cheridito, M. Kupper, Recursiveness of indifference prices and translation-invariant preferences. Math. Finan. Econ. **2**, 173–188 (2009)
19. P. Cheridito, M. Kupper, Composition of time consistent dynamic monetary risk measures in discrete time. Int. J. Theor. Appl. Financ. **14**(1), 137–162 (2011)
20. P. Cheridito, M. Stadje, Time-inconsistency of VaR and time consistent alternatives. Finance Res. Lett. **6**(1), 40–46 (2009)
21. P. Cheridito, F. Delbaen, M. Kupper, Dynamic monetary risk measures for bounded discrete-time processes. Electron. J. Probab. **11**, 57–106 (2006)
22. J. Claro, J.P. de Sousa, A multiobjective metaheuristic for a mean-risk multistage capacity investment problem with process flexibility. Comput. Oper. Res. **39**(4), 838–849 (2012)
23. R. Collado, D. Papp, A. Ruszczyński, Scenario decomposition of risk-averse multistage stochastic programming problems. Ann. Oper. Res. **200**, 147–170 (2012)
24. G. Consigli, M.A.H. Dempster, Dynamic stochastic programming for asset-liability management. Ann. Oper. Res. **81**, 131–161 (1998)

25. G.M. Constantinides, A.G. Malliaris, Portfolio theory, in *Handbooks on Operations Research and Management Science*, vol. 9, ed. by R. Jarrow et al. (Elevier Science, Amsterdam, 1995)
26. X.Y. Cui, D. Li, S.Y. Wang, S.S. Zhu, Better than dynamic mean-variance: time inconsistency and free cash flow stream. Math. Financ. **22**(2), 346–378 (2012)
27. X.Y. Cui, J.J. Gao, X. Li, D. Li, Optimal multi-period mean-variance policy under no-shorting constraint. Eur. J. Oper. Res. **234**, 459–468 (2014)
28. D. Cuoco, H. He, S. Isaenko, Optimal dynamic trading strategies with risk limits. Oper. Res. **56**(2), 358–368 (2008)
29. C. Czichowsky, Time-consistent mean-variance portfolio selection in discrete and continuous time. Financ. Stoch. **17**, 227–271 (2013)
30. F. Delbaen, The structure of m-stable sets and in particular of the set of risk-neutral measures. Preprint, ETH Zürich (2003)
31. F. Delbaen, W. Schachermayer, *The Mathematics of Arbitrage* (Springer, Berlin, 2006)
32. K. Detlefsen, G. Scandolo, Conditional and dynamic convex risk measures. Financ. Stoch. **9**(4), 539–561 (2005)
33. D. Duffie, *Security Markets: Stochastic Models* (Academic Press, Boston, 1998)
34. J. Dupacŏvá, V. Kozmík, Structure of risk-averse multistage stochastic programs. OR Spectr. **37**, 559–582 (2015)
35. L.G. Epstein, S.E. Zin, Substitution, risk aversion, and the temporal behavior of consumption and asset returns: a theoretical framework. Econometrica **57**(4), 937–969 (1989)
36. C.I. Fábián, Handling CVaR objectives and constraints in two-stage stochastic models. Eur. J. Oper. Res. **191**(3), 888–911 (2008)
37. H. Föllmer, I. Penner, Convex risk measures and the dynamics of their penalty functions. Stat. Decis. **24**(1), 61–96 (2006)
38. H. Föllmer, A. Schied, Convex risk measures, in *Encyclopedia of Quantitative Finance*, ed. by R. Cont (Wiley, New York, 2010), pp. 355–363
39. M. Frittelli, E. Rosazza Gianin, Dynamic convex risk measures, in *Risk Measures for the 21st Century*, ed. by G. Szegö (Wiley, Chichester, 2004), pp. 224–248
40. M. Frittelli, G. Scandolo, Risk measure and capital requirements for processes. Math. Financ. **18**(4), 589–612 (2006)
41. H. Geman, S. Ohana, Time-consistency in managing a commodity portfolio: a dynamic risk measure approach. J. Bank. Financ. **32**(10), 1991–2005 (2008)
42. N. Gülpinar, B. Rustem, Worst-case robust decisions for multi-period mean-variance portfolio optimization. Eur. J. Oper. Res. **183**, 981–1000 (2007)
43. N. Gülpinar, B. Rustem, R. Settergren, Multistage stochastic programming in computational finance, in *Computational Methods in Decision-Making, Economics and Finance*, ed. by E.J. Kontoghiorghes, B. Rustem, S. Siokos (Springer, New York, 2002), pp. 35–47
44. S. Gülten, A. Ruszczyński, Two-stage portfolio optimization with higher-order conditional measures of risk. Ann. Oper. Res. **229**, 409–427 (2015)
45. T. Homem-de-Mello, B.K. Pagnoncelli, Risk aversion in multistage stochastic programming: a modeling and algorithmic perspective. Eur. J. Oper. Res. (2015). doi:doi:10.1016/j.ejor.2015.05.048
46. L. Jobert, L.C.G. Rogers, Valuations and dynamic convex risk measures. Math. Financ. **18**, 1–22 (2008)
47. P.R. Kleindorfer, L. Li, Multi-period VaR-constrained portfolio optimization with applications to the electric power sector. Energy J. **26**(1), 1–26 (2005)
48. S. Kloppel, M. Schweizer, Dynamic indifference valuation via convex risk measures. Math. Financ. **17**(4), 599–627 (2007)
49. R. Kovacevic, G.C. Pflug, Time consistency and information monotonicity of multiperiod acceptability functionals. Radon Ser. Comput. Appl. Math. **8**, 1–24 (2009)
50. V. Kozmík, D.P. Morton, Evaluating policies in risk-averse multi-stage stochastic programming. Math. Prog. **152**, 275–300 (2015)
51. M. Kupper, W. Schachermayer, Representation results for law invariant time consistent functions. Math. Financ. Econ. **2**, 189–210 (2009)

52. M. Leippold, F. Trojani, P. Vanini, Equilibrium impact of value-at-risk regulation. J. Econ. Dyn. Control **30**, 1277–1313 (2006)
53. D. Li, W.L. Ng, Optimal dynamic portfolio selection: multi-period mean-variance formulation. Math. Financ. **10**, 387–406 (2000)
54. D. Li, T.F. Chan, W.L. Ng, Safety-first dynamic portfolio selection. Dyn. Contin. Discrete Impuls. Syst. **4**(4), 585–600 (1998)
55. H. Markowitz, Portfolio selection. J. Financ. **7**, 77–91 (1952)
56. Y. Matmoura, S. Penev, Multistage optimization of option portfolio using higher order coherent risk measures. Eur. J. Oper. Res. **227**, 190–198 (2013)
57. R.C. Merton, Lifetime portfolio selection under uncertainty: the continuous time case. Rev. Econ. Stat. **51**, 247–257 (1969)
58. A.F. Perold, Large-scale portfolio optiminization. Manag. Sci. **30**, 1143–1160 (1984)
59. G.C. Pflug, A. Pichler, Time-inconsistent multistage stochastic programs: martingale bounds. Eur. J. Oper. Res. (2015). doi:10.1016/j.ejor.2015.02.033
60. G.C. Pflug, W. Römisch, Modeling, in *Measuring and Managing Risk* (World Scientific, Singapore, 2007)
61. A.B. Philpott, V.L. de Matos, Dynamic sampling algorithms for multi-stage stochastic programs with risk aversion. Oper. Res. Lett. **218**(2), 470–483 (2012)
62. A. Prékopa, *Stochastic Programming* (Akademiai Kiado, Budapest, 1995)
63. F. Riedel, Dynamic coherent risk measures. Stoch. Proc. Appl. **112**, 185–200 (2004)
64. R.T. Rockafellar, Uryasev, S.P.: Conditional value-at-risk for general loss distributions. J. Bank. Financ. **26**, 1443–1471 (2002)
65. B. Roorda, J.M. Schumacher, Time consistency conditions for acceptability measures, with an applications to Tail value at risk. Insur. Math. Econ. **40**, 209–230 (2007)
66. B. Roorda, J.M. Schumacher, J.C. Engwerda, Coherent acceptability measures in multiperiod models. Math. Financ. **15**, 589–612 (2005)
67. B. Rudloff, A. Street, D. Valladão, Time consistency and risk averse dynamic decision models: Definition, interpretation and practical consequences. Eur. J. Oper. Res. **234**, 743–750 (2014)
68. A. Ruszczyński, Risk-averse dynamic programming for Markov decision processes. Math. Prog. Ser. B **125**, 235–261 (2010)
69. A. Ruszczyński, A. Shapiro, Conditional risk mappings. Math. Oper. Res. **31**, 544–561 (2006)
70. A. Shapiro, On a time consistency concept in risk averse multistage stochastic programming. Oper. Res. Lett. **37**, 143–147 (2009)
71. A. Shapiro, A dynamic programming approach to adjustable robust optimization. Oper. Res. Lett. **39**, 83–87 (2011)
72. A. Shapiro, Minimax and risk averse multistage stochastic programming. Eur. J. Oper. Res. **219**, 719–726 (2012)
73. A. Shapiro, D. Dentcheva, A. Ruszczyński, *Lectures on Stochastic Programming: Modeling and Theory* (SIAM, Philadelphia, 2009)
74. R.J. Shen, S.Z. Zhang, Robust portfolio selection based on a multi-stage scenario tree. Eur. J. Oper. Res. **191**, 864–887 (2008)
75. N.L. Stokey, R.E. Lucas, *Recursive Methods in Economic Dynamics* (Harvard University, Cambridge, 1989)
76. N. Topaloglou, H. Vladimirou, S.A. Zenios, A dynamic stochastic programming model for international portfolio management. Eur. J. Oper. Res. **185**, 1501–1524 (2008)
77. T. Wang, A class of dynamic risk measure. Working paper, University of British Columbia (1999). http://web.cenet.org.cn/upfile/57263.pdf
78. T. Wang, Conditional preferences and updating. J. Econ. Theory **108**(2), 286–321 (2003)
79. J. Wang, P.A. Forsyth, Continuous time mean variance asset allocation: a time consistent strategy. Eur. J. Oper. Res. **209**, 184–201 (2011)
80. S. Weber, Distribution-invariant risk measures, information, and dynamic consistency. Math. Financ. **16**, 419–441 (2006)
81. S.Z. Wei, Z.X. Ye, Multi-period optimization portfolio with bankruptcy control in stochastic market. Appl. Math. Comput. **186**, 414–425 (2007)

82. L.W. Xin, D.A. Goldberg, A. Shapiro, Distributionally robust multistage inventory models with moment constraints. Working paper, Georgia Institute of Technology (2013). http://arxiv.org/abs/1304.3074
83. L.W. Xin, A. Shapiro, Bounds for nested law invariant coherent risk measures. Oper. Res. Lett. **40**(6), 431–435 (2012)
84. A. Yoshimoto, The mean-variance approach to portfolio optimization subject to transaction costs. J. Oper. Res. Soc. Jpn. **39**(1), 99–117 (1996)
85. S.S. Zhu, D. Li, S.Y.Wang, Risk control over bankruptcy in dynamic portfolio selection: a generalized mean-variance formulation. IEEE Automat. Control **49**, 447–457 (2004)

Chapter 2
Asset Price Dynamics: Shocks and Regimes

Leonard MacLean and Yonggan Zhao

Abstract Security prices changes are known to have a non-normal distribution, with heavy tails. There are modifications to the standard geometric Brownian motion model which accomodate heavy tails, most notably (1) adding point processes to the Brownian motion or (2) classifying time into regimes. With regimes the prices follow Brownian motion dynamics within regime, but the parameters vary by regime. The unconditional distribution of returns is a mixture of normals, with the mixing coefficients being Markov transition probabilities. The contrasting approaches have a common link—risk factors. In the case of the point processes, the intensity of "shocks" depends on a set of factors, eg. bond-stock yield differential, credit spread, implied volatility, exchange rates. The factors drive shocks, which are a component of the returns. With regimes, the economic state is hidden (latent) and is determined by the period by period observations on factors. The characterization of regimes follows from description in terms of the set of risk factors. In this paper the link between the shocks and regimes is explored. The shocks times defined by risk factors are an alternative method of determining regimes and the classifications by shocks and by the Expectation-Maximization algorithm are examined. The connections *factors* $\rightarrow$ *regimes* $\rightarrow$ *shocks* further justifies a classifiaction of financial markets into homogeneous epochs. The regime structure leads to improved estimates for distribution parameters. The methods are applied to the prediction of returns on Sector Exchange Traded Funds (ETFs).The allocation of investment capital to funds based on predicted returns generates favorable wealth accumulation over a planning horizon.

Keywords Risk processes • Market regimes • Jumps • Portfolio performance • Regime switching • Exchange traded funds

---

L. MacLean (✉) • Y. Zhao
School of Business Administration, Dalhousie University, Halifax, NS, Canada B3H 3J5
e-mail: l.c.maclean@dal.ca; yonggan.zhao@dal.ca

© Springer International Publishing Switzerland 2017
G. Consigli et al. (eds.), *Optimal Financial Decision Making under Uncertainty*,
International Series in Operations Research & Management Science 245,
DOI 10.1007/978-3-319-41613-7_2

2.1 Introduction

There are many factors affecting the trading decisions of investors and correspondingly the trading prices of assets in financial markets. In reaching a decision the investor formulates an expectation for the change in asset prices: will they increase or decrease and by how much? It is intuitive that the factors defining the state of the economy provide important information about investment opportunities and future price movements. In this paper, macro and micro economic indicators are considered in relation to the changes in the trading prices of securities. Large deviations in the indicators from normal levels are a force for a reversal in price movements, with investors more likely to respond to unstable market conditions. One approach to incorporating the economic factors is through nonhomogeneous point processes [7, 19]. The motivation for the point process is the perspective of investor behavior, and the setting of a likelihood or probability of price reversion. The price reversion is viewed as a risk, with the likelihood and size (consequence) of the reversion being economy dependent. Risk is typically defined as the product of the likelihood and consequence of a hazard, so the point process has a natural risk setting. The point process component is in addition to a diffusive component. The idea is that the diffusion reflects the long term market dynamics and large deviations from stable levels can trigger the point processes.

There are alternative modelling approaches to incorporating economic conditions into asset price dynamics. Factor models have been used in a wide variety of situations, but they have certain limitations since all the key parameters, including the interest rate and the stock appreciation/volatility rates, are assumed to be insensitive to the (very likely) drastic changes in the market. The underlying market may have many "modes" or "regimes" that switch among themselves from time to time. The market mode could reflect the state of the underlying economy, the general mood of investors in the market, and so on. For example, the market can be roughly divided as "bullish" and "bearish", while the market parameters can be quite different in the two modes. One could certainly introduce more intermediate states between the two extremes. A system, commonly referred to as the regime switching model, can be formulated as a set of stochastic differential equations whose coefficients are modulated by a continuous-time Markov chain.

Active research focuses on the relationship between asset returns and common risk factors. The literature has considered some common factors such as lagged returns [9, 22], the dividend-to-price ratio [4, 10, 14], the earnings-to-price ratio [5], the book-to-market ratio [18], the dividend payout ratio [16], the share of equity in new finance [2, 21], yield and credit spreads [3, 11, 15], recent changes in short-term interest rates [3, 14], and the level of consumption relative to income and wealth [17]. Many of these variables are related directly or indirectly to various stages of the business cycle and are used to predict a counter-cyclical variation in stock returns [11, 17].

The presence of stages or regimes in the financial market suggests a regime switching structure. The Markov regime-switching model has been applied to economic and financial modeling for decades. Hamilton [12, 13] applied a Markov switching model for the US GDP data and identified the various regimes in the US economy based on the observed data. Schwert [23] considered that asset returns may be associated with market volatility which switches over time. Ang and Bekaert [1] studied an international asset allocation model with regime shifts. This modeling approach is very flexible in addressing a variety of interesting questions about capital markets. For example: (1) What are plausible market regimes? (2) How frequently do these regimes switch? (3) When do these regimes change and what drives them to change over time?

In this paper, alternative models of market returns are defined: (1) a model with a diffusion augmented by non-homogeneous point processes depending on market factors; (2) a model with hidden regimes characterized by the market factors. The models have the same structure, with the point process approach being directly linked to the market factors and providing a mechanism for understanding when these regimes change and what drives them to change over time. The approach is applied to the prediction of returns on Sector Exchange traded Funds and the accumulation of capital through investment decisions based on improved estimates of returns.

2.2 Risk Factors in Financial Markets

We assume the economic situation is driven by a set of risk factors which include both micro and macro market indicators, such as realized stock and bond market returns, Currency strength, market volatility indicators, corporate dividend yield, interest rate policies, yield spread, and credit spread. Let F_t be the vector of these indicators at time t.

Micro Economic Factors

- F_1 = Current stock market return (SR_t): Log return from time t-1 to t.
- F_2 =Current bond market return (BR_t): Log return from time t-1 to t.
- F_3 =Current currency return (CR_t): Log return from time t-1 to t.
- F_4 =Current level of the implied volatility (IV_t): VIX index divided by 100.

Macro Economic Factors

- F_5 =Dividend yield (DY_t): Aggregate Dividend Yield at time t.
- F_6 =Interest rate (RF_t): U.S. Interbank offer rate at time t.
- F_7 =Yield spread (YS_t): 10 year U.S. treasury bond—3 month t-Bill at time t.
- F_8 =Credit spread (CS_t): U.S. Corporate BAA—U.S. Corporate AAA at time t.

Weekly values of the macro and micro economic factors from January 4, 1999 to November 7, 2009 are shown in Figs. 2.1 and 2.2.

The dynamics of the factors characterize the dynamics of the financial market. It is assumed that the market is composed of epochs. Furthermore, the market states over time $\{S(t), t > 0\}$ follow a discrete state continuous time Markov

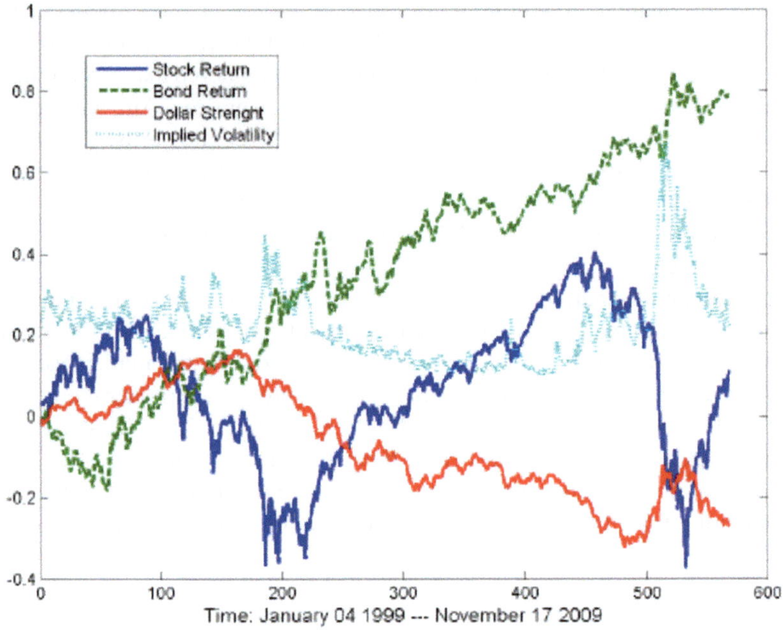

Fig. 2.1 Factors—micro

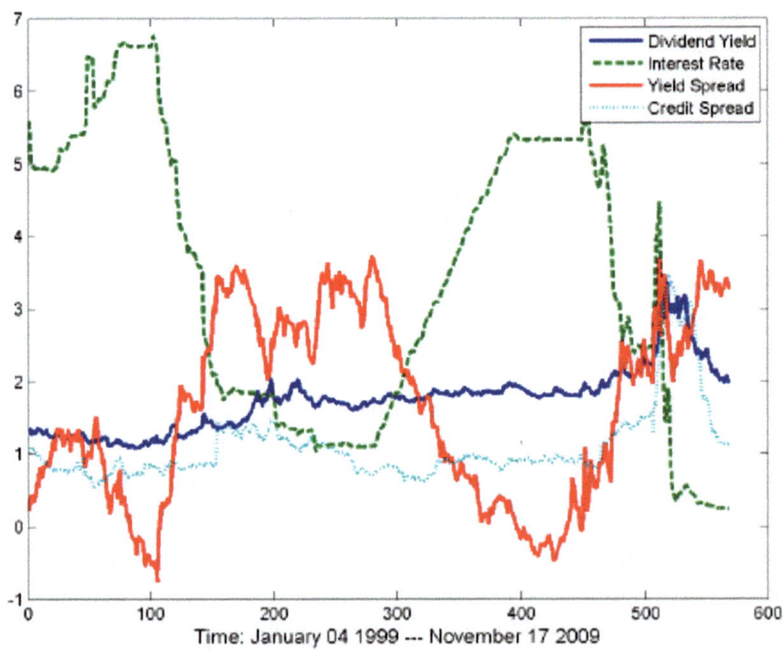

Fig. 2.2 Factors—macro

process. The state space is finite $S = \{S_1, \ldots, S_m\}$, and states will be referred to as regimes: $\{i = 1, \ldots, m\}$. The dynamics of the Markov process are driven by the intensity g_{ij}, which is the rate of transitioning from regime i to regime j. The rate of switching from regime i at time t to another regime j at time $t + h$ is $P[S(t + h) = j|S(t) = i] = g_{ij} \cdot h + o(h)$, where $\frac{o(h)}{h} \to 0$ as $h \to 0$. If the process is in regime i it transitions out of i to another regime with rate $g_i = \sum_{j \neq i} g_{ij}$, and $P[S(t + h) = i|S(t) = i] = (1 - q_i)h + o(h)$. Then $p_{ij} = \frac{g_{ij}}{g_i}$ is the probability that the process moves to regime j from regime i. For regimes i, j the transition probability function $P_{ij}(t) = Pr[S(t) = j|X(0) = i]$ is a continuous function of t. The function satisfies the Chapman-Kolmogorov equations $P_{ij}(t + s) = \sum_{k \in S} P_{ik}(t) P_{kj}(s)$. This relation allows us to consider transitions at discrete time points: $t_1 = 0, t_2 = d, t_3 = 2d, \ldots, t_n = (n - 1)d$. For the transition probability function $P_{ij}(d)$ the interval time d is fixed, so we will drop the time d and simply refer to the fixed matrix $P = (P_{ij})$. If the distribution over regimes at time t is $\pi(t) = (\pi_1(t), \ldots, \pi_m(t))$, then $\pi(t) = \pi(t - 1)P$.

Given the Markov process for regimes, we assume that the factors F_t follow a regime-switching vector autoregressive VAR(1) model.

$$F_t = \alpha_{S_t} + F_{t-1}\beta_{S_t} + \epsilon_{S_t}, \tag{2.1}$$

where α_{S_t} and β_{S_t} are regime-dependent coefficients and ϵ_{S_t} is an iid process with standard multivariate normal distribution having covariance Σ_{S_t}. Within a regime, ie a type of market, the factors are a stochastic dynamic process,where dynamics of the factors are defined by a first order autoregressive model. The regimes follow a Markov process with the dynamics defined by the transition matrix P. The following assumptions are made for the dynamics between time points:

1. There is at most one regime transition in the time interval $(t, t + d)$. Given regime i at time t, then τ_i = the time to switch from regime i to another regime is Exponential with parameter q_i, and $Pr[\tau_i \leq d] = 1 - e^{-q_i d} \approx q_i d$, which is small for a short time interval. For two transitions $Pr\left[\tau_i + \tau_j \leq d\right] \approx q_i q_j d^2$, a negligible quantity.
2. If there is a transition it occurs at the start of the interval $(t, t + d)$. The probability that there is one transition in the interval and it is from i to j is $P_{ij}(d) = Pr[S(t + d) = j|S(t) = i] = Pr[S(d) = j|S(0) = i] \approx q_i d \times p_{ij}$. Then the chance of remaining in regime i is $P_{ii}(d) \approx 1 - q_i d$.
3. The probability of remaining in a regime is high, with the chance of moving to a neighboring regime decreasing as the neighboring regime is further away. Although the state is hidden, it is reflected in the market factors. It is assumed there are expected or stable levels on the factors and as the actual level deviates form the expected there is market instability and the probability of a regime switch increases. So a regime is characterized by ranges/boundaries on the factors, with the switching probability high near a boundary.

It is interesting to identify how these common risk factors are related. As evidenced in Figs. 2.1 and 2.2, the correlations within subsets factors (F_1, F_4, F_6, F_7),

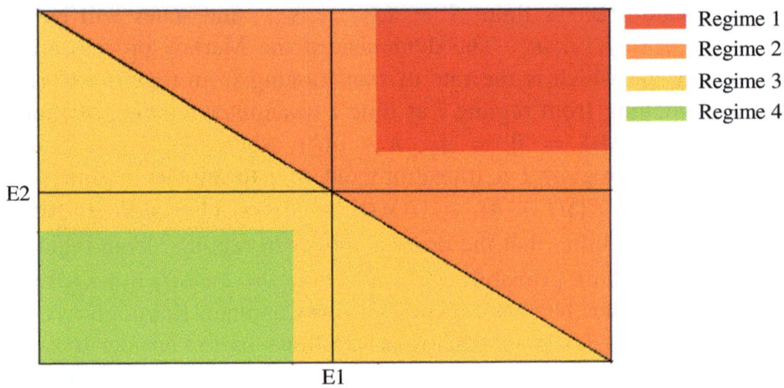

Fig. 2.3 Regimes

(F_2, F_3, F_5, F_8) are high and between subsets the cross-correlations are low. This indicates that these subsets of factors represent different forces in the financial market.

From the market factors the set of independent indices are defined for the factors

$$E_{it} = \sum a_{ij} F_{jt}, i = 1, 2. \tag{2.2}$$

The coefficients may be determined from an orthogonal factor analysis. The dynamics on the factors define the dynamics on the indices. Deviation in factors generates extreme values on the indices, which in turn determine regimes. A simplified representation of boundaries on indices and regimes is depicted in Fig. 2.3.

The use of indices composed from factors which characterize economic regimes has been employed by Vliet and Blitz [24]. In that work a single index is used to determine regimes.

2.2.1 *Regimes from Factor Thresholds*

One of the issues with regimes is the drivers of regime switching. With the autoregressive model it is assumed that the probability of switching regimes is a function of the deviation of the factors from stable levels. A direct approach to regime estimation based on observations for the factors would provide valuable insight to the triggers for switching. Furthermore, the stochastic dynamic model for factors is a mechanism for anticipating regime changes and that is a foundation for decisions on investment in assets driven by the market factors. Vliet and Blitz [24] used the prediction of regimes from a single index to derive a successful dynamic asset allocation strategy.

Based on the joint distribution for factor indices we can define upper and lower thresholds (contours) so that values above/below the thresholds are in a space for a different regime—anologous to clusters. In this section a search method in the space of (lower, upper) thresholds is proposed. A grid is formed and the optimal thresholds over the grid are identified. The thresholds are determined from factor index deviations. The criterion for optimality of thresholds is the fitting error for observed factor values. That is, the fitted vector $\hat{F}_t(k)$ is calculated for the estimated parameters with regimes defined by threshold (grid) index $k = (k_1, k_2)$, and the mean fitting error $MFE(k) = \frac{1}{T}\sum_{t=1}^{T}(F_t - \hat{F}_t(k))'(F_t - \hat{F}_t(k))$ is minimized.

For each factor index consider the index deviations $\left\{\hat{\psi}_{jt} = e_{jt} - \bar{e}_j\right\}$, with mean $\bar{e}_j$, covariance $\hat{\sigma}_j$. The threshold grid points $(\psi_{j1}^*, \psi_{j2}^*)$ for low and high deviations are determined as the number of deviations from the mean. With extreme values for the differentials below and above the mean, respectively, the times with $\left\{\hat{\psi}_{j1,t} < \psi_{j1}^*\right\}$ and/or $\left\{\hat{\psi}_{j2,t} > \psi_{j2}^*\right\}$ identify different regimes in the market.

The method proceeds as follows:

1. Specify a grid size $\omega\hat{\sigma} > 0$, and for each factor index, $j = 1, 2$, the minimum thresholds (w_{j1}, w_{j2}). Set $k_{j1} = k_{j2} = 0$.
2. With integer grid point pair $k = \{k_1, k_2\} = \{(k_{11}, k_{21}), (k_{12}, k_{22})\}$ and extreme values $\psi_{j1}^* = w_{j1} + k_{j1}\omega\hat{\sigma}_j, \psi_{j2}^* = w_{j2} - k_{j2}\omega\hat{\sigma}_j$, identify times/indices

$$T_{1k} = \left\{t \mid [\hat{\psi}_{12,t} > \psi_{12}^*] \cap [\hat{\psi}_{22,t} > \psi_{22}^*]\right\}$$

$$T_{2k} = \left\{t \mid \hat{\psi}_{1t} + \hat{\psi}_{1t} > 0, [\hat{\psi}_{12,t} < \psi_{12}^*] \cup [\hat{\psi}_{22,t} < \psi_{22}^*]\right\}$$

$$T_{3k} = \left\{t \mid \hat{\psi}_{1t} + \hat{\psi}_{1t} < 0, [\hat{\psi}_{11,t} > \psi_{j11}^*] \cup [\hat{\psi}_{21,t} > \psi_{21}^*]\right\}$$

$$T_{4k} = \left\{t \mid [\hat{\psi}_{11,t} < \psi_{11}^*] \cap [\hat{\psi}_{21,t} < \psi_{21}^*]\right\}$$

3. Assume there are distinct regimes at time sets $\{T_{1k}, T_{1k}, T_{1k}, T_{1k}\}$. For this sequence of regime times, calculate the conditional maximum likelihood estimates for $VAR(1)$ model parameters.
4. For the regime times and estimated parameter values, compute the fitted factor values and the mean fitting error $MFE(k) = \frac{1}{T}\sum_{t=1}^{T}(F_t - \hat{F}_t(k))'(F_t - \hat{F}_t(k))$.
5. Select another grid point pair k and return to [2].
6. After evaluating all grid point pairs, select the thresholds and shock times which have the smallest $MFE(k)$, that is determine $k^* = argmin_k MFE(k)$.

The threshold method identifies the regime at each time point: $\{S_t, t = 1, \ldots, T\}$, where $S_t = i$ if $t \in T_{ik^*}$. Adjacent time points identify transitions, with t_{ij} = number of transitions from regime i to regime j. Then $\hat{p}_{ij} = \frac{t_{ij}}{t}$, and $\hat{P} = (\hat{p}_{ij})$.

The methods in this section provide estimates for the parameters in the VAR(1) model: $\left\{\hat{\alpha}_{S_t}, \hat{\beta}_{S_t}, \hat{\Sigma}_{S_t}\right\}$ and the transition matrix $\hat{P}$ for factor dynamics.

2.2.2 Regimes from Hidden States

The threshold approach determines regimes directly from the values on the factors. It is usually considered that the actual regimes are hidden and the observable factors are a reflection of the unobservable regimes. The unobservable regimes can be defined as parameters which need to be estimated in addition to the coefficients in the VAR(1) model. The standard estimation method is an adaptation of the EM algorithm [8], which consists of two steps, the E-Step (estimation of the unknown states/regimes) and the M-Step (maximization of the likelihood conditional on the estimated regimes). Given an initial condition, the two steps alternate in updating parameters. The algorithm is modified to accommodate the structure implied by the regime-switching model.

To describe the algorithm used to estimate a regime-switching model, we provide a generic version of the expectation-maximization algorithm. Let Θ be the set of parameters $\{\alpha_{S_t}, \beta_{S_t}, \Sigma_{S_t}, P\}$ for the model, X the sequence of observations of the factors $\{F_t\}$ over time, and Y the sequence of unobservable regimes $\{S_t\}$ over time. Denote $\mathcal{Y}$ the space of all possible regime sequences for the time period. The marginal maximum log-likelihood is expressed as:

$$max_{\Theta}\left\{ln\sum_{Y\in\mathcal{Y}}P(X,Y;\Theta)\right\},$$

where $P(X, Y; \Theta)$ is the joint probability distribution function of X and Y.

An iterative algorithm can be designed as follows:

1. Set the number of regimes at m. This determines the number of parameters in the regime switching VAR(1) model.
2. E-step: Set an initial value Θ^0 for the true parameter set Θ, calculate the conditional distribution function, $Q(Y) = P(Y|X; \Theta^0)$, and determine the expected log-likelihood, $E^Q[lnP(X, Y; \Theta)]$.
3. M-step: Maximize the expected log-likelihood with respect to the conditional distribution of the hidden variable to obtain an improved estimate of Θ. The improved estimate is:

$$\Theta^1 = argmax_{\Theta}\left\{E^Q[lnP(X,Y;\Theta)]\right\}.$$

4. With Θ^1 as the new initial value for Θ, return to the E-Step.

In the E-step, given the observed data and current estimate of the parameter set, the hidden data are estimated using the conditional expectation. After estimating the parameters, a dynamic programming algorithm is applied to characterize the prevailing regime in each period by maximizing the joint probability of regimes given the observed data.

If it is assumed that the probability of regime switching is monotone increasing in the deviations from mean factor levels, then there is a sequence of thresholds on factor indices which maximizes the joint probability of regimes. So depending on the granulatity of the grid in the thresholds approach, the parameter estimates from the alternative approaches will align. The direct approach with factor thresholds provides the signaling information on regime changes which would be valuable in investment decisions.

2.2.3 Regime Fitting

The suitability of the regime switching VAR(1) model will be illustrated with data on the factors ($F_1, \ldots, F_8$) for weeks from January 4, 1999 to November 7, 2009. The data were analyzed to obtain estimates for the parameters in the regime switching factor model, with results provided in the tables below. The number of regimes was set to 4, consistent with the thresholds on factor indices.

The following factor dynamics by regime are observed.

- In regime 1 (Bear Market), future stock return is positively related to realized stock return, bond return, strength of the currency, and implied volatility, while it is negatively related to the observed interest rate, yield spread, and credit spread (Table 2.1).
- In regime 4 (Bull Market), future stock return is negatively related to the realized stock return, interest rate, yield spread, and credit spread, while it is positively related to the realized bond return, strength of the currency, implied volatility, and dividend yield.
- In regimes 2 & 3 (Transit markets), future stock return is negatively related to the realized stock return and bond returns, while it is positively related to interest rate and yield spread (Tables 2.2 and 2.3).
- Strength of the currency, dividend yield, and credit spread, are negatively (positively) related to future stock returns, in Transit markets (Table 2.4).

The estimated transition matrix based on the prevailing regime in each periods provided in Table 2.5. For the weeks from January 4, 1999 to November 7, 2009, the time spent in each regime were: Regime 1 (7 %); Regime 2 (22 %); Regime 3 (35 %); Regime 4 (36 %).

Table 2.1 Factors: regime 1

Factor	Cons	F_1	F_2	F_3	F_4	F_5	F_6	F_7	F_8
F_1	0.09	0.41	0.29	0.63	0.40	0.00	−0.02	−0.03	−0.06
F_2	−0.06	0.24	−0.01	0.00	0.13	0.02	0.00	−0.00	−0.02
F_3	−0.04	−0.23	−0.26	−0.13	−0.06	0.00	0.01	0.00	0.01
F_4	−0.09	−0.64	0.09	0.02	0.19	0.01	0.03	0.04	0.11
F_5	0.04	−2.05	−1.31	1.48	0.04	0.93	0.02	0.01	0.06
F_6	−0.43	−5.30	1.71	−3.36	−3.45	0.66	1.02	0.08	−0.06
F_7	0.89	−6.22	2.93	−3.98	−2.75	−0.15	−0.01	0.89	0.40
F_8	−0.45	0.63	1.19	2.06	0.38	0.21	0.03	0.11	0.78

Table 2.2 Factors: regime 2

Factor	Cons	F_1	F_2	F_3	F_4	F_5	F_6	F_7	F_8
F_1	−0.07	−0.20	−0.13	−0.56	0.19	−0.01	0.01	0.01	−0.00
F_2	−0.05	−0.02	−0.10	−0.78	0.05	0.01	0.00	0.01	−0.00
F_3	0.05	−0.01	−0.02	0.02	−0.02	−0.01	−0.00	−0.00	−0.01
F_4	0.14	0.31	0.19	0.20	0.71	−0.01	−0.01	−0.02	−0.01
F_5	0.08	−1.23	0.62	0.39	−0.25	0.99	−0.00	−0.01	0.02
F_6	0.67	−0.36	−0.63	0.86	−0.69	−0.17	0.94	−0.07	0.04
F_7	0.40	−0.19	0.31	3.35	0.32	0.04	−0.06	0.92	−0.13
F_8	−0.01	−0.27	0.29	−0.17	−0.01	0.04	0.00	−0.01	0.96

Table 2.3 Factors: regime 3

Factor	Constant	F_1	F_2	F_3	F_4	F_5	F_6	F_7	F_8
F_1	−0.01	−0.09	−0.05	0.06	−0.10	0.01	0.00	0.00	0.00
F_2	−0.04	−0.06	0.08	0.27	0.02	0.01	0.00	0.01	−0.01
F_3	−0.04	−0.02	0.05	−0.01	0.01	0.01	0.00	0.00	0.01
F_4	0.02	0.06	0.04	−0.19	1.03	−0.01	−0.00	−0.00	0.00
F_5	0.09	−1.17	0.03	−0.06	−0.02	0.97	−0.01	−0.01	0.00
F_6	−0.04	−0.19	−0.23	−0.23	−0.12	0.06	1.00	0.01	−0.04
F_7	0.43	0.18	−0.19	−1.46	−0.14	−0.16	−0.04	0.94	0.07
F_8	−0.05	−0.04	−0.01	0.33	0.11	0.03	0.00	0.00	0.98

Table 2.4 Factors: regime 4

Factor	Constant	F_1	F_2	F_3	F_4	F_5	F_6	F_7	F_8
F_1	−0.02	−0.18	0.09	0.01	0.02	0.02	−0.00	−0.00	−0.01
F_2	−0.07	0.07	0.13	0.12	0.03	0.02	0.01−001		−0.01
F_3	0.02	−0.00	0.02	−0.04	0.01	−0.00	−0.00	−0.00	−0.00
F_4	0.04	0.36	−0.11	−0.10	0.92	−0.02	−0.00	−0.00	0.02
F_5	0.08	−1.18	−0.05	0.31	0.00	0.98	−0.01	−0.01	−0.00
F_6	0.01	−0.02	−0.04	−0.08	−0.01	−0.00	1.00	−0.00	−0.00
F_7	0.64	−0.35	−1.07	−0.72	−0.24	−0.16	−0.08	0.89	0.11
F_8	0.17	−0.53	−0.05	−0.53	0.04	−0.02	−0.01	−0.02	0.91

Table 2.5 Transition matrix

	S_1	S_2	S_3	S_4
S_1	0.46	0.46	0.08	0.00
S_2	0.11	0.64	0.14	0.11
S_3	0.01	0.07	0.82	0.10
S_4	0.03	0.06	0.08	0.83

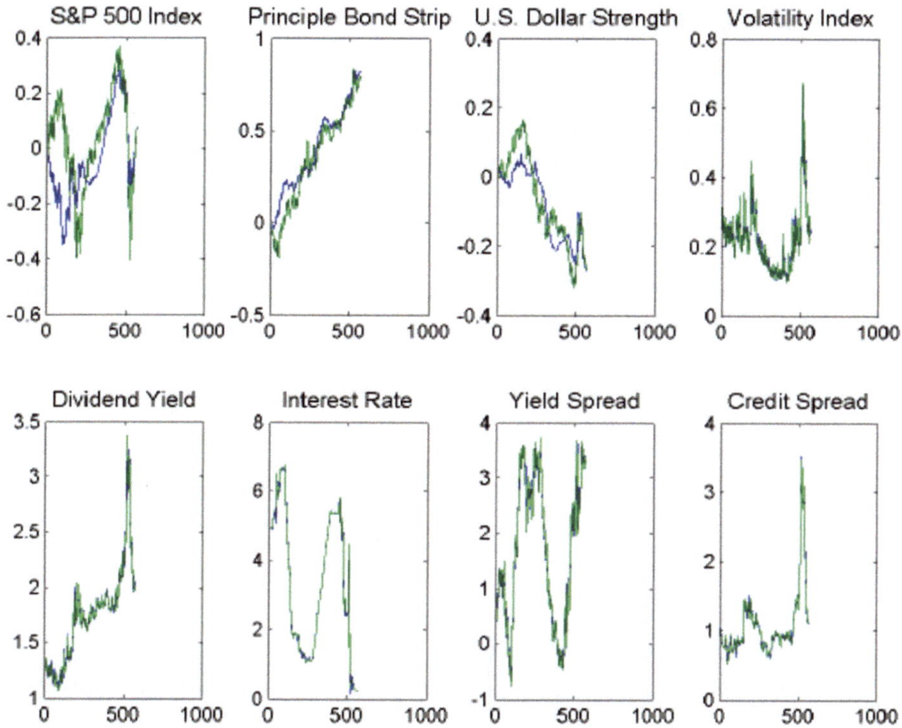

Fig. 2.4 Fitted factors

Using the estimated model, the predicted factor scores $\hat{F}_t = \hat{\alpha}_{\hat{S}_t} + \hat{F}_{t-1}\hat{\beta}_{\hat{S}_t}$ for the weeks in the time window were calculated. A comparison of the predicted (blue) and the observed (green) factor scores are shown on Fig. 2.4.

Clearly in this time window the model is an accurate depiction of the factor dynamics.

2.3 Discrete Time Asset Pricing Model

Consider a competitive financial market with n assets whose prices are stochastic dynamic processes. Let the vector of prices at time t be

$$P_t = (P_{0t}, P_{1t}, \ldots, P_{nt})', \tag{2.3}$$

where P_{0t} is the price of the risk free asset, with rate of return r_t at time t. Let $Y_{it} = \ell n P_{it}, i = 0, \ldots, n$ be the log-prices.

2.3.1 Model with Jumps

Consider that asset prices are identified at equally spaced points in time, $t = 1, \ldots, T$. Assume that the dynamics of price movements between times are defined by a geometric random walk with drift plus point processes. The idea with the point processes is to include the effect of market factors in the price dynamics. Following the discussion of factor thresholds and regimes, when the factors deviate substantially from benchmark values the asset prices react. The conditional price dynamics, given parameter values and initial values Y_{i0}, are defined by the equations for $i = 1, \ldots, n$ and $t = 1, \ldots, T$

$$Y_{it} = Y_{i0} + R_{i1} + \ldots + R_{it}, \tag{2.4}$$

where for $s = 1, \ldots, t$

$$R_{is} = \left[\alpha_i + \sum_{j=1}^{n} \delta_{ij} Z_{js}\right] + \left[\sum_{j=1}^{2} \vartheta_{ijs} \Delta N_j(\lambda_{js})\right]. \tag{2.5}$$

In these equations, $Z_s = \begin{pmatrix} Z_{1s} \\ \vdots \\ Z_{ns} \end{pmatrix}$, $s = 1, \ldots, t$ are independent multivariate normal variables and $N_j(\lambda)$ are counting processes with intensities $\lambda_j, j = 1, 2$. The intensities/chance of a jump depend on the independent factor indices $E = \{E_1, E_2\}$. If the marginal distributions for the indices at time s are $\{G_{1s}, G_{2s}\}$, then the intensities are $\lambda_{1s} = \frac{g_{1s}(\cdot)}{1-G_{1s}(\cdot)}, \lambda_{2s} = \frac{g_{2s}(\cdot)}{1-G_{2s}(\cdot)}$, respectively. As the deviations increase the chance of a jump driven by the factor indices increases, with a high probability of a jump in the regimes with extreme (beyond the thresholds) scores.

The parameters $\vartheta_{ij}, j = 1, 2$ are independent random variables capturing the size of the jumps to asset prices, and it is assumed they also depend on the deviation in factors. To have the jump size reflecting extreme returns, it is assumed that size at time t depends linearly on the factor deviation $\psi_j(s)$. If there is a shock at time s, the size is assumed to be

$$\vartheta_{ijs} = \varphi_{ij} + \theta_{ij}\psi_{js} + \eta_i W_{is},\ j = 1, 2, \tag{2.6}$$

where W_i, $i = 1, 2$, are independent, standard Gaussian variables. The sign of ψ_{js} determines the direction of the jump (UP or DOWN).

In Eq. (2.5) the counting processes augment the random walk, generating more extreme price movements. If the intensities are such that jumps occur in clusters of the same type (UP or DOWN), then the trajectory of cumulative price movements drifts away from the random walk. If the deviations are in opposite directions then the jumps could cancel, and then dynamics would be closer to the random walk. It is instructive, therefore, to consider the effect of the point processes in the regions/regimes defined by the factors. In doing so, let $\Delta N_j(\lambda_{js}) = 0$ below the thresholds and $\Delta N_j(\lambda_{js}) = 1$ above the thresholds. Then in Regime 1, for example,

$$R_{is} = \left[\alpha_i + \sum_{j=1}^{n} \delta_{ij} Z_{js}\right] + \left[\sum_{j=1}^{2} \vartheta_{ijs} \Delta N_j(\lambda_{js})\right] =$$

$$\left[\alpha_i + \sum_{j=1}^{n} \delta_{ij} Z_{js}\right] + \left[\sum_{j=1}^{2} \vartheta_{ijs}\right] =$$

$$\left[\alpha_i + \sum_{j=1}^{n} \delta_{ij} Z_{js}\right] + \sum_{j=1}^{2} \left[\varphi_{ij} + \theta_{ij} \psi_{js} + \eta_i W_{is}\right].$$

If the deviation in factor indices is written in terms of the factors, the return model has the matrix form

$$R_s = A_1 + F_s B_1 + \Gamma_1 \epsilon_s.$$

In the same way for each regime with $S(t) = j, j = 1, \ldots, 4$, the model for returns is

$$R_s = A_j + F_s B_j + \Gamma_j \epsilon_s. \tag{2.7}$$

Although the components are combined in (2.7) the separation into a random walk and non-homogenous point process is important to understanding market forces. Furthermore it is possible to separate the components in the estimation.

It is important to note that the predicted prices in the next period are the basis of investment decisions for that period. The regime switching VAR(1)model is used to forecast $\hat{F}_s$, one period ahead factors, from observations on F_{s-1}. Then the predicted returns in (2.7) are linear functions of $\hat{F}_s$.

2.3.2 Model with Regimes

The consolidation of the pricing model with random walk and jump components into a linear factor model establishes the setup for a regime based model. Having presented a regime-switching model for risk factors, it is assumed that returns of all primary investment assets follow a linear model with regime-dependent risk sensitivity. Our interest is in developing a predictive model for asset returns based on the forecast of the risk factors. Explicitly, the following structure is specified:

$$R_t = A_{S_t} + \hat{F}_t B_{S_t} + \Gamma_{S_t} \epsilon_t,$$

where $\hat{F}_t = \hat{\alpha}_{\hat{S}_t} + F_{t-1}\hat{\beta}_{\hat{S}_t}$ is the predicted value of risk factors from the risk factor model, $(A_{S_t}, B_{S_t}, \Gamma_{S_t})$ are regime-dependent coefficients and ϵ_t is an iid process with standard multivariate normal distribution.

Thus, the dynamics of asset returns are linearly related to the prediction of risk factors. Given the predicted factor and the state at time t, the one-period return vector is conditionally multivariate normal with conditional mean return vector and covariance matrix as

$$\mu_{S_t} = A_{S_t} + \hat{F}_t B_{S_t}$$
$$\Sigma_{S_t} = \Gamma_{S_t}\Gamma_{S_t}^{\top}.$$

Unlike conventional Brownian models, this model provides time-varying and state-dependent returns driven by risk factors.

2.4 Application: Exchange Traded Funds

The regime dependent VAR(1) model provided accurate predictions of risk factor dynamics. A linear factor model is now estimated for the returns on Sector Exchange Traded Funds.

2.4.1 Predicting Asset Returns

The regime dependent linear factor model will be estimated for the returns on exchange traded funds—the Standard & Poor's Depositary Receipts, SPDRs. The Sector Select ETFs are chosen because they represent major sectors of the Standard & Poor's entire US stock market and they have a slightly longer trading history (started December 23, 1998) than other sector ETFs. Out of the ten sectors, the telecommunication sector ETF is not included because this sector consists of only nine companies in the Standard & Poor's 500 stocks. The remaining nine sectors are represented by the following sector ETFs:

1. Consumer Discretionary
2. Consumer Staples
3. Energy
4. Financials
5. Health
6. Industrials
7. Materials
8. Technology
9. Utilities.

Weekly returns on the SPDRs from January 4, 1999 to November 7, 2009 provide the data. The regimes are those determined from the previously described factors.

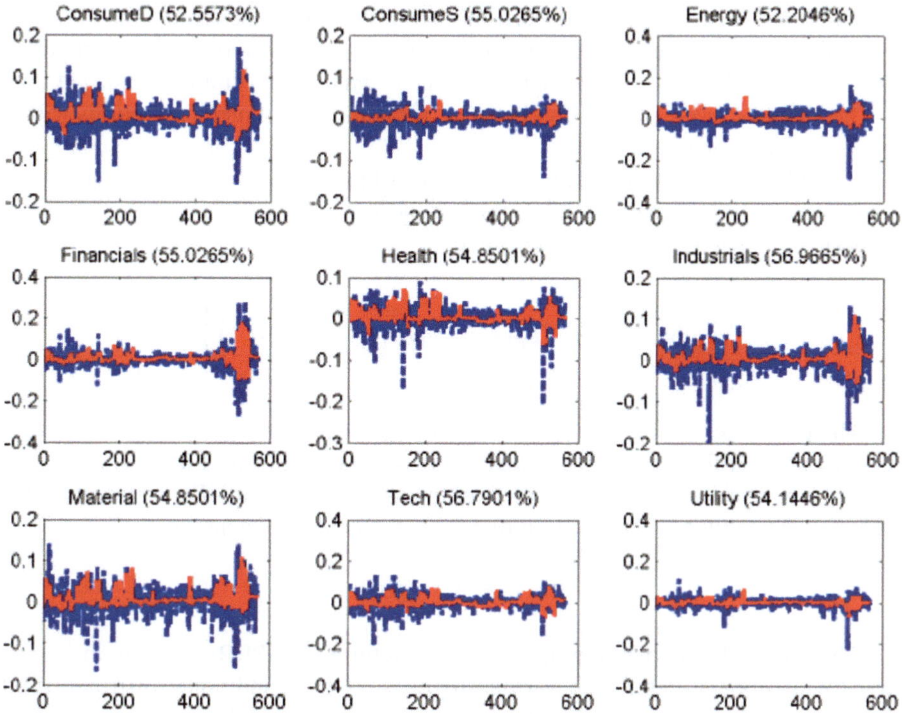

Fig. 2.5 Predicted returns

The maximum likelihood estimates for the factor model parameters can be determined from standard multivariate regression of the ETF returns on the predicted factor values. The predicted weekly returns from the fitted model are $\hat{R}_t$,and the prediction error is $R_t - \hat{R}_t$. The prediction errors are compared to the error from the standard mean predictor. The percent error is $PE = \frac{(R_t-\hat{R}_t)'(R_t-\hat{R}_t)}{(R_t-\bar{R})'(R_t-\bar{R})} \times 100$. In Fig. 2.5 the fits and actuals for the 9 SPDRs are displayed. The accuracy of predictions (PE) is included with the display.

The movements in prices are picked up by regime switching and the model is a good basis for anticipating price changes and for making investment decisions. (See MacLean et al. [20].) The improved prediction of returns should translate into successful portfolio strategies since there is evidence of the significant impact of estimation errors on portfolio returns [6].

2.4.2 *Portfolio Performance*

To illustrate the advantage of accurate predictions of price movements, investments in the SPDRS will be considered for the time January 4, 1999 to November 17, 2009. Let x_t be the vector of portfolio weights in the SPDRs at the beginning of time period t. The return on the portfolio in period t given the regime j is $R_{tj}(x_t) = r_t + (R_{tj} - r_t)'x_t$.The expected return on the portfolio in period t given the regime j is $E_{tj}(x_t) = r_{ft} + (\mu_{tj} - r_{ft})'x_t$, where μ_{tj} is the mean return vector in regime j at time t. The variance of the portfolio is $V_{tj}(x) = x_t' \Sigma_{tj} x_t$. It is assumed that returns within a regime are normal.

Consider that the current regime (at the end of period $t-1$) is known (or implied) and the transition probability to regime j in period t is π_t. The investment strategy in period t is determined by the objective

$$max_{x_t} \left\{ \sum_j \pi_j \left(E_{tj}(x_t) - \lambda(E_t \left[b_t - R_{tj}(x_t) \right]^+ \right) \right\}.$$

The parameter b_t is a benchmark return level against which the portfolio return is compared. If the return falls below the benchmark (shortfall) there is a penalty. The risk parameter λ is chosen to reflect the aversion to shortfalls, so the expected shortfall is penalized. The shortfall rate is controlled by the constraint

$$Pr\left[(b_t - R_t(x_t)) \leq \rho_t \right] \leq \alpha_t.$$

The probability is over the distribution of returns on SPDRs and within regime transition probabilities to the various regimes. If the conditional distribution of returns R_{tj}, given the regime, are normal, then the unconditional return R_t is a mixture of normals. In addition to the shortfall constraint limits are placed on the investment fractions

$$l_t \leq x_t \leq u_t.$$

This investment model is analogous to a mean-variance setup, with the modification that the downside (falling below the benchmark) is controlled. The problem has a complex formulation, but the assumption of normality within regimes enables the objective and constraint functions to be expressed as deterministic equivalents. The equivalent deterministic problem is non-convex and is solved using a Monte Carlo approach.

In the implementation of the investment model, starting wealth is set at $w_0 = \$1000$, the Penalty size is $\lambda = 5$, the benchmark is 99 % of the current wealth level $b_t = 0.99 w_{t-1}$,and the maximum shortfall is $\rho_t = 0.03$. The shortfall rate is $\alpha_t = 0.05$. As well maximum short sales $l_t = -0.05$ and maximum investment percentage $u_t = 0.15$ are applied. Portfolio rebalance frequency is weekly.

The optimal one period strategy is almost a fixed fraction. That is the fraction of invested capital in the individual SPDRs is very stable. Figure 2.6 displays the fraction in the last 20 weeks of the study period. For the SPDRs which have positive returns the fractions are basically at the upper limit.

The allocation of capital to investment does vary depending on the state of the financial market as shown in Fig. 2.7.

Period	Materials	Health	Cons Stple	Cons Disy	Energy	Financials	Industrial	Technology	Utility
1	0.1495	0.0020	0.1489	0.1496	0.1476	0.1493	0.1492	0.1493	-0.0472
2	0.1500	0.1500	0.1500	0.1500	-0.0000	0.1500	0.1500	0.1500	-0.0500
3	0.1500	0.0000	0.1500	0.1500	0.1500	0.1500	0.1500	0.1500	-0.0500
4	0.1500	0.1143	0.1500	0.1500	0.0357	0.1500	0.1500	0.1500	-0.0500
5	0.1500	0.1500	0.1500	0.1500	0.0000	0.1500	0.1500	0.1500	-0.0500
6	0.1493	0.0416	0.1488	0.1492	0.1090	0.1490	0.1488	0.1490	-0.0461
7	0.1494	0.0100	0.1489	0.1493	0.1324	0.1491	0.1491	0.1494	-0.0412
8	0.1493	0.0989	0.1485	0.1492	0.0527	0.1489	0.1492	0.1493	-0.0472
9	0.1500	0.1500	0.1500	0.1500	0.0000	0.1500	0.1500	0.1500	-0.0500
10	0.1500	0.1500	0.1500	0.1500	-0.0000	0.1500	0.1500	0.1500	-0.0500
11	0.1493	0.1464	0.1492	0.1492	-0.0208	0.1491	0.1491	0.1493	-0.0221
12	0.1490	0.1329	0.1490	0.1488	0.0166	0.1485	0.1490	0.1487	-0.0439
13	0.1500	0.1500	0.1500	0.1500	-0.0000	0.1500	0.1500	0.1500	-0.0500
14	0.1500	0.1500	0.1500	0.1500	-0.0000	0.1500	0.1500	0.1500	-0.0500
15	0.1500	0.1500	0.1500	0.1500	-0.0000	0.1500	0.1500	0.1500	-0.0500
16	0.1500	0.1500	0.1500	0.1500	-0.0000	0.1500	0.1500	0.1500	-0.0500
17	0.1500	0.1500	0.1500	0.1500	0.0000	0.1500	0.1500	0.1500	-0.0500
18	0.1500	0.1500	0.1500	0.1500	-0.0303	0.1500	0.1500	0.1500	-0.0496
19	0.1495	0.1483	0.1493	0.1493	-0.0331	0.1493	0.1493	0.1495	-0.0157
20	0.1500	0.1500	0.1500	0.1500	-0.0500	0.1500	0.1500	0.1500	0.0000

Fig. 2.6 Investment weights: July 7, 2009 to November 17, 2009

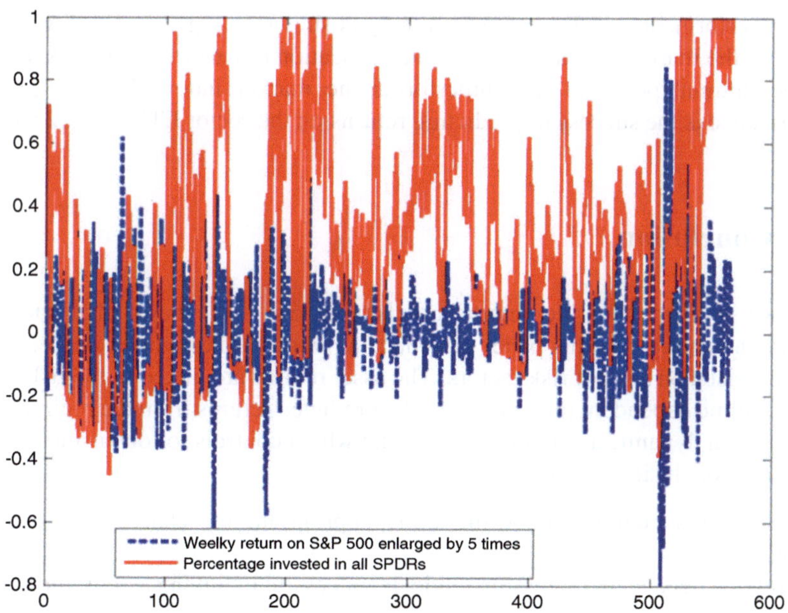

Fig. 2.7 Capital allocation: Jan 4, 1999 to Nov 17, 2009

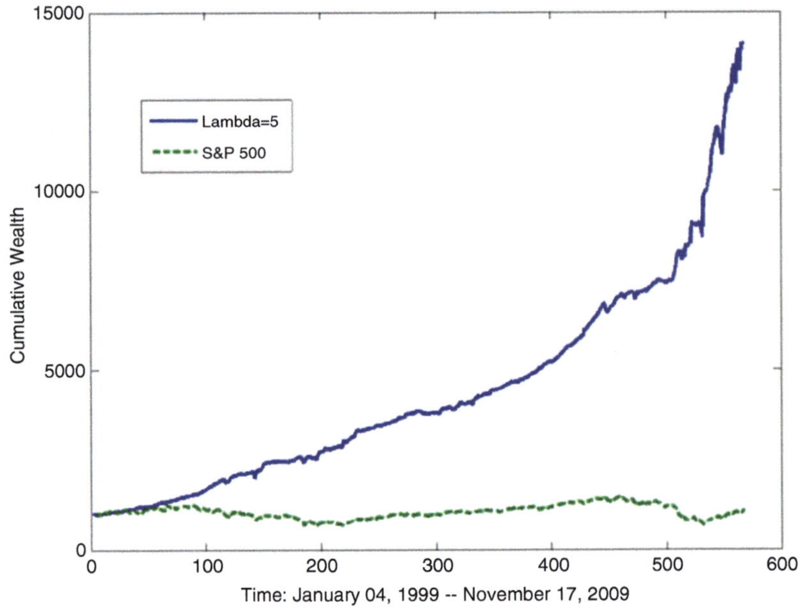

Fig. 2.8 Wealth trajectory

The performance of the optimal strategy is shown in Fig. 2.7. The S&P 500 is shown as the market return and the capital allocation moves with the market.

The optimal portfolio far outperforms the market index. The key for this performance is the success in predicting returns on the sector ETFs (Fig. 2.8).

2.5 Conclusion

Accurate prediction of the returns on risky assets is the basis of a successful investment strategy. In this paper the financial market is segmented into economic regimes based on a set of risk factors. The asset returns are linked to the risk factors with a regime dependent linear model. The predicted returns from the regime model are used in a dynamic asset allocation model which controls for downside risk. The following conclusion can be reached:

1. A regime switching factor model is appropriate for characterizing market regimes.
2. A regime dependent regression model is successful in linking asset returns to risk factors.
3. Based on a model test with data on Sector Exchange Traded Funds, the estimation methods provide a solid basis for sound investment strategies.

References

1. A. Ang, G. Bekaert, International asset allocation with regime shifts. Rev. Financ. Stud. **15**(4), 1137–1187 (2002)
2. M. Baker, J. Wurgler, The equity share in new issues and aggregate stock returns. J. Financ. **55**(5), 2219–2257 (2000)
3. J.Y. Campbell, Stock returns and the term structure. J. Financ. Econ. **18**(2), 373–399 (1987)
4. J.Y. Campbell, R.J. Shiller, The dividend-price ratio and expectations of future dividends and discount factors. Rev. Financ. Stud. **1**(3), 195–228 (1988)
5. J.Y. Campbell, R.J. Shiller, Stock prices, earnings, and expected dividends. J. Financ. **43**(3), 661–676 (1988)
6. V. Chopra, W.T. Ziemba, The effect of errors in means, variances, and covariances on optimal portfolio choice. J. Portf. Manag. Winter 1993 **19**(2), 6–11 (1993)
7. G. Consigli, L. MacLean, Y. Zhao, W. Ziemba, The bond-stock yield differential as a risk indicator in financial markets. J. Risk **11**(3), 1–22 (2009)
8. A.P. Dempster, N.M. Laird, D.B. Rubin, Maximum likelihood from incomplete data via the EM algorithm. J. R. Stat. Soc. (Ser. B) **39**, 1–38 (1977)
9. E.F. Fama, K.R. French, Permanent and temporary components of stock prices. J. Polit. Econ. **96**(2), 246–273 (1988)
10. E.F. Fama, K.R. French, Dividend yields and expected stock returns. J. Financ. Econ. **22**(1), 3–27 (1988)
11. E.F. Fama, K.R. French, Business conditions and expected returns on stocks and bonds. J. Financ. Econ. **25**(1), 23–49 (1989)
12. J.D. Hamilton, A new approach to the economic analysis of non-stationary time series and the business cycle. Econometrica **57**(2), 357–384 (1989)
13. J.D. Hamilton, Analysis of time series subject to changes in regime. Econometrica **45**, 39–70 (1989)
14. R.J. Hodrick, Dividend yields and expected stock returns: alternative procedures for inference and measurement. Rev. Financ. Stud. **5**(3), 357–386 (1992)
15. D. Keim, R. Stambaugh, Predicting returns in stock and bond markets. J. Financ. Econ. **17**(2), 357–390 (1986)
16. O. Lamont, Earnings and expected returns. J. Financ. **53**(5), 1563–1587 (1998)
17. M. Lettau, S. Ludvigson, Consumption, aggregate wealth, and expected stock returns, working paper, Federal Reserve Bank of New York, New York, NY (1999)
18. J. Lewellen, The time-series relations among expected return, risk, and book-to-market. J. Financ. Econ. **54**(1), 5–43 (1999)
19. L. MacLean, G. Consigli, Y. Zhao, W. Ziemba, Risk Indicators in Financial Markets. Math. Meth. Econ. Financ. **3**(2), 101–118 (2008)
20. L. MacLean, E. Thorp, W. Ziemba, *The Kelly Capital Growth Investment Criterion: Theory and Practice* (World Scientific Publishing Co., Singapore, 2010)
21. W.R. Nelson, Three essays on the ability of the change in shares outstanding to predict stock returns. Ph.D. dissertation, Yale University, Newhaven, CT, 1999
22. J. Poterba, L.H. Summers, Mean reversion in stock returns: evidence and implications. J. Financ. Econ. **22**(1), 27–60 (1988)
23. G.W. Schwert, Business cycles, financial crises, and stock volatility, in *Carnegie-Rochester Conference Series on Public Policy*, vol. 31 (Elsevier, Barking, 1989), pp. 83–125
24. P. Vliet, D. Blitz, Dynamic Strategic Asset Allocation: Risk Return Across Economic Regimes. J. Asset Manage. **12**, 360–375 (2011)

Chapter 3
Scenario Optimization Methods in Portfolio Analysis and Design

Giuseppe Carlo Calafiore

Abstract This chapter discusses techniques for analysis and optimization of portfolio statistics, based on direct use of samples of random data. For a given and fixed portfolio of financial assets, a classical approach for evaluating, say, the value-at-risk (V@R) of the portfolio is a *model-based* one, whereby one first assumes some stochastic model for the component returns (e.g., Normal), then estimates the parameters of this model from data, and finally computes the portfolio V@R. Such a process hinges upon critical assumptions (e.g., the elicited return distribution), and leaves unclear the effects of model estimation errors on the computed quantity of interest. Here, we propose an alternative *direct* route that bypasses the assumption and estimation of a model for the returns, and provides the estimated quantity of interest (together with its out-of-sample reliability tag) directly from data generated by a *scenario generation oracle*. This idea is then extended to the situation where one simultaneously optimizes over the portfolio composition, in order to achieve an optimal portfolio with a guaranteed level of expected shortfall probability. Such a scenario-based portfolio design approach is here developed for both single-period and multi-period allocation problems. The methodology underpinning the proposed computational method is that of random convex programming (RCP). Besides the described data-driven problems, we show in this chapter that the RCP paradigm can also be employed alongside more standard mean-variance portfolio optimization settings, in the presence of ambiguity in the statistical model of the returns, providing a viable technique to address robust portfolio optimization problems.

Keywords Data-driven portfolio optimization • Scenario methods • Random convex programs • Empirical quantiles • Multi-period asset allocation

---

G.C. Calafiore (✉)
Dipartimento di Automatica e Informatica, Politecnico di Torino, Torino, Italy
e-mail: giuseppe.calafiore@polito.it

© Springer International Publishing Switzerland 2017

G. Consigli et al. (eds.), *Optimal Financial Decision Making under Uncertainty*, International Series in Operations Research & Management Science 245,
DOI 10.1007/978-3-319-41613-7_3

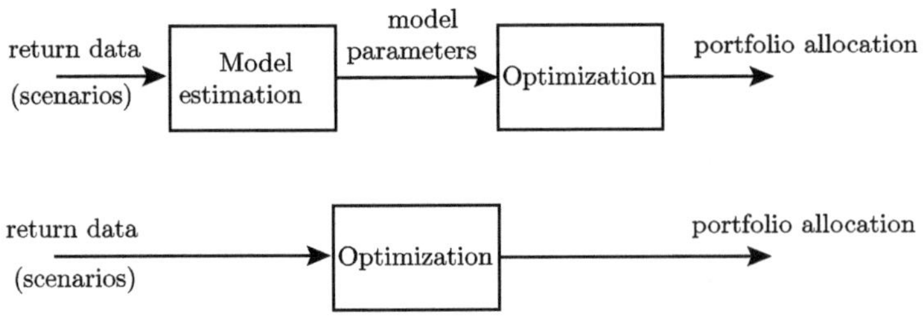

Fig. 3.1 Model based approach (*top*) vs. data-driven approach (*bottom*)

3.1 Introduction

In this chapter we present numerical methods for analysis and design of investment portfolios by means of data-driven techniques based on the scenario optimization technology [6, 8]. The idea behind data-driven approaches to asset allocation is to use directly return data to numerically compute the optimal portfolios. Mainstream model-based approaches, derived from the classical Markowitz setup [18], focus on parameters of the return distribution (such as expected returns and covariance) that need be somehow estimated, and then derive the optimal portfolios on the basis of these parameters. Data-driven methods focus instead on *data*, and aim at determining the optimal allocations using the data directly, as illustrated in Fig. 3.1. Data-driven approaches, hence, do not necessarily or explicitly require the intermediate step of estimating a statistical model of the returns, although some model of the returns may still be necessary for generating scenarios beyond the cardinality available from historical data.

In both the model-based and the data driven approaches, a key issue relates to assessing the *reliability* of the resulting portfolio allocation. It is for instance well known (see, e.g., [1, 11]) that the allocation resulting from "classical" approaches is quite sensitive to the estimated model parameters (e.g., expected returns and covariances, or other parameters of the elicited return distribution). As a consequence, due to model estimation errors, a portfolio x^* designed to have, say, a certain value γ at risk level ϵ, may well fail to provide the expected performance "out of sample," that is on new, future, scenarios that have not been accounted for at the model estimation stage. Similarly, most literature on data-driven methods (see, e.g., [12, 14, 16, 17, 19]) focuses on computational issues in the data-driven optimization problem, but still lacks a rigorous analysis of the out-of-sample reliability of the optimal data-driven portfolios.

The objective of this chapter is to describe classes of portfolio allocation problems that can be cast and solved in a data-driven framework, and to provide for these problems a rigorous analysis of their out-of-sample reliability, based on recent advances in the theory of random convex programming [6]. This chapter is organized as follows: in Sect. 3.2 we study an analysis problem on a given portfolio over a single period. As a data-driven risk measure we consider the empirical value at risk of the portfolio, that is the empirical quantile of a stream of random portfolio returns of cardinality N. The main results in Sect. 3.2.2 show that the empirical quantile is, with arbitrarily large probability, a reliable approximator of the real quantile, provided that N satisfies an explicit lower cardinality bound (see, e.g., Eq. (3.13)). In Sect. 3.3 we extend this idea to the single-period *design* problem, that is we compute optimal data-driven portfolios by optimizing an empirical quantile, and we provide in Sect. 3.3.3 an assessment of their out-of-sample reliability as a function of the scenarios cardinality N. Section 3.3 is based on original results appeared in [7]; a similar approach has also been recently proposed in [20]. In Sect. 3.4 we further extend the proposed data-driven approach to multi-period problems. The idea is to exploit linear reaction policies, as proposed in [4]; however, the approach in [4] was a model-based one, whereas we here propose an original route, based on oracle-generated return paths. In Sect. 3.5 we show instead how scenario optimization techniques can also be employed in the context of model-based approaches, in the presence of uncertainty in the statistical model description of the returns (ambiguous models). Finally, Sect. 3.6 illustrates the concept presented in Sect. 3.3 via a numerical example concerning allocation over seven investment sectors.

3.1.1 Definitions and Preliminaries

We denote with $a_1, \ldots, a_n$, a collection of assets, and with $p_i(k)$ the market price of a_i at time $k\Delta$, where k is an integer, and Δ is a fixed period of time. The rate of return (or, return, for brevity) of an investment in asset i over the k-th period from $(k-1)\Delta$ to $k\Delta$ is

$$r_i(k) \doteq \frac{p_i(k) - p_i(k-1)}{p_i(k-1)}, \; i = 1, \ldots, n; \; k = 1, 2, \ldots,$$

and the corresponding *gain* (or total return) is defined as

$$g_i(k) \doteq 1 + r_i(k), \; i = 1, \ldots, n; \; k = 1, 2, \ldots.$$

We denote with $r(k) \doteq [r_1(k) \; \cdots \; r_n(k)]^\top$ the vector of assets' returns over the k-th period, and with $g(k)$ the corresponding vector of gains.

The return and gain vectors are assumed to be random quantities, and we denote with $\mathbb{P}_{1,2,\ldots}$ the probability distribution of $\{r(1), r(2), \ldots\}$, given the past $\{\ldots, r(-1), r(0)\}$, where $k = 0$ denotes the current time, at which the portfolio decision is to be taken. In this chapter we shall consider both single-period and multi-period allocation problems. If $T \geq 1$ denotes the number of forward periods over which the allocation decisions need be taken, then we let $\mathbb{P}$ simply denote the joint marginal probability distribution of $\{r(1), \ldots r(T)\}$ given the past. We do not assume that $\mathbb{P}$ is known. We only assume that there is available a *scenario-generating oracle* which is capable of generating independent and identically distributed (iid) samples of the forward returns $\{r(1), \ldots r(T)\}$, according to $\mathbb{P}$.

A *portfolio* of assets $a_1, \ldots, a_n$ is defined by a vector $x(k) \in \mathbb{R}^n$ whose entry $x_i(k)$, $i = 1, \ldots, n$, describes the (signed) amount of an investor's wealth invested in asset a_i at time $k = 0, \ldots, T-1$, where $x_i(k) \geq 0$ denotes a "long" position, and $x_i(k) < 0$ denotes a "short" position. In portfolio design, the portfolio vector $x(k)$ is typically subject to various constraints, reflecting the investor's a-priori policies and bindings. For example, short-selling might be forbidden, in which case the components of $x(k)$ must be nonnegative (which we write as $x(k) \geq 0$, with element-wise inequality), or the portfolio should be self financing (the sum of portfolio entries must be equal to a constant), or yet constraints may include minimum and maximum exposure in an individual asset, or limits in the exposure over classes of assets, etc. In this paper, we shall treat the problem in reasonable generality by assuming that the portfolio vector is constrained in a polytope (a bounded polyhedron) $\mathcal{X}(k)$. In single-period allocation problems, where the portfolio composition is only set at time $k = 0$, we denote the portfolio simply with $x \doteq x(0)$, and the composition constraints simply with $\mathcal{X} \doteq \mathcal{X}(0)$. The classical Markowitz case is given by the conditions $\mathbf{1}^\top x = 1$, $x \geq 0$ (no short-selling), in which case $\mathcal{X}$ is the standard simplex.

3.2 Single-Period Analysis of Portfolio Shortfall Probability

In this section we consider the portfolio $x \in \mathcal{X}$ to be fixed and held for one single period forward. We let

$$z \doteq \varrho(x) = r^\top x,$$

represent the random return of the given portfolio at the end of the period, where $r \in \mathbb{R}^n$ is the vector of random returns of the component assets over the forward period. We denote with $\mathbb{P}$ the probability distribution of r, and with $\mathbb{P}_z$ the distribution of z, having support $Z \in \mathbb{R}$. We assume that the scenario generating oracle provides N iid random observations $\{r^{(1)}, \ldots, r^{(N)}\}$ of r and, correspondingly, N iid observations $\mathbf{z} = \{z_1, \ldots, z_N\}$ of the portfolio return $\varrho(x)$. We next study the probabilistic properties of the k-th smallest return in this sequence of N returns.

3.2.1 The Shortfall Probability of the k-th Order Sample

Let $\mathbf{z} = \{z_1, \ldots, z_N\}$ be a sequence of N independent and identically distributed (iid) samples extracted according to $\mathbb{P}_z$. We denote with $z_{[1]} \leq z_{[2]} \leq \cdots \leq z_{[N]}$ the observations in $\mathbf{z}$ arranged in nondecreasing order, i.e., the order statistics: $z_{[1]} = \min\{\mathbf{z}\}$, $z_{[2]} = \min\{\mathbf{z} \setminus z_{[1]}\}$, $z_{[3]} = \min\{\mathbf{z} \setminus z_{[1]}, z_{[2]}\}$, etc.

Note that events related to $\mathbf{z}$ are measured by the product probability $\mathbb{P}_z^N = \mathbb{P}_z \times \mathbb{P}_z \times \cdots \times \mathbb{P}_z$ (N times) on Z^N, and that $z_{[k]}$ is a function of $\mathbf{z}$, hence also events related to $z_{[k]}$ are measured by $\mathbb{P}_z^N$. For $\zeta \in \mathbb{R}$, let

$$F(\zeta) \doteq \mathbb{P}_z\{z \in Z : \; z \leq \zeta\}$$

denote the cumulative probability distribution function of z. $F(\zeta)$ is nondecreasing and right continuous, that is $F(\zeta) = \lim_{\xi \to \zeta_+} F(\xi)$. Let further

$$V(\zeta) \doteq \lim_{\xi \to \zeta_-} F(\xi) = \mathbb{P}_z\{z \in Z : \; z < \zeta\}.$$

Obviously, $V(x)$ is nondecreasing and left continuous. Moreover, $F(x) - V(x) = \mathbb{P}_z\{z = x\}$, and this latter quantity is zero if $F(x)$ is continuous. We define the η-quantile of z as follows:

$$q_\eta \doteq \sup\{\zeta : V(\zeta) \leq \eta\} \equiv \inf\{\zeta : F(\zeta) \geq \eta\}.$$

It is then a standard fact that

$$V(\zeta) \leq \eta \quad \Leftrightarrow \quad \zeta \leq q_\eta, \tag{3.1}$$

and that

$$F(q_\eta) \geq \eta, \quad (F(q_\eta) = \eta, \text{ if } F(x) \text{ is continuous}). \tag{3.2}$$

We now evaluate $V(\zeta)$ at $z_{[k]}$: since this point is random, $V(z_{[k]})$ is itself a random variable taking values in $[0, 1]$. We call this random quantity the *shortfall probability* of the k-th smallest observation, since it represents the probability with which the random return z may fall below the value $z_{[k]}$. We are thus interested in determining the cumulative probability distribution

$$\mathbb{P}_z^N\{\mathbf{z} \in Z^N : \; V(z_{[k]}) \leq \epsilon\}, \quad \epsilon \in [0, 1].$$

The following key result holds.

Theorem 1. *For $k = 1, \ldots, N$, it holds that*

$$\mathbb{P}_z^N\{\mathbf{z} \in Z^N : V(z_{[k]}) \leq \epsilon\} = \mathbb{P}_z^N\{\mathbf{z} \in Z^N : z_{[k]} \leq q_\epsilon\} \geq \Phi_{N,k}(\epsilon), \tag{3.3}$$

where

$$\Phi_{N,k}(\epsilon) \doteq \sum_{j=k}^{N} \binom{N}{j} \epsilon^j (1-\epsilon)^{N-j}.$$

Moreover, equality holds in (3.3)*, if* $F(x)$ *is continuous.*

Proof. First note that from (3.1) with $\zeta = z_{[k]}$ it follows that

$$\mathbb{P}_z^N\{\mathbf{z} \in Z^N : V(z_{[k]}) \leq \epsilon\} = \mathbb{P}_z^N\{\mathbf{z} \in Z^N : z_{[k]} \leq q_\epsilon\}.$$

We thus concentrate on this latter probability:

$$\begin{aligned}
\mathbb{P}_z^N\{\mathbf{z} \in Z^N : z_{[k]} \leq q_\epsilon\} &= \mathbb{P}_z^N\{\mathbf{z} \in Z^N : \text{at least } k \text{ of the } z_i\text{'s are } \leq q_\epsilon\} \\
&= \sum_{j=k}^{N} \mathbb{P}_z^N\{\mathbf{z} \in Z^N : \text{exactly } j \text{ of the } z_i\text{'s are } \leq q_\epsilon\} \\
&= \sum_{j=k}^{N} \binom{N}{j} (F(q_\epsilon))^j (1 - F(q_\epsilon))^{N-j} \\
&\doteq \Phi_{N,k}(F(q_\epsilon)),
\end{aligned}$$

where the last passage follows from the fact that the z_i's are independent, and the probability that $z_i \leq q_\epsilon$ is, by definition, $F(q_\epsilon)$. Note that $\Phi_{N,k}(\xi)$ represents the probability of having k or more "successes" in N Bernoulli trials, each trial having success probability ξ, and that this expression is obviously increasing in ξ (increasing the success probability ξ in a single trial increases the probability of having at least k successes in N trials). Therefore, it follows from (3.2) that $\Phi_{N,k}(F(q_\epsilon)) \geq \Phi_{N,k}(\epsilon)$, with equality holding whenever F is continuous, which permits to conclude (3.3). □

Remark 1. Notice that, for continuous F, the distribution of $V(z_{[k]})$ is precisely the upper tail of a cumulative Binomial distribution, and that this distribution holds irrespective of the distribution F originally assumed on the z samples. Considering the complementary event in (3.3) we obtain that

$$\begin{aligned}
\mathbb{P}_z^N\{\mathbf{z} \in Z^N : V(z_{[k]}) > \epsilon\} &\leq 1 - \Phi_{N,k}(\epsilon) \\
&\doteq \bar{\Phi}_{N,k}(\epsilon) = \sum_{j=0}^{k-1} \binom{N}{j} \epsilon^j (1-\epsilon)^{N-j},
\end{aligned}$$

with equality holding for continuous distributions. We shall show in Corollary 1 that, by appropriate selection of N and k, the above residual probability can be rendered arbitrarily small. ◊

A result analogous to Theorem 1 holds for the k-th largest sample $z_{[N-k+1]}$, as formalized in the next theorem. Let

$$\tilde{V}(\zeta) \doteq \mathbb{P}_z\{z \in Z : z > \zeta\} = 1 - F(\zeta)$$
$$\tilde{V}_-(\zeta) \doteq \lim_{\xi \to \zeta_-} \tilde{V}(\zeta) = \mathbb{P}_z\{z \in Z : z \geq \zeta\}$$

and let $\tilde{q}_\eta = \inf\{x : \tilde{V}(x) \leq \eta\}$.

Theorem 2. *For $k = 1, \dots, N$, it holds that*

$$\mathbb{P}_z^N\{\mathbf{z} \in Z^N : \tilde{V}(z_{[N-k+1]}) \leq \epsilon\} = \mathbb{P}_z^N\{\mathbf{z} \in Z^N : z_{[N-k+1]} \geq \tilde{q}_\epsilon\} \geq \Phi_{N,k}(\epsilon). \quad (3.4)$$

Moreover, equality holds in (3.4)*, if $F(x)$ is continuous.*

Proof. Note that $\tilde{V}$ is right continuous, hence it holds that $\tilde{V}(x) \leq \eta \Leftrightarrow x \geq \tilde{q}_\eta$, and

$$\tilde{V}_-(\tilde{q}_\eta) \geq \eta, \text{ with equality holding if } F \text{ is continuous.} \quad (3.5)$$

Therefore, $\mathbb{P}_z^N\{\tilde{V}(z_{[N-k+1]}) \leq \epsilon\} = \mathbb{P}_z^N\{z_{[N-k+1]} \geq \tilde{q}_\epsilon\}$. For this latter probability, we have that

$$\begin{aligned}
\mathbb{P}_z^N\{\mathbf{z} \in Z^N : z_{[N-k+1]} \geq \tilde{q}_\epsilon\} &= \mathbb{P}_z^N\{\mathbf{z} \in Z^N : \text{at least } k \text{ of the } z_i\text{'s are } \geq \tilde{q}_\epsilon\} \\
&= \sum_{j=k}^{N} \mathbb{P}_z^N\{\mathbf{z} \in Z^N : \text{exactly } j \text{ of the } z_i\text{'s are } \geq \tilde{q}_\epsilon\} \\
&= \sum_{j=k}^{N} \binom{N}{j} (\tilde{V}_-(\tilde{q}_\epsilon))^j (1 - \tilde{V}_-(\tilde{q}_\epsilon))^{N-j} \\
&\doteq \Phi_{N,k}(\tilde{V}_-(\tilde{q}_\epsilon)),
\end{aligned}$$

where the last passage follows from the fact that the z_i's are independent, and the probability that $z_i \geq \tilde{q}_\epsilon$ is precisely $\tilde{V}_-(\tilde{q}_\epsilon)$. Note that $\Phi_{N,k}(\xi)$ is increasing in ξ, therefore it follows from (3.5) that $\Phi_{N,k}(V_-(\tilde{q}_\epsilon)) \geq \Phi_{N,k}(\epsilon)$, with equality holding whenever F is continuous, which permits to conclude the proof. □

Corollary 1. *Let $\beta \in (0, 1)$ and $k \in \{1, \dots, N\}$. If N is an integer such that*

$$N \geq \frac{1}{\epsilon}\left(k + \ln \beta^{-1} + \sqrt{\ln^2 \beta^{-1} + 2k \ln \beta^{-1}}\right), \quad (3.6)$$

then

$$\mathbb{P}_z^N\{\mathbf{z} \in Z^N : V(z_{[k]}) > \epsilon\} \leq \beta. \quad (3.7)$$

Moreover, in the special case when $k = 1$, (3.7) holds if and only if

$$N \geq \frac{\ln \beta^{-1}}{\ln(1-\epsilon)^{-1}}. \tag{3.8}$$

Proof. First consider the case when $k = 1$. In such a case, we simply have $\bar{\Phi}_{N,1}(\epsilon) = (1-\epsilon)^N$, hence $\bar{\Phi}_{N,1}(\epsilon) \leq \beta$ holds if and only if $N \ln(1-\epsilon)^{-1} \geq \ln \beta^{-1}$, form which (3.8) follows (notice that the bound is tight in this case).

For the general case $k > 1$, the classical Chernoff's inequality [10] for the lower Binomial tail yields the bound

$$\bar{\Phi}_{N,k}(\epsilon) \leq \exp\left(-\frac{(N\epsilon - k + 1)^2}{2N\epsilon}\right), \quad \text{for } N\epsilon \geq k-1.$$

Therefore, for $k > 1$ and $\beta \in (0,\ 1)$, the following implications hold:

$$\begin{aligned}
\bar{\Phi}_{N,k}(\epsilon) \leq \exp\left(-\frac{(N\epsilon - k + 1)^2}{2N\epsilon}\right) &\leq \beta \\
&\Updownarrow \\
-\frac{(N\epsilon - k + 1)^2}{2N\epsilon} &\leq \ln \beta \\
&\Updownarrow \\
N^2\epsilon^2 - 2N\epsilon(k - 1 + \ln \beta^{-1}) + (k-1)^2 &\geq 0 \\
&\Updownarrow \\
\frac{1}{\epsilon}\left(k - 1 + \ln \beta^{-1} + \sqrt{\ln^2 \beta^{-1} + 2(k-1)\ln \beta^{-1}},\right) &\leq N
\end{aligned}$$

which concludes the proof. □

Remark 2 (Working Under "Near Certainty" Conditions). When the probability in (3.7) is bounded from above by a very small β, say $\beta = 10^{-6}$ or smaller, to most practical purposes we may assume that the event $V(z_{[k]}) > \epsilon$ is negligible. Equivalently, we shall say that $V(z_{[k]}) \leq \epsilon$ holds with near certainty. Setting for instance $\beta = 8.3153 \times 10^{-7}$ (this was chosen just in order to make $\ln \beta^{-1}$ an integer number), bound (3.6) reads

$$N \geq \frac{14 + k - 1 + \sqrt{196 + 28k}}{\epsilon}.$$

This formula gives a readily computable lower bound on N such that (3.7) holds. However, the exact value of N can be obtained numerically, by computing the least integer N such that

$$\sum_{j=0}^{k-1} \binom{N}{j} \epsilon^j (1-\epsilon)^{N-j} \leq \beta.$$

Some tabulated values obtained by numerical inversion of the above formula are reported in Tables 3.1 and 3.2. ♢

Table 3.1 Minimum number N of extractions guaranteeing (3.7), with $\beta = 8.3153 \times 10^{-7}$

	$\epsilon = 0.4$	$\epsilon = 0.3$	$\epsilon = 0.2$	$\epsilon = 0.15$	$\epsilon = 0.1$
$k = 1$	28	40	63	87	133
$k = 1 + \lceil 0.02N \rceil$	34	56	88	135	246
$k = 1 + \lceil 0.05N \rceil$	44	69	144	251	719
$k = 1 + \lceil 0.1N \rceil$	58	108	336	1096	–
$k = 1 + \lceil 0.15N \rceil$	85	199	1417	–	–

Table 3.2 Minimum number N of extractions guaranteeing (3.7), with $\beta = 10^{-5}$

	$\epsilon = 0.4$	$\epsilon = 0.3$	$\epsilon = 0.2$	$\epsilon = 0.15$	$\epsilon = 0.1$
$k = 1$	23	33	52	71	110
$k = 1 + \lceil 0.02N \rceil$	29	41	76	117	217
$k = 1 + \lceil 0.05N \rceil$	34	54	113	208	592
$k = 1 + \lceil 0.1N \rceil$	47	88	268	885	–
$k = 1 + \lceil 0.15N \rceil$	71	160	1125	–	–

3.2.2 The k-th Order Sample as an Approximator of the ϵ-Quantile

In this section we show that, with an appropriate choice of N and k, the k-th order sample $z_{[k]}$ can approximate arbitrarily well the ϵ quantile q_ϵ of the random return z. We preliminarily state the following corollary.

Corollary 2. *Given N, $k \leq N$, let $\epsilon_-, \epsilon_+ \in (0,1)$ be such that $\epsilon_+ > \epsilon_-$, and let the probability distribution on z be continuous. Then,*

$$\mathbb{P}_z^N\{\mathbf{z} \in Z^N : q_{\epsilon_-} < z_{[k]} \leq q_{\epsilon_+}\} = \Phi_{N,k}(\epsilon_+) - \Phi_{N,k}(\epsilon_-). \tag{3.9}$$

Moreover, given $\beta \in (0, 1)$, *if*

$$N\epsilon_- + \sqrt{\frac{N}{2}\ln(2/\beta)} \leq k \leq N\epsilon_+ + 1 - \sqrt{\frac{N}{2}\ln(2/\beta)} \tag{3.10}$$

then it holds that

$$\mathbb{P}_z^N\{\mathbf{z} \in Z^N : q_{\epsilon_-} < z_{[k]} \leq q_{\epsilon_+}\} \geq 1 - \beta.$$

Proof. Inspecting the proof of Theorem 1 we see that, if the distribution on z is continuous, then

$$\mathbb{P}_z^N\{\mathbf{z} \in Z^N : z_{[k]} \leq q_\eta\} = \Phi_{N,k}(\eta),$$

from which (3.9) immediately follows. Define next $A \doteq \{\mathbf{z} \in Z^N : q_{\epsilon_-} < z_{[k]} \leq q_{\epsilon_+}\}$, and let $\bar{A}$ denote the complementary event. We next establish a lower bound on the probability of A. From (3.9), we have that

$$\mathbb{P}_z^N\{\bar{A}\} = 1 - \mathbb{P}_z^N\{A\} = \bar{\Phi}_{N,k}(\epsilon_+) + \Phi_{N,k}(\epsilon_-), \tag{3.11}$$

where we recall that

$$\bar{\Phi}_{N,k}(\epsilon_+) = \sum_{j=0}^{k-1} \binom{N}{j} \epsilon_+^j (1 - \epsilon_+)^{N-j},$$

$$\Phi_{N,k}(\epsilon_-) = \sum_{j=k}^{N} \binom{N}{j} \epsilon_-^j (1 - \epsilon_-)^{N-j}.$$

Using the Chernoff bound for, respectively, the lower and the upper Binomial tails, we obtain that, for $0 < k \leq N$

$$\bar{\Phi}_{N,k}(\epsilon_+) \leq \exp\left(-2(N\epsilon_+ + 1 - k)^2/N\right), \quad k \leq N\epsilon_+ + 1,$$

$$\Phi_{N,k}(\epsilon_-) \leq \exp\left(-2(k - N\epsilon_-)^2/N\right), \quad k \geq N\epsilon_-.$$

Let now $\beta \in (0, 1)$ be given. Then, it can be readily verified that

$$\exp\left(-2(N\epsilon_+ + 1 - k)^2/N\right) \leq \beta/2$$

$$\exp\left(-2(k - N\epsilon_-)^2/N\right) \leq \beta/2$$

hold if

$$N\epsilon_- + \sqrt{\frac{N}{2}\ln(2/\beta)} \leq k \leq N\epsilon_+ + 1 - \sqrt{\frac{N}{2}\ln(2/\beta)}. \tag{3.12}$$

Therefore, (3.12) implies that $\bar{\Phi}_{N,k}(\epsilon_+) + \Phi_{N,k}(\epsilon_-) \leq \beta$, hence, from (3.11), $\mathbb{P}_z^N\{A\} \geq 1 - \beta$, which concludes the proof. □

Finally, we observe that, for the k-th order sample $z_{[k]}$, the empirical probability on the observed extraction z_i's being no larger than $z_{[k]}$ is precisely k/N, hence $z_{[k]}$ can be viewed as an "empirical ϵ-quantile" of the underlying distribution, for $\epsilon = k/N$. We can then specialize the previous corollary to assess precisely how close this empirical quantile is to the actual ϵ-quantile. This is established in the next corollary, which also provides an explicit, non-asymptotic condition on N guaranteeing a desired approximation of the quantile to arbitrary confidence.

Corollary 3. *Given N, $k \leq N$, $\alpha > 0$, let the probability distribution on z be continuous. Then,*

$$\mathbb{P}_z^N\{\mathbf{z} \in Z^N : q_{k/N-\alpha/2} < z_{[k]} \leq q_{k/N+\alpha/2}\} = \Phi_{N,k}(k/N + \alpha/2) - \Phi_{N,k}(k/N - \alpha/2).$$

Moreover, given $\beta \in (0,1)$, if

$$N \geq \frac{2}{\alpha^2} \ln(2/\beta) \tag{3.13}$$

then it holds that

$$\mathbb{P}_z^N\{\mathbf{z} \in Z^N : q_{k/N-\alpha/2} < z_{[k]} \leq q_{k/N+\alpha/2}\}\} \geq 1 - \beta.$$

Proof. Let $\epsilon_- \doteq \epsilon - \alpha/2$, $\epsilon_+ \doteq \epsilon + \alpha/2$, with $\epsilon = k/N$, and use (3.10): the k variable gets eliminated, and we obtain

$$N\frac{\alpha}{2} + 1 \geq \sqrt{\frac{N}{2} \ln(2/\beta)}$$

$$N\frac{\alpha}{2} \geq \sqrt{\frac{N}{2} \ln(2/\beta)}.$$

From the second of these conditions (which implies the first), we readily have that $N \geq \frac{2}{\alpha^2} \ln(2/\beta)$, which concludes the proof. □

Remark 3 (Practical Use of the Bounds). In practice, one may typically wish to use the k-th order sample to approximate the ϵ-quantile of the return distribution, for given ϵ. In this case, ϵ is given, and so are the accuracy $\alpha > 0$ and confidence β: integer k is then fixed as $k = \lceil \epsilon N \rceil$, hence the problem amounts to determining a suitable N such that

$$\mathbb{P}_z^N\{\mathbf{z} \in Z^N : q_{\epsilon-\alpha/2} < z_{[\lceil \epsilon N \rceil]} \leq q_{\epsilon+\alpha/2}\} \geq 1 - \beta. \tag{3.14}$$

Using Corollary 2, and recalling that $N\epsilon \leq \lceil \epsilon N \rceil \leq N\epsilon + 1$, one can easily check that conditions (3.10) simply boil down to $N \geq \frac{2}{\alpha^2} \ln(2/\beta)$, i.e., to the same condition as (3.13). Note however that (3.13) is a conservative bound on the required N: in some applications it is important to actually determine the smallest possible sample size N such that (3.14) is satisfied. This can be done by direct numerical evaluation of the exact expression from (3.9)

$$\mathbb{P}_z^N\{\mathbf{z} \in Z^N : q_{\epsilon-\alpha/2} < z_{[\lceil \epsilon N \rceil]} \leq q_{\epsilon+\alpha/2}\}\} = \Phi_{N,k}(\epsilon + \alpha/2) - \Phi_{N,k}(\epsilon - \alpha/2),$$

as a function of N, for $k = \lceil \epsilon N \rceil$. ◊

3.3 Single-Period Scenario Design

In Sect. 3.2 we discussed how to reliably estimate the ϵ-quantile of the return distribution of a fixed portfolio, by direct analysis of a stream of N portfolio returns generated by a scenario-generation oracle. In this section, we make a key leap forward, moving into the area of data-driven portfolio *design*; the results in this section have been originally exposed in [7]. The portfolio composition x in now unknown, and our objective is to determine x so that the random portfolio return has some desired properties, such as a guaranteed level of expected shortfall. To this end, we shall consider a stream of N oracle-generated iid random returns, collected by rows in a matrix R_N:

$$R_N^\top = \left[r^{(1)} \; r^{(2)} \; \cdots \; r^{(N)} \right] \in \mathbb{R}^{n,N}.$$

Notice that R_N is a random matrix, with each row independently distributed according to the (possibly unknown) distribution $\mathbb{P}$; events related to R_N are measured by the product probability measure $\mathbb{P}^N$, having support Δ^N. If $x \in \mathcal{X}$ is a portfolio vector, then the product

$$\rho_N(x) = R_N x = [\varrho_1(x) \; \varrho_2(x) \; \cdots \; \varrho_N(x)]^\top \in \mathbb{R}^N$$

is a vector of iid portfolio returns, where

$$\varrho_i(x) \doteq r^\top(i) x, \quad i = 1, \ldots, N.$$

Our approach is to select x so to maximize a return level γ that is exceeded by at least $N - q$ of the returns, where $q \leq N - n - 1$ is a given nonnegative integer. In other words, we select x so that γ is the empirical q/N-quantile of the portfolio distribution. This selection procedure is described next.

3.3.1 The Return Selection Rule

Let $q \leq N-n-1$ be a given nonnegative integer. We introduce a *rule* $\mathcal{S}_q$ for selecting a subset of cardinality $N-q$ of the returns in R_N. Rule $\mathcal{S}_q$ takes as input the matrix R_N and returns a partition $\mathcal{I}_q$, $\mathcal{I}_{N-q}$ of the set of indices $\mathcal{I} = \{1,\ldots,N\}$, such that, with probability one, the following properties are satisfied:

(a) $|\mathcal{I}_q| = q$, $|\mathcal{I}_{N-q}| = N-q$, and $\mathcal{I}_q \cup \mathcal{I}_{N-q} = \mathcal{I}$, $\mathcal{I}_q \cap \mathcal{I}_{N-q} = \emptyset$;
(b) let γ^*, x^* denote the optimal solutions of the following optimization problem:

$$\max_{x\in\mathcal{X},\gamma} \gamma \text{ subject to: } \varrho_i(x) \geq \gamma,\ i \in \mathcal{I}_{N-q}. \tag{3.15}$$

Then, it must be $\varrho_i(x^*) < \gamma^*$, for all $i \in \mathcal{I}_q$.

In words, the rule selects q returns in the sequence $\{\varrho_i(x)\}$ such that γ^* is the largest lower bound over a (suitably selected, see Remark 4) subset of $N-q$ returns, while q of the returns fall below γ^*.

As it will be made rigorously clear in the next section, we are in the presence of a fundamental tradeoff here: while level γ^* increases by increasing q, intuitively this level also becomes less and less *reliable* that is, informally, the probability of the actual portfolio return $\varrho(x^*)$ being larger than γ^* decreases. This fact should not come too much as a surprise, since level γ^* can be interpreted as the empirical (q/N)-quantile of the return sequence $\{\varrho_i(x^*)\}_{i=1,\ldots,N}$.

Remark 4 (On the Implementation of the Selection Rule). All results in this section hold for *any* selection rule that fulfills the requirements (a), (b) above. There are indeed several ways to define a suitable selection rule; some of these possibilities are briefly described next.

(i) *Optimal selection rule.* One possibility is to remove those q returns that provide the best possible improvement of the γ^* level in problem (3.15). We call this rule the *optimal* selection rule. From a computational point of view, implementing the optimal selection rule may be hard numerically, since it corresponds in principle to a combinatorial problem: among all subsets of $\{1,\ldots,N\}$ of cardinality q, select the one subset that provides the largest value in γ^*. Finding the optimal portfolio x^* and the corresponding level γ^* under the optimal selection rule may be cast in the form of a mixed-integer linear program as follows:

$$\max_{x\in\mathcal{X},\gamma,s_i\in\{0,1\}} \quad \gamma \tag{3.16}$$

$$\text{subject to: } Ms_i + \varrho_i(x) \geq \gamma,\ i = 1,\ldots,N$$

$$\textstyle\sum_{i=1}^N s_i = q,$$

where M is some "large" positive number (e.g., one may take $M = 1$, if all $\varrho_i(x)$ are known to be all smaller than one), and s_i, $i = 1, \ldots, N$, are additional 0/1 variables. Such problems can usually be solved quite effectively, for moderate sizes, using numerical packages for mixed-integer linear optimization, such as IBM CPLEX.

(ii) m-at-a-time rule. An alternative, suboptimal, rule for return removal can be implemented as described next. The idea is that although problem (3.16) is theoretically hard, it turns out in practice that it can be solved quite rapidly, if the number of suppressed constraints is small. Therefore, while it can be prohibitive to remove all q constraints at once (as it is prescribed by the optimal rule), it is usually doable to remove $m \ll q$ of them iteratively. The "m-at-a-time rule" rule thus simply prescribes to suppress iteratively $1 \leq m \ll q$ returns at a time, by solving repeatedly a problem of the form (3.16) with m instead of q, until all q constraints have been removed. In principle, this approach is suboptimal, and may not yield the same result as the optimal rule. However, it usually gives good results in practice.

(iii) Lagrange multiplier-based rule. Another possibility (useful if one does not have a mixed-integer solver available) is to prune the returns sequentially (one by one, or $m \geq 1$ at a time) according to their impact on objective sensitivity. With this approach, one first solves the LP with all returns in place, then removes the $1 \leq m \leq \min(n, q)$ returns that yield the best local improvement in the objective, then solves again the LP, and so on, until all q returns are removed. Suitable implementation of such a technique provides a valid selection rule. At each iteration, the returns to be removed can be determined by looking at the values λ_i of the Lagrange multipliers (dual variables) associated with the surviving constraints $\varrho_i(x) \geq \gamma$. It is indeed well known (see, e.g., [3]) that a positive Lagrange multiplier λ_i represents the sensitivity of the optimal objective value to variations in the i-th inequality constraint, hence the locally-best choice is to remove the m constraints corresponding to the m largest λ_i, since this would induce (to first order approximation) the largest improvement in objective value. A distinctive advantage of this selection rule is that the optimal portfolio allocation problem in (3.15) is solved efficiently by solving a sequence of standard linear programming problems.

Other constraint removal heuristics may be devised, besides the described ones. It is important, however, to stress again the fact that the theory that is presented in this paper *does not depend* on the specific selection rule and, in particular, it does not need implementation of the optimal rule (which may be hard to compute). The results in this paper hold for *any* selection rule that satisfies the requirements (a), (b) specified at the beginning of Sect. 3.3.1. $\Diamond$

3.3.2 The Shortfall Probability

Given a selection rule $\mathcal{S}_q$, the optimal portfolio allocation strategy x^* that we propose is a solution of the following LP

$$\gamma^* = \max_{x \in \mathcal{X}, \gamma} \quad \gamma \tag{3.17}$$
$$\text{subject to: } r^{(i)\top} x \geq \gamma, \; i \in \mathcal{I}_{N-q},$$

where, by the definition of the selection rule, it holds that $\varrho_i(x^*) = r^{(i)\top} x^* < \gamma^*$, for all $i \in \mathcal{I}_q$. It is important to underline that we here take an a-priori point of view that is, a priori, the return vectors $r^{(i)}$, $i = 1, \ldots, N$, are random variables, hence also the optimal solutions γ^*, x^* of (3.17) are random variables, which are functions of the random data of the problem (i.e., of $r^{(i)}$, $i = 1, \ldots, N$, which are collected in the random matrix R_N). Events involving γ^*, x^* are thus measured by the product probability $\mathbb{P}^N$. The problem under study belongs to the class of so-called random convex programs (RCP), or scenario-based optimization, see [6, 8, 9]. In particular, we here build upon the technique of random programs with violated constraints described in [6] in order to derive the desired probabilistic bounds.

If we *observe* an actual *realization* of the returns (for example, by looking a posteriori at the stream of N returns generated by the oracle), then the observed return sequence becomes deterministic, and (3.17) would return a deterministic vector x^* and a deterministic level γ^*. However, *before* we look at the actual realization, these two variables remain uncertain and random. We are interested in providing an a-priori probabilistic characterization on the optimal solution of (3.17). To this end, we introduce a further assumption and a definition.

Assumption 1 (Uniqueness). *$\mathcal{X}$ is a nonempty polytope and, with probability one, the optimal solution x^*, γ^* of (3.17) is unique.* ⋆

Remark 5. The assumption that $\mathcal{X}$ is a nonempty polytope guarantees that the set $\mathcal{X}$ is compact and nonempty, which implies that problem (3.17) is feasible and it attains an optimal solution; this assumption is generally fulfilled in portfolio optimization problems, hence it is not restrictive in practice. Assumption 1 further postulates that the optimal solution is uniquely identified, i.e., that the optimum of the LP is attained at a vertex. This is usually the case for LP constrains in "general position" (e.g., excluding cases of two or more identical returns, which happen with zero probability under continuous distributions). Moreover, an infinitesimal perturbation of the constraints, or introduction of a strictly convex regularization term in the objective would always make the optimal solution unique. Assumption 1 is thus made for technical reasons, and it is not restrictive from a practical point of view, in the present context. ♢

For a *fixed* portfolio $x \in \mathcal{X}$ and return level $\gamma \in \mathbb{R}$, we define the shortfall probability as

$$V(x, \gamma) = \mathbb{P}\{r : r^\top x < \gamma\}.$$

Such a probability is a *number* in [0, 1]. However, if we now ask about the shortfall probability relative to the optimal solution of (3.17), we have

$$V^* \doteq V(x^*, \gamma^*) = \mathbb{P}\{r : r^\top x^* < \gamma^*\}, \tag{3.18}$$

and this is, a priori, a random variable, since x^*, γ^* are so. Indeed, for each different realization of the random returns R_N we will get different x^*, γ^*, hence a different V^*. Therefore, V^* is a random variable with support [0, 1], and events related to V^* are measured by the product probability $\mathbb{P}^N$. It is then natural to consider as a measure of "riskiness" of the optimal portfolio the expected value (with respect to $\mathbb{P}^N$) of the shortfall probability V^*. This leads to the following definition.

Definition 1 (Expected Shortfall Probability). The expected shortfall probability of the optimal portfolio resulting from (3.17) is defined as

$$\mathbb{E}_{\mathbb{P}^N}\{V^*\} = \mathbb{E}_{\mathbb{P}^N}\{\mathbb{P}\{r : r^\top x^* < \gamma^*\}\}.$$

Our key result concerns a quantification of an upper bound on the expected shortfall probability of the optimal portfolio. This is developed in the next section.

3.3.3 *Shortfall Probability of the Optimal Data-Driven Portfolio*

The first result we report concerns an upper bound on the upper tail of the distribution of V^*. This results follows directly from Theorem 4.1 and Corollary 4.2 in [6], considering that the problem (3.17) we are dealing with is precisely a random convex program with Helly's dimension upper bounded by $n + 1$, which is the number of decision variables in problem (3.17); see [6] for further details and definitions.

Lemma 1 (Upper-Tail Bound on V^*). *Let Assumption 1 hold, and let x^*, γ^* be the optimal solution of problem (3.17), under any given selection rule satisfying properties (a)–(c) specified in Sect. 3.3.1. Let V^* be defined as in (3.18). Then it holds that*

$$\mathbb{P}^N\{V^* > z\} \le \binom{q+n}{q} \bar{\Phi}(z; q+n, N), \tag{3.19}$$

where

$$\bar{\Phi}_{N,q+n+1}(z) \doteq \sum_{j=0}^{q+n} \binom{N}{j} z^j (1-z)^{N-j}, \tag{3.20}$$

and

$$\binom{q+n}{q} = \frac{(q+n)!}{q!n!} = \prod_{i=1}^{q} \frac{n+i}{i}.$$

★

Note that $\bar{\Phi}$ in (3.20) can be expressed in terms of the regularized incomplete beta function $I(z; a, b)$ as follows

$$\begin{aligned} \bar{\Phi}_{b,a+1}(z) &= I(1-z; b-a, a+1) \\ &= 1 - I(z; a+1, b-a). \end{aligned}$$

An important consequence of Lemma 1 is that, for given level z and a suitable choice of the time window N and of the removal cardinality q, we can make the upper tail bound (3.19) as small as desired, so that with practical certainty the optimization will provide a V^* such that $V^* \leq z$.

We next state a result which provides an explicit and efficient upper bound on the expected shortfall probability.

Lemma 2 (Upper Bound on the Expected Shortfall Probability). *Let Assumption 1 hold, and let x^*, γ^* be the optimal solution of problem (3.17), under any given selection rule satisfying properties (a)–(c) specified in Sect. 3.3.1. Let V^* be defined as in (3.18). Then it holds that*

$$\mathbb{E}_{\mathbb{P}^N}\{V^*\} \leq \frac{q}{N} + \left(\frac{n}{N} + \frac{\omega(n,q)}{2\sqrt{N}}\right), \tag{3.21}$$

where $\omega(n,q) = \mathcal{O}(\sqrt{2n\ln(q+n)})$ and, more precisely,

$$\omega(n,q) = \frac{2n(1+\ln(q+n)-\ln n) - 2\ln 2 + 1}{\sqrt{2n(1+\ln(q+n)-\ln n) - 2\ln 2}}.$$

★

A proof for Lemma 2 is given in Appendix 1 of [7].

Remark 6. Equation (3.21) has the following interpretation: $\eta \doteq q/N$ is the *empirical* shortfall probability, i.e., the shortfall probability of the optimal data-driven portfolio on the data R_N that are used for the optimization. In other words, η is the in-sample shortfall empirical probability. The extra term in (3.21)

$$\epsilon(N,q,n) = \frac{n}{N} + \frac{\omega(n,q)}{2\sqrt{N}}$$

represents the *excess* shortfall probability due to the fact that the data upon which the optimal portfolio is built (the R_N) are themselves random. ◊

Lemma 2 provides us with an explicit upper bound on the expected shortfall probability, for given data length N and removal cardinality q. The formula in (3.21), as well as the formula in (3.19), can be "inverted," at least numerically, in order to find suitable N and q, given assigned levels of tolerable shortfall probability or of expected shortfall probability. These fundamental tradeoffs are highlighted in the following corollary, whose proof is reported in Appendix 3 of [7].

Corollary 4 (Explicit Conditions on N and q). *Let $\beta \in (0,1)$ be a very small probability level chosen by the user (e.g., set $\beta = 10^{-6}$, or lower, for "practical certainty"). Let $z_{\mathrm{tol}} \in (0,1)$ be a desired tolerable shortfall probability level, let $z_{\mathrm{exp}} \in (0,1)$ be a desired expected shortfall probability level, and let $q \leq N-n-1$. Then, the following statements are true:*

(1) If

$$N \geq \frac{2}{z_{\mathrm{tol}}} \ln \beta^{-1} + \frac{4}{z_{\mathrm{tol}}}(q+n), \tag{3.22}$$

then $\{V^* \leq z_{\mathrm{tol}}\}$ holds with probability larger than $1-\beta$ (i.e., with practical certainty).

(2) If

$$N \geq 4\frac{q+n}{z_{\mathrm{exp}}} + \frac{(c+1/c)^2}{4{z_{\mathrm{exp}}}^2}, \tag{3.23}$$

with $c \doteq \sqrt{2n + 2n\ln\frac{n+q}{q} - 2\ln 2}$, then it holds that $\mathbb{E}_{\mathbb{P}^N}\{V^*\} \leq z_{\mathrm{exp}}$. For "large" q, bound (3.23) simplifies approximately to

$$N \geq 4\frac{q+n}{z_{\mathrm{exp}}} + \frac{2n + 2n\ln\frac{n+q}{q} - 2\ln 2}{4{z_{\mathrm{exp}}}^2}.$$

★

Remark 7. Equations (3.22), (3.23) provide a rigorous quantification of the tradeoff between acceptable risk and the cardinality of scenarios used for optimization. The practical use of these equations is illustrated next; to fix ideas we concentrate on design based on the expected shortfall probability (Eq. (3.23)), the discussion on (3.22) being analogous. We make three observations.

(*i*) For fixed N and for a given desired level of expected shortfall probability z_{exp}, it is the investor's interest to make γ^* as large as possible. On a given realization of the returns, level γ^* increases if we increase the number q of suppressed returns, hence we want to make q as large as possible. However, if one increases q too much, then the resulting portfolio will fail to satisfy the expected shortfall probability requirement. The right-hand-side of (3.23) is increasing in q, hence this term tells us precisely how large q can be made, while satisfying the requirement $\mathbb{E}_{\mathbb{P}^N}\{V^*\} \leq z_{\text{exp}}$: we choose q such that the right-hand-side of (3.23) is the largest integer that does not exceed N.

(*ii*) If N is not fixed (e.g., we are free to decide what the scenario cardinality to be used for optimization should be), then Eq. (3.23) can be used to plot a tradeoff set on an (N, q) plane, for the given desired level z_{exp}. Any pair N, q in the admissible set is a valid pair guaranteeing that the result of the optimization will satisfy $\mathbb{E}_{\mathbb{P}^N}\{V^*\} \leq z_{\text{exp}}$.

(*iii*) An interesting feature that is captured by the present theory is that, all the other parameters being the same, the expected shortfall probability bound increases as n (the number of securities in the portfolio) increases. The reason for this lies at the fundamental tradeoff between the complexity of the random optimization model (here, the number of variables, $n + 1$) and the out-of-sample *reliability* of the model: the more complex the model is (i.e., the larger n is), the more training data we need for achieving a given reliability level (i.e., the larger N needs to be). A financial interpretation of this phenomenon is that high diversification of a portfolio (large n) needs a large number N of data in order to provide meaningful portfolios.

$\diamond$

3.4 Multi-Period Scenario Design

In this section, we outline a multi-period extension of the idea developed in the previous section. We consider a decision problem over T periods (or *stages*), where at each period we have the opportunity of rebalancing our portfolio allocation, with the objective of obtaining a maximal level of return at the final stage, while guaranteeing a desired level of expected shortfall.

We start in Sect. 3.4.1 by considering, for simplicity of exposition, an "open-loop" strategy with no recourse. This model can be extended to include conditional decisions with affine recourse policies, along the path described in [4, 5]; such extension is outlined in Sect. 3.4.2.

3.4.1 Open-Loop Strategy

Consider a decision horizon of T periods, where the k-th period starts at time $k-1$ and ends at time k, see Fig. 3.2.

We denote with $x_i(k)$ the Euro value of the portion of the investor's total wealth invested in security a_i at time k. The portfolio at time k is the vector

$$x(k) \doteq \left[x_1(k) \cdots x_n(k)\right]^\top.$$

The investor's total wealth at time k is

$$w(k) \doteq \sum_{i=1}^{n} x(k) = \mathbf{1}^\top x(k),$$

where $\mathbf{1}$ denotes a vector of ones. Let $x(0)$ be the given initial portfolio composition at time $k = 0$. For instance, we may assume that $x(0)$ is all zeros, except for one entry representing the initial available amount of cash. At $k = 0$, we have the opportunity of conducting transactions on the market and therefore adjusting the portfolio by increasing or decreasing the amount invested in each asset. Just after transactions, the adjusted portfolio is

$$x^+(0) = x(0) + u(0),$$

where $u_i(0) > 0$ if we increase the position on the i-th asset, $u_i(0) < 0$ if we decrease it, and $u_i(0) = 0$ if we leave it unchanged. Suppose now that the portfolio is held unchanged for the first period of time Δ. At the end of this first period, the portfolio composition is

$$x(1) = G(1)x^+(0) = G(1)x(0) + G(1)u(0),$$

where $G(1) = \mathrm{diag}(g_1(1), \ldots, g_n(1))$ is a diagonal matrix of the asset gains over the period from time 0 to time 1. At time $k = 1$, we perform again an adjustment of the portfolio: $x^+(1) = x(1) + u(1)$, and then hold the updated portfolio for another period of duration Δ. At time $k = 2$ the portfolio composition is hence

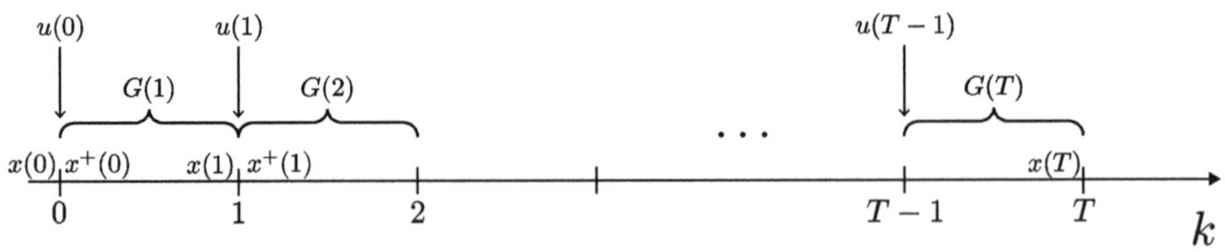

Fig. 3.2 Portfolio dynamics and periods

$$x(2) = G(2)x^+(1) = G(2)x(1) + G(2)u(1).$$

Proceeding in this way for $k = 0, 1, 2, \ldots$, we determine the iterative dynamic equations of the portfolio composition at the end of period $(k+1)$:

$$x(k+1) = G(k+1)x(k) + G(k+1)u(k), \quad k = 0, \ldots, T-1 \tag{3.24}$$

as well as the equations for portfolio composition just after the $(k+1)$-th transaction (see Fig. 3.2)

$$x^+(k) = x(k) + u(k).$$

From (3.24) it results that the (random) portfolio composition at time $k = 1, \ldots, T$ is

$$\begin{aligned} x(k) &= \Phi(1,k)x(0) + \left[\, \Phi(1,k)\ \Phi(2,k) \cdots \Phi(k-1,k)\ \Phi(k,k) \,\right] \begin{bmatrix} u(0) \\ u(1) \\ \vdots \\ u(k-2) \\ u(k-1) \end{bmatrix} \\ &= \Phi(1,k)x(0) + \Omega_k \mathbf{u}, \end{aligned}$$

where we defined $\Phi(\nu,k)$, $\nu \leq k$, as the *compounded gain* matrix from the beginning of period ν to the end of period k:

$$\Phi(\nu,k) \doteq G(k)G(k-1)\cdots G(\nu), \quad \Phi(k,k) \doteq G(k),$$

and

$$\begin{aligned} \mathbf{u} &= \left[\, u(0)^\top \cdots u(T-2)^\top\ u(T-1)^\top \,\right]^\top, \\ \Omega_k &= \left[\, \Phi(1,k) \cdots \Phi(k-1,k)\ \Phi(k,k) \middle| 0 \cdots 0 \,\right]. \end{aligned}$$

We thus have for the total wealth

$$w(k) = \mathbf{1}^\top x(k) = \phi(1,k)^\top x(0) + \omega_k^\top \mathbf{u},$$

where

$$\begin{aligned} \phi(\nu,k)^\top &\doteq \mathbf{1}^\top \Phi(\nu,k), \\ \omega_k^\top &\doteq \mathbf{1}^\top \Omega_k = [\phi(1,k)^\top \ \cdots \ \phi(k-1,k)^\top\ \phi(k,k)^\top \mid 0 \ \cdots \ 0]. \end{aligned}$$

We consider the portfolio to be self-financing, that is

$$\sum_{i=1}^{n} u_i(k) = 0, \quad k = 0, \ldots, T-1,$$

and we include generic linear constraints in the model by imposing that the updated portfolios $x^+(k)$ lie within a given polytope $\mathcal{X}(k)$. The purpose of our decision model is to determine, at time 0, the adjustments $u(0), \ldots, u(T-1)$ so to maximize a level γ of final return under an assigned empirical shortfall constraint.

The cumulative gross return of the investment over the whole horizon is

$$\varrho(\mathbf{u}) \doteq \frac{w(T)}{w(0)} = \frac{\mathbf{1}^\top x(T)}{\mathbf{1}^\top x(0)} = \frac{\phi(1,T)^\top x(0)}{\mathbf{1}^\top x(0)} + \frac{1}{\mathbf{1}^\top x(0)} \omega_T^\top \mathbf{u}.$$

We see that $\varrho(\mathbf{u})$ is an affine function of the decision variables $\mathbf{u}$, with a random vector ω_T of coefficients that depends on the random gains over the T periods.

Suppose now that N iid samples (scenarios) of the period gains $\{G(k),\ k = 1, \ldots, T\}$ are available from a scenario generating oracle. These sample produce in turn N scenarios for each of the Ω_k matrices, $k = 1, \ldots, T$, and hence of the ω_k and $\phi(1,k)$ vectors. We denote such scenarios with $\Omega_k^{(i)}$, $\omega_k^{(i)}$, $\phi^{(i)}(1,k)$, $i = 1, \ldots, N$, and with $x^{(i)}(k)$, $w^{(i)}(k)$, $\varrho^{(i)}(\mathbf{u})$, respectively, the portfolio composition at time k, the total wealth at time k, and the cumulative final return, under the i-th scenario. Consider next a selection rule $\mathcal{S}_q$, analogous to the one introduced in Sect. 3.3.1, that selects $N-q$ of the generated scenarios, such that

$$\begin{aligned}(\mathbf{u}^*, \gamma^*) = \arg\max_{\mathbf{u},\gamma} \quad & \gamma \\ \text{s.t.:} \quad & \varrho^{(i)}(\mathbf{u}) \geq \gamma \quad i \in \mathcal{I}_{N-q} \\ & x^{(i)}(k) \in \mathcal{X}(k),\ k = 1, \ldots, T;\ i \in \mathcal{I}_{N-q} \\ & \mathbf{1}^\top u(k) = 0,\ \ k = 0, \ldots, T-1,\end{aligned} \tag{3.25}$$

and

$$\varrho^{(i)}(\mathbf{u}^*) < \gamma^*, \quad \text{for } i \in \mathcal{I}_q.$$

The selection rule can be implemented in one of the ways described in Remark 4. With the implementation in point *(iii)*, finding the optimal portfolio adjustments $\mathbf{u}^*$ amounts to solving a sequence of linear programming problems. However, regardless of the chosen selection rule, and under the usual assumptions of existence and uniqueness of the optimal solution, we can apply the result in Lemma 2 to quantify an a-priori guarantee on the expected shortfall probability of the final return. To this end, we define

$$V^* = \mathbb{P}\{\varrho^{(i)}(\mathbf{u}^*) < \gamma^*\},$$

and then apply Eq. (3.21) substituting n with the actual number of decision variables nT (i.e., the dimension of vector $\mathbf{u}$), thus obtaining

$$\mathbb{E}_{\mathbb{P}^N}\{V^*\} \leq \frac{q}{N} + \left(\frac{nT}{N} + \frac{\omega(nT,q)}{2\sqrt{N}}\right).$$

Here, probability $\mathbb{P}$ refers to the probability measure on the gain sequences $\{G(k),\ k = 1,\dots,T\}$, and $\mathbb{P}^N$ to the corresponding N-fold product measure.

3.4.2 Closed-Loop Strategy with Affine Policies

As it is well known, the open-loop strategy discussed in the previous section may be suboptimal in an actual implementation, since all adjustment decisions $u(0),\dots,u(T-1)$ are computed at time $k = 0$. While the first decision $u(0)$ must be immediately implemented (here-and-now variable), the future decisions may actually wait-and-see the actual outcomes of the returns in the forward periods, and hence benefit from the uncertainty reduction that comes from these observations. For example, at time $k \geq 1$, when we need to implement $u(k)$, we have *observed* a realization of the asset returns over the periods from 1 to k. Hence, we would like to exploit this information, by considering *conditional* allocation decisions $u(k)$, that may react to the returns observed over the previous periods. This means that, instead of focusing on fixed decisions $u(k)$, we wish to determine suitable *policies* that prescribe what the actual decision should be, in dependence of the observed returns from 1 up to k. In determining the structure of the decision policy one should evaluate a tradeoff between generality and numerical viability of the ensuing optimization problems. In some recent papers, see, e.g., [2, 4, 5] it has been observed that linear or affine policies do provide an interesting tradeoff by allowing reactive policies to be efficiently computed via convex optimization techniques. In this paper, we follow this route, and consider decisions prescribed by affine policies of the following form

$$u(k) = \bar{u}(k) + \Theta(k)\,(g(k) - \bar{g}(k))\,, \quad k = 1,\dots,T-1, \tag{3.26}$$

and $u(0) = \bar{u}(0)$, where $\bar{u}(k) \in \mathbb{R}^n$, $k = 0,\dots,T-1$ are "nominal" allocation decision variables, $g(k)$ is the vector of gains over the k-th period, $\bar{g}(k)$ is a given estimate of the expected value of $g(k)$, and $\Theta(k) \in \mathbb{R}^{n,n}$, $k = 1,\dots,T-1$, are the policy "reaction matrices," whose role is to adjust the nominal allocation with a term proportional to the deviation of the gain $g(k)$ from its expected value. Since the budget conservation constraint $\mathbf{1}^\top u(k) = 0$ must hold for any realization of the gains, we shall impose the restrictions

$$\mathbf{1}^\top \bar{u}(k) = 0, \;\; \mathbf{1}^\top \Theta(k) = 0, \quad k = 0, 1, \dots, T-1.$$

Applying the adjustment policy (3.26) to the portfolio dynamics Eqs. (3.24), (3.4.1), we have

$$x^{+}(k) = x(k) + \bar{u}(k) + \Theta(k)\,(g(k) - \bar{g}(k)) \tag{3.27}$$

$$x(k+1) = G(k+1)x^{+}(k), \quad k = 0, 1 \ldots, T-1, \tag{3.28}$$

with $\Theta(0) \doteq 0$. From repeated application of (3.27), (3.28) we obtain the expression for the portfolio composition at a generic instant $k = 1, \ldots, T$:

$$x(k) = \Phi(1,k)x(0) + \Omega_k\bar{\mathbf{u}} + \sum_{t=1}^{k} \Phi(t,k)\Theta(t-1)\tilde{g}(t-1),$$

where $\Theta(0) = 0$, and

$$\bar{\mathbf{u}} \doteq \left[\bar{u}(0)^\top \cdots \bar{u}(T-2)^\top\ \bar{u}(T-1)^\top\right]^\top,$$
$$\tilde{g}(k) \doteq g(k) - \bar{g}(k), \quad k = 1, \ldots, T.$$

A key observation is that $x(k)$ is an affine function of the decision variables $\bar{u}(k)$ and $\Theta(k)$, $k = 1, \ldots, T-1$. The cumulative gross return of the investment over the whole horizon is then

$$\begin{aligned}\varrho(\bar{\mathbf{u}}, \boldsymbol{\Theta}) &= \frac{w(T)}{w(0)} = \frac{\mathbf{1}^\top x(T)}{\mathbf{1}^\top x(0)} \\ &= \frac{1}{\mathbf{1}^\top x(0)}\left(\phi(1,T)^\top x(0) + \omega_T^\top\bar{\mathbf{u}} + \sum_{t=1}^{T}\Phi(t,T)\Theta(t-1)\tilde{g}(t-1)\right),\end{aligned}$$

which is again affine in the variables $\bar{\mathbf{u}}$ and $\boldsymbol{\Theta} \doteq [\Theta(1) \cdots \Theta(T-1)]$. Given N iid samples (scenarios) of the period gains $\{G(k),\, k = 1, \ldots, T\}$, generated by a scenario generating oracle, and a selection rule $\mathcal{S}_q$, we can determine optimal policies by solving a problem similar to (3.25), that is

$$\begin{aligned}(\bar{\mathbf{u}}^*, \boldsymbol{\Theta}^*, \gamma^*) = \arg\max_{\bar{\mathbf{u}}, \boldsymbol{\Theta}, \gamma}\ & \gamma \\ \text{s.t.: } & \varrho^{(i)}(\bar{\mathbf{u}}, \boldsymbol{\Theta}) \geq \gamma\ i \in \mathcal{I}_{N-q} \\ & x^{(i)}(k) \in \mathcal{X}(k),\ k = 1, \ldots, T;\ i \in \mathcal{I}_{N-q} \\ & \mathbf{1}^\top \mathbf{u}(k) = 0,\ \ k = 0, \ldots, T-1, \\ & \mathbf{1}^\top \Theta(k) = 0,\ \ k = 1, \ldots, T-1,\end{aligned}$$

and

$$\varrho^{(i)}(\bar{\mathbf{u}}^*, \boldsymbol{\Theta}^*) < \gamma^*, \quad \text{for } i \in \mathcal{I}_q.$$

These optimal allocations may be determined in a numerically efficient way by solving, for instance, a sequence of linear programming problems. The results of Lemma 2 apply also to this problem for quantifying an a-priori guarantee on the expected shortfall probability of the final return, considering that the number of optimization variables is now $nT + n^2(T-1)$.

3.4.3 Sliding-Horizon Implementation

We remark that, in practice, both the open-loop and the closed-loop investment models will typically be applied using a *sliding horizon* strategy, whereby only the first adjustment $u(0)$ (or $\mathbf{u}(0)$, in the closed-loop approach) is executed at the current time $k = 0$. Then, the investor waits one period and, at time $k = 1$, solves the whole problem again over a forward-shifted horizon and executes only the first adjustment of the computed sequence, and so on. This approach permits to effectively take full advantage of the information that becomes progressively available as the time moves forward (e.g., at time $k = 0$ the gain $G(1)$ is uncertain and random, while at time $k = 1$ we can see a realization of $G(1)$, and hence profit from this information when computing the adjustment $u(1)$, etc.).

3.5 Scenario Methods for Single-Period Robust Portfolio Design

In this section we outline how random convex programming (RCP) or scenario-based optimization methods can be effectively used also in the context of model-based robust portfolio design. We focus on a single-period mean-variance framework, and consider only one prototype problem of robust variance minimization under a minimal return constraint. Different from the previous sections, we shall assume henceforth that an *ambiguous model* for the asset returns has been obtained, for instance, by historical data analysis, possibly in conjunction with expert knowledge, as detailed next.

3.5.1 Robust Portfolio Allocation Models

We assume that the return vector is a random variable with expected value $\hat{r} \in \mathbb{R}^n$ and covariance matrix $\Sigma \in \mathbb{R}^{n,n}$ (symmetric and positive semidefinite). We assume, however, that these two moments are not known exactly, i.e., that the return distribution is *ambiguous*, that is, there exist uncertainty in the value of these parameters. We model this uncertainty by assuming that $(\hat{r}, \Sigma)$ belongs to some given bounded uncertainty set $\mathcal{U}$, over which we also specify a probability distribution, modeling our prior knowledge on the likelihood of the outcomes of these parameters. For a given portfolio x, the worst-case portfolio variance is given by

$$\sigma_{\rm wc}{}^2 = \sup_{(\hat{r},\Sigma)\in\mathcal{U}} x^\top \Sigma x.$$

A minimum worst-case variance portfolio design problem would then be

$$\begin{aligned} \min_x \quad & \sup\nolimits_{(\hat{r},\Sigma)\in\mathcal{U}} x^\top \Sigma x \\ \text{s.t.:}\ & \inf\nolimits_{(\hat{r},\Sigma)\in\mathcal{U}} \hat{r}^\top x \geq \gamma_{\min} \\ & x \in \mathcal{X}, \end{aligned}$$

where $\gamma_{\min}$ is a given lower bound on the portfolio expected return. The problem can be stated equivalently as follows:

$$\begin{aligned} \min_{x,t} \quad & t \\ \text{s.t.:}\ & x^\top \Sigma x \leq t \ \ \forall\, (\hat{r},\Sigma) \in \mathcal{U} \\ & \hat{r}^\top x \geq \gamma_{\min} \ \forall\, (\hat{r},\Sigma) \in \mathcal{U}, \\ & x \in \mathcal{X}. \end{aligned} \tag{3.29}$$

This problem can be solved efficiently and exactly only in some special cases, where the uncertainty set $\mathcal{U}$ is "simple." For instance, if $\mathcal{U} = \{(\hat{r},\Sigma) : \hat{r} \in \mathcal{U}_r,\ \Sigma \in \mathcal{U}_\Sigma\}$, where $\mathcal{U}_r, \mathcal{U}_\Sigma$ are interval sets

$$\begin{aligned} \mathcal{U}_r &= \{\hat{r} : r_{\min} \leq \hat{r} \leq r_{\max}\}, \\ \mathcal{U}_\Sigma &= \{\Sigma : \Sigma_{\min} \leq \Sigma \leq \Sigma_{\max},\ \Sigma \succeq 0\}, \end{aligned}$$

then, under the further assumptions that $x \geq 0$ and $\Sigma_{\max} \succeq 0$, problem (3.29) is equivalent to (see Tütüncü and Koenig [21])

$$\begin{aligned} \min_{x,t} \quad & x^\top \Sigma_{\max} x \\ \text{s.t.:}\ & r_{\min}{}^\top x \geq \gamma_{\min} \\ & x \in \mathcal{X}. \end{aligned}$$

Another uncertainty model, proposed by Goldfarb and Iyengar [15], assumes that the return vector is described by a factor model

$$r = \hat{r} + V^\top f + \varepsilon,$$

where $V \in \mathbb{R}^{m,n}$ is the matrix of factor loadings, $f \in \mathbb{R}^m$ is the vector of returns of the factors that drive the market, and ε is a vector of residual terms. In the model of [15], it is assumed that f has zero mean and covariance F, and ε has zero mean and covariance $D = \mathrm{diag}(d_1, \ldots, d_n)$; F is assumed to be known exactly, while interval uncertainty is assumed on D (i.e., $d_i \in [\underline{d}_i, \bar{d}_i]$, $i = 1, \ldots, n$) and on $\hat{r}$ (i.e., $\hat{r}_i = \bar{r}_i + \xi_i$, $|\xi_i| \leq \lambda_i$, $i = 1, \ldots, n$), and column-wise uncertainty is assumed on the factor loading matrix, i.e.,

$$V = V_0 + W, \quad \text{where } w_i^\top G w_i \leq \rho_i^2,\ i = 1, \ldots, n,$$

where w_i denotes the i-th column of matrix W, and $G \succ 0$ is a given matrix. Under such hypotheses, problem (3.29) can be reformulated into an equivalent second order cone program (SOCP), and hence solved efficiently.

However, for more generic uncertainty sets, not satisfying the hypotheses of the two particular cases mentioned above, the robust minimum worst-case variance problem (3.29) remains computationally intractable, in general. The same claim can be made for related robust portfolio design problems, such as the robust maximum return problem (in which one maximizes the minimum expected return, under an upper bound on the worst-case portfolio variance), the robust maximum Sharpe ratio problem (in which one maximizes the worst-case ratio of the expected excess return on the portfolio to the standard deviation of the return), or the robust value-at-risk problem [13].

3.5.2 *The Scenario Approach*

In the scenario approach to robust optimization (also known as the random convex programming (RCP) paradigm), we tackle "hard" robustness problems that cannot be solved efficiently via exact approaches, by relaxing the deterministic worst-case paradigm via a probabilistic one, and by resorting to uncertainty randomization. This approach has two aspects, one computational and the other theoretical. From the computational side, the scenario approach is very simple: collect N independent samples—the scenarios—$(\hat{r}^{(i)}, \Sigma^{(i)})$, $i = 1, \ldots, N$, of the uncertain parameters from the set $\mathcal{U}$ according to their probability distribution $\mathbb{P}_u$, and then solve the standard convex optimization problem

$$\begin{aligned} \min_{x,t} \quad & t \\ \text{s.t.:} \quad & x^\top \Sigma^{(i)} x \leq t \;\; i = 1, \ldots, N, \\ & \hat{r}^{(i)\top} x \geq \gamma_{\min} \;\; i = 1, \ldots, N, \\ & x \in \mathcal{X}. \end{aligned} \tag{3.30}$$

From the theoretical side, the RCP theory provides us a guarantee on the level of "probabilistic robustness" of the optimal solution of problem (3.30). Indeed, observe that, a priori, the optimal solution of (3.30), (x^*, t^*), is a random variable that depends on the scenario multi-extraction $(\hat{r}^{(i)}, \Sigma^{(i)})$, $i = 1, \dots, N$, hence events related to (x^*, t^*) are measured by the product probability $\mathbb{P}_u^N$. The probabilistic robustness R of a *given* portfolio x and level t is defined as

$$R(x, t) = \mathbb{P}_u\{(\hat{r}, \Sigma) : x^\top \Sigma x \leq t, \ \hat{r}^\top x \geq \gamma_{\min}\}.$$

For instance, if (x, t) is the solution of a deterministically robust design, one would have $R(x, t) = 1$, since this solution guarantees satisfaction of the constraints, for all possible outcomes of the "uncertainty" $(\hat{r}, \Sigma) \in \mathcal{U}$. We are interested in quantifying the probabilistic robustness of the optimal solution of the scenario problem (3.30), that is in

$$R^* \doteq R(x^*, t^*).$$

However, since (x^*, t^*) is random, $R(x^*, t^*)$ is also a random variable, with support in the interval $[0, \ 1]$. The best we could hope for is thus to determine the probability distribution of $R(x^*, t^*)$, or at least a lower bound on it. Clearly, the more R^* is concentrated towards high values (i.e., close to one), the more "robust" the solution is. Under the assumption that problem (3.30) is feasible with probability one and that the optimal solution is unique, we can apply the result in Corollary 3.4 of [6]. Since the number of variables of problem (3.30) is $n+1$ and the problem is assumed to be feasible almost surely, the Helly's dimension of the problem is $\zeta = n + 1$, and we obtain from Corollary 3.4 of [6] that, for $\epsilon \in (0, 1)$

$$\mathbb{P}_u^N\{R^* < 1 - \epsilon\} \leq \bar{\Phi}_{N,n+1}(\epsilon),$$

where $\bar{\Phi}_{N,n+1}$ is defined in (3.20), namely

$$\bar{\Phi}_{N,n+1}(\epsilon) = \sum_{j=0}^{n} \binom{N}{j} \epsilon^j (1 - \epsilon)^{N-j},$$

In practice, one selects a very small value $\beta \in (0, \ 1)$, say $\beta = 10^{-7}$, and a desired level of probabilistic robustness $1 - \epsilon$ for the portfolio, and determines numerically a value for integer N such that $\bar{\Phi}_{N,n+1}(\epsilon) \leq \beta$. Solving the scenario problem (3.30) with this value of N will guarantee a priori that the resulting portfolio will satisfy $R^* \geq 1 - \epsilon$, with overwhelming probability $1 - \beta$. Since β is chosen so small that the event $R^* < 1 - \epsilon$ is unlikely to occur to all practical purposes, we have a tool for obtaining portfolio designs that are probabilistically robust to level ϵ. The advantage of this approach is that it works for any generic structure of the uncertainty, and it does not require lifting the class of optimization problems one needs to solve, i.e., the "robustified" problem (3.30) remains a convex quadratic optimization problem.

3.6 A Practical Asset Allocation Example

We present a numerical example illustrating the technique described in Sect. 3.3 on a problem of asset allocation, with a comparison with results obtained via a traditional value-at-risk (V@R) minimization approach. We considered $n = 7$ asset classes, corresponding to the following sectors:

1. US Industrial sector equities, represented by proxy *SPDR Dow Jones Industrial ETF*, ticker DIA;
2. US Transportation sector equities, represented by proxy *iShares Dow Jones Transportation ETF*, ticker IYT;
3. US Utilities sector equities, represented by proxy *iShares Dow Jones Utilities ETF*, ticker IDU;
4. US Technology sector equities, represented by proxy *First Trust Nasdaq-100 Ex-Tech EFT*, ticker QQXT;
5. Euro Large Cap equities, represented by proxy *SPDR Euro Stoxx 50 ETF*, ticker FEZ;
6. Long-term Treasury bonds, represented by proxy *iShares Barclays 20+ Year Treas. Bond ETF*, ticker TLT;
7. Corporate bonds, represented by proxy *iShares iBoxx USD High Yid Corp Bond ETF*, ticker HYG.

Our simulation horizon considers daily price data of the component assets from 2-Jan-2011 to 13-Dec-2011 (240 trading days), as shown in Fig. 3.3.

Starting from the initial date of the considered period, we proceed as follows:

- We collect return data for the 7 considered assets, over a look-back period of 100 trading days preceding the current date;
- We use these data to estimate expected returns and the covariance matrix of the assets (to be used for the V@R-optimal portfolio design), and we re-sample the data so to obtain a larger sample of size $N = 500$ to be used for the scenario-optimization method;
- We set expected shortfall level $z_{\text{exp}} = 0.1$, and determine a value for q such that Eq. (3.21) holds, with $n = 7, N = 500$;
- We solve the scenario optimization problem (3.17) using an m-at-a-time return removal rule (with $m = 3$), and determine the corresponding optimal portfolio allocation x_{es}. We applied a threshold rule on portfolio update: if the maximum portfolio variation is above 5 % w.r.t. the previously computed portfolio, then we apply the variation to the updated portfolio, otherwise, we keep the previously computed portfolio for the next period;

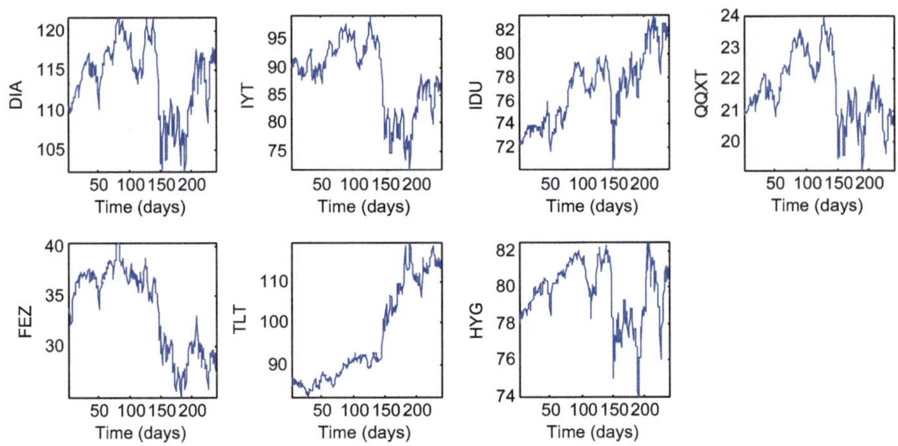

Fig. 3.3 Daily price data (in USD) of the considered assets over the period from 2-Jan-2011 to 13-Dec-2011

- We hold the portfolios fixed for 20 trading days, and simulate the corresponding performance on the actual forward data;
- We update the current date to the end of the simulation period of 20 days;
- We iterate the whole process, until the final date is reached.
- For comparison purposes, we also solve at each step a standard V@R optimization problem with shortfall level z_{exp} and determine the corresponding optimal portfolio allocation x_{var}. Also, we use for comparison a randomly chosen portfolio, as well as a one-over-n fixed portfolio allocation (every asset fixed to $1/n$ proportion in the portfolio).

Figure 3.4 shows the time evolution of a unit initial investment in portfolios managed according to the empirical shortfall approach (bold line). The V@R optimal portfolios, along with random portfolios and one-over-n portfolios are also reported for comparison. In this test, the empirical-shortfall strategy yielded a slightly superior return at the end of the simulation interval, compared to the V@R strategy (16.62 % gross return against 13.95 % of the V@R strategy). The 0.1-empirical, out of sample, quantile of the return streams generated by the two strategies resulted to be almost identical, being −0.44 % for the empirical-shortfall strategy and −0.42 % for the V@R strategy. The random and one-over-n portfolio strategies have larger 0.1-empirical quantiles, resulting respectively in −1.24 % and −1.39 %. The composition of the empirical shortfall portfolios over the 12 considered holding periods is shown in Fig. 3.5.

An analogous simulation, over the period from 14-Dec-2011 to 28-Nov-2012, yielded similar results, showing that the empirical-shortfall portfolios yield a path similar to the one of the V@R portfolios (the out-of-sample 0.1-empirical quantiles were −0.33 % for the empirical-shortfall strategy, −0.34 % for the V@R strategy,

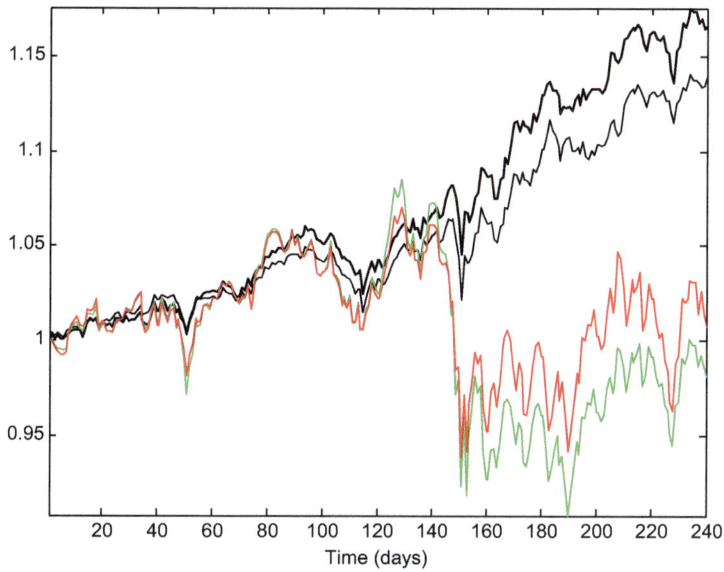

Fig. 3.4 Evolution of the investment according to the empirical shortfall optimization strategy (*bold black line*) and to the V@R optimization strategy (*light black line*), over the period from 2-Jan-2011 to 13-Dec-2011. The *green line* reports the evolution of a random portfolio allocation strategy, and the *red line* the one of a one-over-n fixed portfolio strategy

-0.68 % for the random portfolios, and -0.73 % for the one-over-n fixed portfolio), as shown in Fig. 3.6.

3.7 Conclusions

The mainstream approach to single-period portfolio analysis and design problems (such as mean-variance design, or value-at-risk design) is usually a model-based one, i.e., it strongly relies on an elicited statistical model for the returns. In this chapter, we proposed in Sect. 3.2 an alternative, data-driven, approach for analyzing the shortfall probability of a given portfolio, and we extended this idea in Sect. 3.3 to a design technique that uses samples of the returns directly in the optimization phase for finding an optimal allocation. From the computational side, this approach typically requires the solution of a sequence of linear optimization problems, which can be efficiently performed on modern computing platforms. From the theoretical side, it provides rigorous guarantees relating the finite number of scenarios used in the optimization phase to the out-of-sample reliability of the obtained design.

The data-driven idea can also be employed in multi-stage allocation problems (discussed in Sect. 3.4), especially in conjunction with specific affine reaction policies, where it might be competitive with the more standard sampling approaches

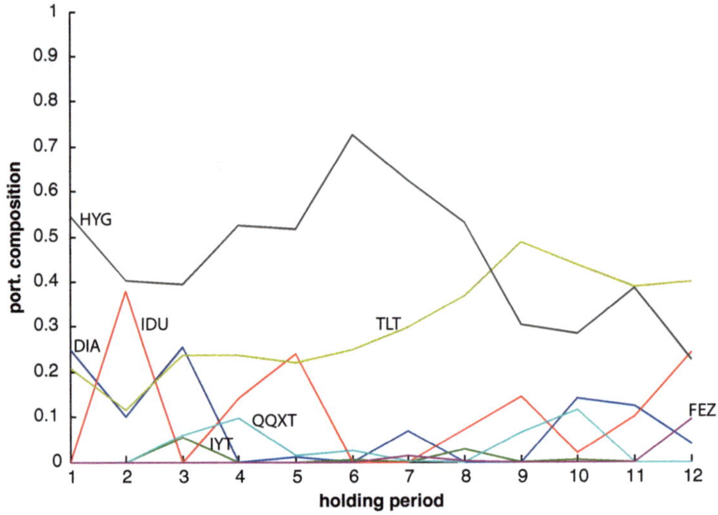

Fig. 3.5 Composition of the empirical shortfall portfolios over the 12 holding periods of 20 days each, from 2-Jan-2011 to 13-Dec-2011

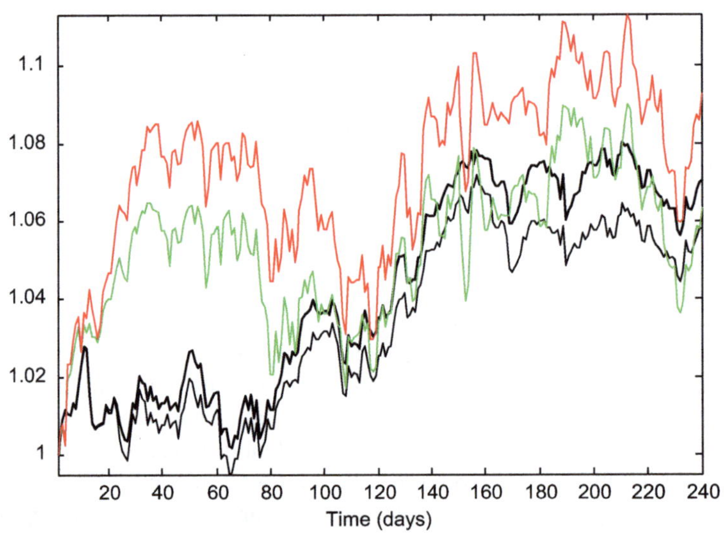

Fig. 3.6 Evolution of the investment according to the empirical shortfall optimization strategy (*bold black line*) and to the V@R optimization strategy (*light black line*), over the period from 14-Dec-2011 to 28-Nov-2012. The *green line* reports the evolution of a random portfolio allocation strategy, and the *red line* the one of a one-over-*n* fixed portfolio strategy

based on scenario trees, who usually have the drawback of rapid combinatorial explosion. In-depth numerical experimentation and comparison of these two types of approaches to multi-period decision problems is still needed, and it should make the object of further research.

References

1. V.S. Bawa, S.J. Brown, R.W. Klein, *Estimation Risk and Optimal Portfolio Choice* (North-Holland, Amsterdam, 1979)
2. D. Bertsimas, A. Thiele, Robust and data-driven optimization: modern decision making under uncertainty, in *INFORMS Tutorials on Operations Research*, pp. 195–122 (2006)
3. D. Bertsimas, J.N. Tsitsiklis, *Introduction to Linear Optimization* (Athena Scientific, Belmont, MA, 1997)
4. G.C. Calafiore, Multi-period portfolio optimization with linear control policies. Automatica **44**(10), 2463–2473 (2008)
5. G.C. Calafiore, An affine control method for optimal dynamic asset allocation with transaction costs. SIAM J. Control Optim. **48**(8), 2254–2274 (2009)
6. G.C. Calafiore, Random convex programs. SIAM J. Optim. **20**(6), 3427–3464 (2010)
7. G.C. Calafiore, Direct data-driven portfolio optimization with guaranteed shortfall probability. Automatica **49**(2), 370–380 (2013)
8. G.C. Calafiore, M.C. Campi, The scenario approach to robust control design. IEEE Trans. Autom. Control **51**(5), 742–753 (2006)
9. M.C. Campi, S. Garatti, The exact feasibility of randomized solutions of uncertain convex programs. SIAM J. Optim. **19**(3), 1211–1230 (2008)
10. H. Chernoff, A measure of asymptotic efficiency for tests of a hypothesis based on the sum of observations. Ann. Math. Stat. **23**, 493–507 (1952)
11. V.K. Chopra, W.T. Ziemba, The effect of errors in means, variances and covariances on optimal portfolio choice. J. Portf. Manag. (Winter) **19**(2), 6–11 (1993)
12. E. Delage, Y. Ye, Distributionally robust optimization under moment uncertainty with application to data-driven problems. J. Oper. Res. **58**(3), 595–612 (2010)
13. L. El Ghaoui, M. Oks, F. Oustry, Worst-case value-at-risk and robust portfolio optimization: a conic programming approach. Oper. Res. **51**(4), 543–556 (2003)
14. M. Gilli, E. Këllezi, H. Hysi, A data-driven optimization heuristic for downside risk minimization. J. Risk **8**(3), 1–19 (2006)
15. D. Goldfarb, G. Iyengar, Robust portfolio selection problems. Math. Oper. Res. **28**(1), 1–38 (2003)
16. C. Jabbour, J.F. Pena, J.C. Vera, L.F. Zuluaga, An estimation-free, robust conditional value-at-risk portfolio allocation model. J. Risk **11**(1), 1–22 (2008)
17. H. Konno, H. Yamazaki, Mean absolute deviation portfolio optimization model and its application to Tokyo stock market. Manag. Sci. **37**, 519–531 (1991)
18. H.M. Markowitz, Portfolio selection. J. Financ. **7**, 77–91 (1952)
19. K. Natarajan, D. Pachamanova, M. Sim, Incorporating asymmetric distributional information in robust value-at-risk optimization. Manag. Sci. **54**(3), 573–585 (2008)
20. B.K. Pagnoncelli, D. Reich, M.C. Campi, Risk-return trade-off with the scenario approach in practice: a case study in portfolio selection. J. Optim. Theory Appl. **155**(2), 707–222 (2012)
21. R.H. Tütüncü, M. Koenig, Robust asset allocation. Ann. Oper. Res. **132**, 157–187 (2004)

Chapter 4
Robust Approaches to Pension Fund Asset Liability Management Under Uncertainty

Dessislava Pachamanova, Nalan Gülpınar, and Ethem Çanakoğlu

Abstract This entry considers the problem of a typical pension fund that collects premiums from sponsors or employees and is liable for fixed payments to its customers after retirement. The fund manager's goal is to determine an investment strategy so that the fund can cover its liabilities while minimizing contributions from its sponsors and maximizing the value of its assets. We develop robust optimization and scenario-based stochastic programming approaches for optimal asset-liability management, taking into consideration the uncertainty in asset returns and future liabilities. Our focus is on computational tractability and ease of implementation under conditions typically encountered in practice, such as asymmetries in the distributions of asset returns. Computational results from tests with real and generated data are presented to illustrate the performance of these models.

Keywords Asset-liability management • Uncertainty • Stochastic programming • Robust optimization • Asymmetry

4.1 Introduction

Asset-liability management (ALM) is one of the classical problems in financial risk management. Typically, ALM involves the management of assets in such a way as to earn adequate return while maintaining a comfortable surplus of assets over existing and future liabilities. This type of problem is faced by a number of financial services companies, such as pension funds and insurance companies. As

D. Pachamanova (✉)
Mathematics and Sciences Division, Babson College, Babson Park, MA 02457, USA
e-mail: dpachamanova@babson.edu

N. Gülpınar
Warwick Business School, University of Warwick, Coventry CV4 7AL, UK
e-mail: Nalan.Gulpinar@wbs.ac.uk

E. Çanakoğlu
Industrial Engineering, Bahcesehir University, Istanbul, Turkey
e-mail: Ethem.Canakoglu@bahcesehir.edu.tr

© Springer International Publishing Switzerland 2017

G. Consigli et al. (eds.), *Optimal Financial Decision Making under Uncertainty*, International Series in Operations Research & Management Science 245,
DOI 10.1007/978-3-319-41613-7_4

we will explain in more detail later, the problem of finding optimal ALM policies is computationally challenging, and many of the approaches for implementation described in the literature can be too intensive computationally to implement in practice.

There is an extensive literature on modeling and optimization of allocation strategies for ALM based on stochastic programming techniques; see, for example, Ziemba and Mulvey [33]. These approaches usually focus on finding optimal investment rules over a set of scenarios for the future returns on the assets and the liabilities of the company. Such methods have been successfully applied in some instances (see, for example, [29], [15], [16], and [21]); however, they are still not widely used in the financial industry for several reasons. First, ALM is inherently a multiperiod problem, and the number of scenarios needed to represent reality satisfactorily increases exponentially with the number of time periods under consideration. Thus, the dimension of the optimization problem, and correspondingly its computational difficulty, increases. Second, the scenario generation itself requires sophisticated statistical techniques, which is a deterrent to practitioners who need to make decisions in a short amount of time. Finally, often little is known about the specific distributions of future uncertainties in the ALM problem, and little data are available for estimating the probability distributions of these uncertainties. In many cases, it may be preferable to provide general information about the uncertainties, such as means, ranges, and directional deviations, rather than generating specific scenarios.

In this entry, we are concerned with ALM models for pension funds. Different ALM models for pension funds have been developed in the literature. For example, Dert [17] analyzes a dynamic model for asset-liability management for defined benefit pension funds. Bogentoft et al. [11] and Hilli et al. [28] consider optimal decision approaches for multiperiod asset-liability management model for a pension fund using Conditional Value-at-Risk (CVaR). Boender [10] introduces a decision support model to sustain management of pension funds in the strategic planning of the available asset and liability policy instruments. He models risk drivers by scenarios rather than by probability distributions, and describes how the process of managerial learning can be improved by a hybrid simulation and optimization method to determine the asset allocations which determine efficient frontiers of contribution rates and downside insolvency risks. Gülpınar and Pachamanova [25] develop a robust ALM model for a pension fund, mapping a time series model for asset returns to the underlying uncertain parameter structure.

We develop stochastic programming and robust optimization-based approaches for handling the classical ALM formulation. As we will discuss in more detail later, robust optimization—a technique for optimization under uncertainty that is concerned with finding the optimal solution when uncertain parameters in the problem take their worst-case values in pre-specified uncertainty sets—is a natural choice for ALM-type problems. In developing these models, we have several priorities in mind. First, we focus on computational tractability and practical implementation. Second, we consider models that can represent features of the investment reality, such as the ability to incorporate observed asymmetries in asset returns in the model.

The rest of this entry is organized as follows. In Sect. 4.2, we present an ALM problem statement for pension funds. Section 4.3 introduces a scenario-based stochastic programming formulation of the pension fund ALM problem. Section 4.4 provides a brief introduction to robust optimization. Robust formulations of ALM models using symmetric and asymmetric uncertainty sets are developed in Sect. 4.5. Computational experiments are presented in Sect. 4.6. Section 4.7 summarizes our findings.

Notation. We use tilde ($\tilde{*}$) to denote randomness; e.g., $\tilde{z}$ denotes random variable z. Vectors are in boldface and matrices are denoted by boldface capital letters. For example, **a** is a vector and **A** is a matrix. The expected value of random variable $\tilde{z}$, $E[\tilde{z}]$, is denoted by $\hat{z}$. Apostrophe ($'$) denotes matrix transpose.

4.2 ALM Model for Pension Funds: Problem Statement

A pension fund receives premiums from sponsors or current employees and supplies fixed payments after their retirement. The fund constructs a portfolio by investing the available funds in the market, and manages the assets so that at each time period the portfolio holdings can cover the fund's liabilities. The fund manager also aims to maximize investment returns while minimizing the contribution rate by the sponsor and active employees of the fund [11]. Therefore, the ALM problem for a pension fund is to determine an optimal contribution rate and an optimal investment strategy during an investment horizon.

We consider a portfolio constructed from M risky assets and a risk-free asset. Securities are denoted by $m = 1, 2, \cdots, M$, and $m = 0$ identifies the risk-free asset. There are $T + 1$ time periods, $t = 0, \cdots, T$, where $t = 0$ represents today. After the initial investment at $t = 0$, the portfolio may be restructured at discrete times $t = 1, \cdots, T-1$ in terms of both return and liability, and redeemed at the end of the investment horizon, $t = T$.

Decision variables h_t^m, s_t^m and b_t^m denote the amount of asset m to be held, sold and bought at time t, respectively. Asset returns and future liabilities are uncertain due to exogenous and endogenous factors. Asset returns $\tilde{r}_t^m$ for $m = 1, \ldots, M$, as well as the risk-free returns $\tilde{r}_t^0$, are represented by random variables. While the liabilities to be paid out at each stage l_t are known at time 0, the total present value at time t of all future liabilities, $\tilde{L}_t$, is unknown because changes in the discount rates over time affect the present value of the cash flows. A full description of the notation is provided in Table 4.1.

The multiperiod ALM optimization model constraints are defined as follows.

A **balance constraint** describes the wealth accumulated in each asset m at time t, h_t^m. The holdings in asset m at time t are updated based on the holdings at the

Table 4.1 Description of notation

Notation	Description
Parameters	
T	Investment horizon
M	Number of investment assets
ψ	Target funding (asset/liability) ratio
c_b, c_s	Transaction costs for buying and selling, respectively
l_t	Amount (liabilities) paid out at time t
W_t	Amount of wages at time t
λ	Penalty for wage contributions
Decision variables	
h_t^m	Holding in asset m at time t
s_t^m	Amount sold of asset m at time t
b_t^m	Amount bought of asset m at time t
γ_t	Contribution as percentage of wages at time t
Random variables	
$\tilde{r}_t^m$	Return on asset m between time t and $t+1$
$\tilde{L}_t$	Present value of the total amount of future outstanding liabilities at time t

previous time period $t-1$, h_{t-1}^m, the return on the asset from $t-1$ to t, $\tilde{r}_{t-1}^m$, as well as trading activity (purchases b_t^m and sales s_t^m) at time t.

$$h_t^m = (1 + \tilde{r}_{t-1}^m) \cdot h_{t-1}^m - s_t^m + b_t^m,\ t = 1, \ldots, T,\ m = 1, \ldots, M \tag{4.1}$$

The **amount of cash** at time t, h_t^0, is updated based on the return, $\tilde{r}_{t-1}^0$, on the amount of cash h_{t-1}^0 from the previous time period $t-1$, the cash received from position changes (positive inflow from sales and negative inflow from purchases, adjusted for the appropriate transaction costs), the deposit payment $\gamma_t \tilde{W}_t$ at time t, as well as the cash outflow to cover the liabilities l_t at time t.

$$h_t^0 = (1 + \tilde{r}_{t-1}^0) \cdot h_{t-1}^0 + \sum_{m=1}^{M} (1 - c_s) s_t^m - \sum_{m=1}^{M} (1 + c_b) b_t^m + \gamma_t \tilde{W}_t - l_t,\ t = 1, \cdots, T \tag{4.2}$$

At each point in time, the **ratio of assets to liabilities** (the so-called *funding ratio*) needs to be maintained above a certain level ψ determined by the fund. Values for ψ typically used in industry are in the range (0.9–1.1), and values of ψ of greater than 1 are used to inject an extra safety margin for meeting outstanding liabilities. The funding ratio constraints at each time period t can be represented in linear form as $Assets_t \geq \psi \cdot Liabilities_t$, or

$$\sum_{m=1}^{M} h_t^m + h_t^0 \geq \psi \tilde{L}_t,\ t = 1, \ldots, T-1. \tag{4.3}$$

$\tilde{L}_t$ is the total value at time t of the outstanding future liabilities between t and T, and can be expressed as

$$\sum_{m=1}^{M} h_t^m + h_t^0 \geq \psi \sum_{\tau=t+1}^{T} \frac{l_\tau}{(1+\tilde{r}_t^0)\ldots(1+\tilde{r}_{\tau-1}^0)}, \quad t = 1, \ldots, T-1$$

The holdings of asset m at time t are restricted to be **nonnegative**; that is, borrowing and short sales are not allowed:

$$h_t^m \geq 0,\ t = 1, \ldots, T,\ m = 0, \ldots, M. \tag{4.4}$$

Finally, the decision variables for the amount of asset m to be bought or sold at time t cannot be negative:

$$s_t^m \geq 0, \quad b_t^m \geq 0,\ t = 1, \ldots, T-1,\ m = 1, \ldots, M. \tag{4.5}$$

The pension fund manager's goal is to maximize the expected portfolio wealth at time T while minimizing the future value of the total contributions from the fund's customers. The objective function describes the wealth at time T and can be stated as

$$OF = \sum_{m=0}^{M} h_T^m - \lambda \sum_{t=0}^{T-1} \gamma_t W_t \cdot (1+\tilde{r}_t^0)(1+\tilde{r}_{t+1}^0)\cdots(1+\tilde{r}_{T-1}^0) \tag{4.6}$$

where λ is a fixed parameter.

The ALM problem statement for a pension fund can be summarized as follows:

$(\mathcal{P}_{PF}^r)$ max OF (4.6)

s.t. Constraints (4.1), (4.2), (4.3), (4.4), and (4.5)

4.3 Scenario-Based ALM Model for Pension Funds

This section describes a possible stochastic programming representation of the ALM model for a pension fund. We walk the reader through the steps in the multiperiod formulation: the specification of a scenario tree and the description of the optimization model.

Constructing a scenario tree. The uncertain returns on the assets and the risk-free rate are random variables. We discretize these random variables and consider a framework in which each random variable can take finitely many values. Given the event history up to a particular time, the uncertainty in the next time period is characterized by finitely many possible outcomes for the next observation. This branching process is represented through a scenario tree. Thus, the possible events are approximated by a discrete set of scenarios, for instance, see [26].

The root node in the scenario tree represents "today" and is immediately observable from available information. The nodes further down the tree stand for events that are conditional on outcomes at prior stages. The arcs linking the nodes represent various realizations of the uncertain variables. Ideally, the generated set of realizations constitutes the whole universe of possible outcomes of the random variable. Realizations should include both optimistic and pessimistic projections. The reader is referred to Gulpinar et al. [27] for more details on scenario tree generation techniques.

A *scenario* is a possible realization of the stochastic variables $\{\boldsymbol{\rho}_1, \ldots, \boldsymbol{\rho}_T\}$, and can be imagined as a path in the tree. Hence, the set of scenarios corresponds to the set of leaves of the scenario tree, $\mathcal{N}_T$, and nodes of the tree at level $t \geq 1$ (the set $\mathcal{N}_t$) correspond to possible realizations of $\boldsymbol{\rho}^t$. Given the event history up to time t, $\boldsymbol{\rho}^t$, the uncertainty in the next period is characterised by finitely many possible outcomes for the next observation $\boldsymbol{\rho}_{t+1}$.

A *node* of the tree (or an *event*) is denoted by $e = (s, t)$, where s is a scenario (path from root to leaf), and time period t specifies a particular node on that path. The root of the tree is $\mathbf{0} = (s, 0)$, where s can be any scenario because the root node is common to all scenarios. The *ancestor* (or *parent*) of event $e = (s, t)$ is denoted by $a(e) = (s, t-1)$, and the branching probability p_e is the conditional probability of event e given its parent event $a(e)$. The path to event e is a partial scenario with probability $P_e = \prod p_e$ along that path; since probabilities p_e must sum to one at each individual branching, probabilities P_e will sum up to one across each layer of tree-nodes $\mathcal{N}_t$; $t = 0, 1, \ldots, T$.

Scenario-based optimization model. Let h_e^m, s_e^m and b_e^m denote the amount of asset m to be held, sold and bought in event $e \in \mathcal{N}_t$ at time t, respectively. Asset returns $\tilde{r}_e^m$ for $m = 1, \ldots, M$ and the risk-free return $\tilde{r}_e^0$ in event $e \in \mathcal{N}_t$ at time $t = 0, 1, \ldots, T$ are realizations of random variables. Recall that we assume that current liabilities, l_t and the amount of wages W_t at time t are certain and remain the same under each realization of the uncertainty. Future liabilities $\tilde{L}_e$ are random variables but they do not need to be represented separately in the tree because they are determined by future values of the risk-free rate. Further, γ_e denotes the contribution as a percentage of wages in scenario $e \in \mathcal{N}_t$ at time t. The optimization model constraints are defined as follows.

Balance constraints determine trading at each node $e \in \mathcal{N}_t$ on each asset m:

$$h_e^m = (1 + r_{a(e)}^m) \cdot h_{a(e)}^m - s_e^m + b_e^m, \; e \in \mathcal{N}_t, \quad t = 1, \ldots, T, \; m = 1, \ldots, M. \quad (4.7)$$

The amount of cash for any scenario $e = (s, t) \in \mathcal{N}_t$ at t is expressed as

$$h_e^0 = (1 + r_{a(e)}^0) \cdot h_{a(e)}^0 + \sum_{m=1}^{M} (1 - c_s) s_e^m - \sum_{m=1}^{M} (1 + c_b) b_e^m + \gamma_t W_t - l_t,$$
$$e \in \mathcal{N}_t, \quad t = 1, \ldots, T. \quad (4.8)$$

Asset-liability (funding ratio) constraints are formulated as

$$\sum_{m=1}^{M} h_e^m + h_e^0 \geq \psi L_e, \; e \in \mathcal{N}_t, \quad t = 1, \ldots, T - 1 \quad (4.9)$$

where L_e for $e \in \mathcal{N}_t$ is the realization of random variable $\tilde{L}_t$ at time t of the liabilities between t and T. This constraint can be equivalently expressed as

$$\sum_{m=1}^{M} h_e^m + h_e^0 \geq \psi \sum_{\tau=t+1}^{T} \frac{l_\tau}{(1 + r_{e' \in \mathcal{N}_t}^0) \ldots (1 + r_{e' \in \mathcal{N}_{\tau-1}}^0)}, \; e \in \mathcal{N}_t, \quad t = 1, \ldots, T - 1$$
$$(4.10)$$

Short sales are not allowed. Nonnegativity constraints are imposed on the holdings of asset m at time t

$$h_e^m \geq 0, \; e \in \mathcal{N}_t, \quad t = 1, \ldots, T, \; m = 0, \ldots, M. \quad (4.11)$$

There are also **nonnegativity** constraints on the amount of asset m to be bought or sold at time t:

$$s_e^m \geq 0, \quad b_e^m \geq 0, \; e \in \mathcal{N}_t, \quad t = 1, \ldots, T - 1, \; m = 1, \ldots, M. \quad (4.12)$$

The **objective function** (OF) describes the expected terminal wealth and can be stated as follows.

$$OF = \sum_{e \in \mathcal{N}_T} P_e \left(\sum_{m=0}^{M} h_e^m - \lambda \sum_{t=0}^{T-1} \gamma_t W_t \cdot (1 + r_e^0)(1 + r_{e' \in \mathcal{N}_{t+1}}^0) \ldots (1 + r_{e' \in \mathcal{N}_{T-1}}^0) \right)$$
$$(4.13)$$

where $e' \in \mathcal{N}_{t+1}$ represents the realizations of the risk-free rate starting at node $e \in \mathcal{N}_t$ at time t of the scenario tree.

The ALM problem statement for a pension fund is to maximize the net expected profit at the end of investment horizon subject to balance, cash, funding ratio, no short-sale and nonnegativity constraints, and can be summarized as follows:

$$
\begin{array}{lll}
(\mathcal{S}^r_{PF}) & \max & OF\,(4.13) \\
 & \text{s.t.} & \text{Constraints } (4.7), (4.8), (4.10), (4.11), (4.12)
\end{array}
$$

4.4 Robust Investment Decisions

This section is a basic introduction to robust optimization - the technique we will use in Sect. 4.5 to derive robust formulations for the ALM pension fund problem given descriptive statistics for the uncertain inputs of the optimization problem. Robust optimization was independently developed by Ben-Tal and Nemirovski [2] and El Ghaoui and Lebret [19], and has experienced tremendous growth in the last decade.

The robust optimization approach provides robust solutions that are "adequate" even if there is error in the estimates of the input parameters, in the following sense. It solves an optimization problem assuming that the uncertain input data belong to an uncertainty set, and looks for an optimal solution that remains feasible if the uncertainties take any values within that uncertainty set. This reformulation of the problem is referred to as the "robust counterpart" of the original optimization problem. In some special cases, the robust counterpart of the original problem involves the worst-case outcome of the stochastic data within the uncertainty set, and is a tractable optimization problem with no random parameters.

The selection of the uncertainty set is often based on statistical estimates and probabilistic guarantees for the solution. Ellipsoidal, box and polyhedral (e.g., *D*-norm) are the most commonly used uncertainty sets, but more recently, asymmetric uncertainty sets have been used as well in order to capture the probability distribution characteristics of the uncertainties better (see, for example, Natarajan et al. (2008) for an application in estimating the value-at-risk of a portfolio). Thus far, in industry robust optimization has been used only in asset management, and primarily to incorporate the uncertainty introduced by estimation errors into the mean-variance portfolio allocation framework. Goldfarb and Iyengar [23] consider robust mean-variance portfolio allocation strategies under various ellipsoidal and interval uncertainty sets for the input parameters (means and covariance matrices) derived from regression analysis. Ceria and Stubbs [13] introduce the zero-net alpha-adjustment robust framework to reduce the conservativeness of robust mean-variance strategies under ellipsoidal uncertainty sets for the input parameters. Robust investment strategies in a multiperiod setting have been studied by Ben-Tal et al. [5], Bertsimas and Pachamanova [7], and Gulpinar and Rustem [26].

A brief introduction to the main ideas of robust *linear* optimization (the type of problem with which we are dealing in this paper) is provided next; for further information, the reader is referred to Ben-Tal and Nemirovski [1–4] as well as Ben-Tal et al. [6].

Consider a linear program

$$
\max \left\{ \mathbf{c}'\mathbf{x} | f(\mathbf{x}, \tilde{\mathbf{z}}) \leq 0, \;\; \mathbf{x} \in V \right\} \tag{4.14}
$$

where $\mathbf{c} \in R^{n\times 1}$ and V consists of all constraints whose parameters are certain. The vector $\mathbf{x} \in R^{n\times 1}$ represents decision variables and $\tilde{\mathbf{z}} \in R^{J\times 1}$ is a vector of uncertain parameters. The relevant constraint function $f(\mathbf{x}, \tilde{\mathbf{z}})$ in our context is bilinear in $\mathbf{x}$ and $\tilde{\mathbf{z}}$ and can be written as

$$f(\mathbf{x}, \tilde{\mathbf{z}}) = f_0(\mathbf{x}) + \sum_{j=1}^{J} f_j(\mathbf{x})\tilde{z}_j, \tag{4.15}$$

where functions $f_j(\mathbf{x})$ for $j = 1, \ldots, J$ are linear in $\mathbf{x}$. We will also assume without loss of generality that the uncertain factors $\tilde{\mathbf{z}}$ satisfy the normalized distributional conditions $E(\tilde{\mathbf{z}}) = 0$ and $E(\tilde{\mathbf{z}} \cdot \tilde{\mathbf{z}}) = \mathbf{I}$. This can be achieved by a suitable linear transformation [see Natarajan et al. (2008)]. For instance, uncertain problem input parameters such as portfolio returns $\tilde{\mathbf{R}} \in \mathcal{R}^N$ with known means vector $\hat{\mathbf{R}} \in \mathcal{R}^N$ and invertible covariance matrix $\mathbf{\Sigma} \in \mathcal{R}^{N\times N}$ can be expressed as $\tilde{\mathbf{R}} = \hat{\mathbf{R}} + \mathbf{\Sigma}^{1/2}\tilde{\mathbf{z}}$ for some uncertain factors $\tilde{\mathbf{z}} \in \mathcal{R}^N$ satisfying the normalized distributional conditions. Hence, $\tilde{\mathbf{z}} = \mathbf{\Sigma}^{-1/2}(\tilde{\mathbf{R}} - \hat{\mathbf{R}})$.

Robust optimization transfers the original constraint with random parameters into its robust counterpart, defined as

$$f(\mathbf{x}, \mathbf{z}) \leq 0, \quad \forall\, \mathbf{z} \in \mathcal{U}(\tilde{\mathbf{z}})$$

where $\mathcal{U}(\tilde{\mathbf{z}})$ is an uncertainty set specified by the modeler.

The size of the uncertainty set is often related to guarantees on the probability that the constraint with uncertain coefficients will not be violated. There is a tradeoff between optimality and the amount of protection against uncertainty that is desired—the smaller the probability that the constraint will be violated, the lower the value of the objective function of the robust counterpart. (We should clarify, however, that this does not mean that the *realized* objective function value will be always lower on average for the robust counterpart with a large uncertainty set than for the robust counterpart with a smaller uncertainty set.)

The shape of the uncertainty set defines a risk measure on the constraints with uncertain coefficients (Natarajan et al. 2009). In practice, the shape is selected to reflect the modeler's knowledge of the probability distributions of the uncertain parameters, while at the same time making the robust counterpart problem efficiently solvable. The ellipsoidal uncertainty set, for example, defines a standard-deviation-like risk measure on the constraint with uncertain parameters, and in the case of linear optimization, results in a robust counterpart to the original problem that is a second order cone problem—a tractable optimization problem. Specifically, the ellipsoidal set on the risk factors can be expressed as

$$\mathcal{U}_{\Omega}(\tilde{\mathbf{z}}) = \{\mathbf{z} : \|\mathbf{z}\|_2 \leq \Omega\}, \tag{4.16}$$

where the constant Ω is specified by the modeler, and determines the size of the uncertainty set. It is sometimes referred to as the "robustness budget" or the "price of robustness."

For $\tilde{\mathbf{z}}$ in uncertainty set (4.16), the robust counterpart of a constraint of the form

$$f(\mathbf{x}, \tilde{\mathbf{z}}) \geq 0$$

can be derived using convex duality, and is

$$f_0(\mathbf{x}) - \Omega \cdot \sigma(f(\mathbf{x}, \tilde{\mathbf{z}})) \geq 0,$$

where $f_0(\mathbf{x})$ is the expected value and $\sigma(f(\mathbf{x}, \tilde{\mathbf{z}}))$ is the standard deviation of $f(\mathbf{x}, \tilde{\mathbf{z}})$ for any given $\mathbf{x}$. We will discuss the ellipsoidal set in more detail in the next section.

The parameter Ω can be selected in such a way that the probability that the constraint is satisfied is at least $1 - \epsilon$ for some small $\epsilon \in (0, 1)$. Specifically, this is true if $\Omega \geq \sqrt{-2 \ln \epsilon}$ [6]. This imposes a value-at-risk-type risk measure on the constraint containing uncertain coefficients [20] (Natarajan et al. 2008). Further results on probability bounds related to the size and the shape of uncertainty sets can be found, for example, in Ben-Tal and Nemirovski [4], Bertsimas and Sim [8], Bertsimas et al. [9], and Chen et al. [14].

Symmetric uncertainty sets such as the ellipsoidal uncertainty set for the parameters in optimization problem may be overly conservative when the underlying probability distributions are asymmetric, as is often the case with asset return distributions. In other words, the probability distributions of the underlying uncertainties may not be represented sufficiently well. Using additional information about the asymmetries in the underlying probability distributions in defining the uncertainty set helps overcome this issue. Chen et al. [14] introduce the "forward" and "backward" deviations of a random variable, and develop a convex asymmetric uncertainty set based on these variability measures. In the case of a normal distribution, the forward and the backward deviations both equal the standard deviation. In the case of asymmetric probability distributions, one of them is greater than the other depending on the skew of the distribution.

To obtain the uncertainty set introduced by Chen et al. [14], decompose the vector of random variables $\tilde{\mathbf{z}}$ into two vectors of random variables $\tilde{\mathbf{v}}$ and $\tilde{\mathbf{w}}$ such that $\tilde{\mathbf{z}} = \tilde{\mathbf{v}} - \tilde{\mathbf{w}}$ where $\tilde{\mathbf{v}} = \max\{\tilde{\mathbf{z}}, 0\}$ and $\tilde{\mathbf{w}} = \max\{-\tilde{\mathbf{z}}, 0\}$. Both $\mathbf{v}$ and $\mathbf{w}$ are positive and at least one of them is zero. Let $p_j > 0$ and $q_j > 0$, $j = 1, \ldots, J$, represent the forward and backward deviations of random variable $\tilde{z}_j$, respectively. Define diagonal matrices $\mathbf{P} = diag(p_1, \ldots, p_J)$ and $\mathbf{Q} = diag(q_1, \ldots, q_J)$. The asymmetric uncertainty set is

$$\mathcal{U}(\tilde{\mathbf{z}}) = \left\{\mathbf{z} : \ \exists \mathbf{v}, \mathbf{w} \in R_+^J, \ \mathbf{z} = \mathbf{v} - \mathbf{w}, \ ||\mathbf{P}^{-1}\mathbf{v} + \mathbf{Q}^{-1}\mathbf{w}|| \leq \Omega, \ -\underline{\mathbf{z}} \leq \mathbf{z} \leq \bar{\mathbf{z}}\right\}$$

where Ω is the desired degree of robustness. In the case of a symmetric distribution, the uncertainty set above is ellipsoidal, i.e., this asymmetric uncertainty set includes the ellipsoidal uncertainty set as a special case. If the factors $\tilde{\mathbf{z}}$ are independent and

Ω is selected so that $\Omega \geq \sqrt{-2 \ln \epsilon}$, then, similarly to the case of the ellipsoidal uncertainty set, the constraint will be satisfied with probability of at least $1 - \epsilon$ (Natarajan et al. 2008).

The robust counterpart of the constraint $f(\mathbf{x}, \tilde{\mathbf{z}}) \geq 0$ in (4.14) under finite distribution support $[-\underline{\mathbf{z}}, \overline{\mathbf{z}}]$ is equivalent to the following set of inequalities:

$$\begin{aligned}\{f_0(\mathbf{x}) \geq \Omega||\mathbf{u}||_2 + \mathbf{r}'\overline{\mathbf{z}} + \mathbf{s}'\underline{\mathbf{z}},\ u_j \geq -p_j\left(f_j(\mathbf{x}) + r_j - s_j\right),\\ u_j \geq q_j\left(f_j(\mathbf{x}) + r_j - s_j\right),\ j = 1, \ldots, J,\ \ \mathbf{r}, \mathbf{s} \geq 0\}\end{aligned} \tag{4.17}$$

for some $\mathbf{u}, \mathbf{r}, \mathbf{s} \in R^J$.

Given the fact that ALM is concerned with ensuring a level of minimum guaranteed performance to meet future liabilities, robust-optimization-based strategies that place special emphasis on the worst-case realizations of uncertainties are particularly appealing in the ALM context. In the next section we introduce a tractable robust approach to ALM for pension funds and derive the robust counterparts of the ALM problem with two kinds of uncertainty sets: symmetric (ellipsoidal) and asymmetric [the uncertainty set suggested by Chen et al. [14]; see also Natarajan et al. (2008)]. The results of computational experiments designed to judge the performance of the robust formulations derived here are presented in Sect. 4.6.

4.5 Robust ALM Models for Pension Funds

This section introduces the robust counterparts of the ALM model when asset returns are assumed to vary in symmetric and asymmetric uncertainty sets. It also includes a discussion of how such sets may be constructed from data.

When formulating the robust counterparts of the ALM problems, it will be preferable to have a particular uncertain parameter in only one constraint (as opposed to multiple constraints). This avoids the problem of dealing with cross-constraint correlations of uncertain parameters, and reduces the conservativeness of the solution. For example, the uncertain returns currently appear in all balance constraints. We can reduce the number of constraints in which the uncertain returns appear by using a transformation suggested in Ben-Tal et al. [5].

In order to formulate the ALM model in terms of cumulative returns, let us define *cumulative* gross returns, $\tilde{R}_t^m$ for each asset $m = 1, \ldots, M$ at $t = 0$ and $t = 1, \ldots, T$, respectively, as

$$\begin{aligned}R_0^m &= 1, \text{and}\\ \tilde{R}_t^m &= (1 + \tilde{r}_0^m)(1 + \tilde{r}_1^m) \ldots (1 + \tilde{r}_{t-1}^m),\ t = 1, \ldots, T.\end{aligned} \tag{4.18}$$

Introducing new decision variables for assets $m = 1, \ldots, M$ and time $t = 1, \ldots, T$

$$\xi_t^m = \frac{h_t^m}{\tilde{R}_t^m}, \quad \eta_t^m = \frac{s_t^m}{\tilde{R}_t^m}, \quad \zeta_t^m = \frac{b_t^m}{\tilde{R}_t^m},$$

and a free variable ν for the OF, we can rewrite problem $(\mathcal{P}^r)$ as follows: $(\mathcal{P}^R)$

$$\begin{aligned}
\max \quad & \nu \\
\text{s.t.} \quad & \nu \leq \sum_{m=0}^{M} \tilde{R}_T^m \xi_T^m - \lambda \sum_{t=0}^{T-1} \gamma_t W_t \cdot \frac{\tilde{R}_T^0}{\tilde{R}_t^0} \\
& \xi_t^m = \xi_{t-1}^m - \eta_t^m + \zeta_t^m, \quad t = 1, \ldots, T, m = 1, \ldots, M \\
& \xi_t^0 = \xi_{t-1}^0 + \sum_{m=1}^{M} (1 - c_s) \frac{\tilde{R}_t^m}{\tilde{R}_t^0} \eta_t^m - \sum_{m=1}^{M} (1 + c_b) \frac{\tilde{R}_t^m}{\tilde{R}_t^0} \zeta_t^m + \frac{\gamma_t W_t}{\tilde{R}_t^0} - \frac{l_t}{\tilde{R}_t^0}, \; t = 1, \ldots, T \\
& \sum_{m=0}^{M} \xi_t^m \tilde{R}_t^m \geq \psi \sum_{\tau=t+1}^{T} l_\tau \cdot \frac{\tilde{R}_t^0}{\tilde{R}_\tau^0}, \; t = 1, \ldots, T-1 \\
& \xi_t^m \geq 0, \quad t = 1, \ldots, T, \quad m = 1, \ldots, M \\
& \eta_t^m \geq 0, \zeta_t^m \geq 0, \quad t = 1, \ldots, T-1, \quad m = 1, \ldots, M
\end{aligned}$$

Note that after the transformation of the decision variables, the uncertain (cumulative) returns appear only in the cash constraints, as opposed to all balance constraints. They also appear in the objective function and the funding ratio constraint, as they did before the transformation.

The two sets of uncertain parameters in the ALM formulation are the asset returns $\tilde{\mathbf{R}}_t$ (including the return on the riskless asset $\tilde{R}_t^0$) and the value of the future liabilities $\tilde{L}_t$ at each point in time t. The latter depends on the realized changes in interest rates between time 0 and time t. In order to simplify notation, let us use the following notation for the vectors $\tilde{\boldsymbol{\alpha}}$, $\tilde{\boldsymbol{\rho}}_t$ and $\tilde{\boldsymbol{\mu}}_t$ in the objective function, balance constraints for $t = 1, \cdots, T$ and funding ratio constraints for $t = 1, \cdots, T-1$, respectively:

$$\tilde{\boldsymbol{\alpha}} = \left(\tilde{R}_T^0, \cdots, \tilde{R}_T^M, -\lambda W_1 \frac{\tilde{R}_T^0}{\tilde{R}_1^0}, \cdots, -\lambda W_{T-1} \frac{\tilde{R}_T^0}{\tilde{R}_{T-1}^0} \right),$$

$$\tilde{\boldsymbol{\rho}}_t = \left((1 - c_s) \cdot \frac{\tilde{R}_t^1}{\tilde{R}_t^0}, \cdots, (1 - c_s) \cdot \frac{\tilde{R}_t^M}{\tilde{R}_t^0}, -(1 + c_b) \cdot \frac{\tilde{R}_t^1}{\tilde{R}_t^0}, \cdots, -(1 + c_b) \cdot \frac{\tilde{R}_t^M}{\tilde{R}_t^0}, W_t \frac{\cdot 1}{\tilde{R}_t^0}, -l_t \cdot \frac{1}{\tilde{R}_t^0} \right),$$

$$\tilde{\boldsymbol{\mu}}_t = \left(\tilde{R}_t^0, \cdots, \tilde{R}_t^M, -\psi l_{t+1} \frac{\tilde{R}_t^0}{\tilde{R}_{t+1}^0}, \cdots, -\psi l_T \frac{\tilde{R}_t^0}{\tilde{R}_T^0} \right).$$

Let vectors $\hat{\boldsymbol{\alpha}}$, $\hat{\boldsymbol{\rho}}_t$, and $\hat{\boldsymbol{\mu}}_t$ denote the expected values of the random vectors $\tilde{\boldsymbol{\alpha}}$, $\tilde{\boldsymbol{\rho}}_t$, and $\tilde{\boldsymbol{\mu}}_t$, respectively. For instance,

$$\hat{\boldsymbol{\alpha}} = \left(E\left[\tilde{R}_T^0\right], \cdots, E\left[\tilde{R}_T^M\right], -\lambda \gamma_1 W_1 E\left[\frac{\tilde{R}_T^0}{\tilde{R}_1^0}\right], \cdots, -\lambda \gamma_{T-1} W_{T-1} E\left[\frac{\tilde{R}_T^0}{\tilde{R}_{T-1}^0}\right] \right).$$

Similarly, let us define vectors

$$\begin{aligned}
\boldsymbol{\kappa} &= \left(\xi_T^0, \cdots, \xi_T^M, \gamma_1, \cdots, \gamma_{T-1}\right)', \\
\boldsymbol{\pi}_t &= \left(\eta_t^1, \cdots, \eta_t^M, \zeta_t^1, \cdots, \zeta_t^M, \gamma_t, 1\right)', \text{and} \\
\boldsymbol{\tau}_t &= \left(\xi_t^0, \cdots, \xi_t^M, 1, \cdots, 1\right)'.
\end{aligned}$$

4.5.1 Robust ALM Model Formulation with Symmetric Uncertainty Sets

We consider ellipsoidal uncertainty sets involving the uncertain future asset returns $\tilde{R}_t^m, m = 1, \ldots, M$, and riskless returns $\tilde{R}_t^0$ at each point in time $t, t = 1, \ldots, T$. The uncertainty sets are determined as follows:

$$\begin{aligned}
\mathcal{SU}^o &= \left\{\tilde{\boldsymbol{\alpha}} \mid \left\|(\Xi^\alpha)^{-\frac{1}{2}} (\tilde{\boldsymbol{\alpha}} - \hat{\boldsymbol{\alpha}})\right\|_2 \leq \theta^o\right\}, \\
\mathcal{SU}_t^h &= \left\{\tilde{\boldsymbol{\rho}}_t \mid \left\|\left(\Xi_t^\rho\right)^{-\frac{1}{2}} (\tilde{\boldsymbol{\rho}}_t - \hat{\boldsymbol{\rho}}_t)\right\|_2 \leq \theta_t^h\right\}, \quad t = 1, \ldots, T \quad \textit{and} \\
\mathcal{SU}_t^f &= \left\{\tilde{\boldsymbol{\mu}}_t \mid \left\|\left(\Xi_t^\mu\right)^{-\frac{1}{2}} (\tilde{\boldsymbol{\mu}}_t - \hat{\boldsymbol{\mu}}_t)\right\|_2 \leq \theta_t^f\right\}, \quad t = 1, \ldots, T-1.
\end{aligned}$$

where Ξ^α, Ξ_t^ρ and Ξ_t^μ are the covariance matrices of the random vectors $\tilde{\boldsymbol{\alpha}}, \tilde{\boldsymbol{\rho}}_t$, and $\tilde{\boldsymbol{\mu}}_t$, respectively. (We will provide intuition on how such uncertainty sets can be determined in Sect. 4.5.3.)

Theorem 1. *Given uncertainty sets $\mathcal{SU}^o$, $\mathcal{SU}_t^h$ and $\mathcal{SU}_t^f$ for the uncertain parameters, the robust counterpart of the ALM problem ($\mathcal{P}^R$) is*

$$\begin{aligned}
(\mathcal{P}_{sym}^R): \ & \max \ \nu \\
& s.t. \ \nu \leq \hat{\boldsymbol{\alpha}}'\boldsymbol{\kappa} - \theta^o \sqrt{\boldsymbol{\kappa}'\Xi^\alpha \boldsymbol{\kappa}} - \lambda W_T \gamma_T \\
& \quad \xi_t^m = \xi_{t-1}^m - \eta_t^m + \zeta_t^m, && t = 1, \ldots, T, m = 1, \ldots, M \\
& \quad \xi_t^0 \leq \xi_{t-1}^0 + \hat{\boldsymbol{\rho}}_t'\boldsymbol{\pi}_t - \theta_t^h \sqrt{\boldsymbol{\pi}_t'\Xi_t^\rho \boldsymbol{\pi}_t}, && t = 1, \ldots, T \\
& \quad 0 \leq \hat{\boldsymbol{\mu}}_t'\boldsymbol{\tau}_t - \theta_t^f \sqrt{\boldsymbol{\tau}_t'\Xi_t^\mu \boldsymbol{\tau}_t}, && t = 1, \ldots, T-1 \\
& \quad \xi_t^m \geq 0, && t = 1, \ldots, T, \ m = 1, \ldots, M \\
& \quad \eta_t^m \geq 0, \zeta_t^m \geq 0, && t = 1, \ldots, T-1, \ m = 1, \ldots, M
\end{aligned}$$

A formal proof of this result is shown in the appendix, but let us provide intuition about how the robust counterpart of each constraint is obtained. We separate the expressions in the constraints into expressions with uncertain coefficients and expressions with certain coefficients. We then solve inner optimization problems that find the worst-case values of the terms involving uncertain coefficients when these uncertain coefficients vary in the given uncertainty sets. When the uncertainty sets are ellipsoidal and are defined in terms of the means and the covariance matrices of the uncertain coefficients, the robust counterparts of the constraints include the expected values of the expressions with uncertain coefficients, as well as penalty-like terms that are related to their standard deviations.

4.5.2 Robust ALM Model Formulation with Asymmetric Uncertainty Sets

Symmetric uncertainty sets can represent uncertainties well when these uncertainties follow symmetric probability distributions such as the normal distribution. Theoretically, the assumption that asset returns follow normal distributions is not unreasonable because the Central Limit Theorem implies that over the long horizon, returns should be approximately Gaussian as long as short-horizon returns are sufficiently independent (see, for example [12]). Empirically, however, there is evidence that both short- and long-horizon stock returns can be skewed and highly leptokurtic (see, for example [18, 22]).

Let us consider the vectors of uncertain coefficients $\tilde{\boldsymbol{\alpha}}$, $\tilde{\boldsymbol{\rho}}_t$, and $\tilde{\boldsymbol{\mu}}_t$ as defined at the beginning of Sect. 4.5. In order to use the result from Chen et al. [14], we need to write the constraints containing uncertain coefficients in the form (4.15). Let $\tilde{\mathbf{z}}^\alpha \in \mathcal{R}^{G^\alpha}$, $\tilde{\mathbf{z}}_t^\rho \in R^{G_t^\rho}$ and $\tilde{\mathbf{z}}_t^\mu \in \mathcal{R}^{G_t^\mu}$ be sets of independent factors, and let $\boldsymbol{\Delta}^\alpha \in \mathcal{R}^{(M+T)\times G^\alpha}$, $\boldsymbol{\Delta}_t^\rho \in \mathcal{R}^{(2M+2)\times G_t^\rho}$, and $\boldsymbol{\Delta}_t^\mu \in \mathcal{R}^{(M+T-t)\times G_t^\mu}$ be matrices such that

$$\tilde{\boldsymbol{\alpha}} = \hat{\boldsymbol{\alpha}} + \boldsymbol{\Delta}^\alpha \cdot \tilde{\mathbf{z}}^\alpha,$$

$$\tilde{\boldsymbol{\rho}}_t = \hat{\boldsymbol{\rho}}_t + \boldsymbol{\Delta}_t^\rho \cdot \tilde{\mathbf{z}}_t^\rho,$$

$$\tilde{\boldsymbol{\mu}}_t = \hat{\boldsymbol{\mu}}_t + \boldsymbol{\Delta}_t^\mu \cdot \tilde{\mathbf{z}}_t^\mu.$$

Let $\mathbf{P}^o$, $\mathbf{P}_t^h$, and $\mathbf{P}_t^f$ be the diagonal matrices with backward deviations and $\mathbf{Q}^o$, $\mathbf{Q}_t^h$, and $\mathbf{Q}_t^f$ be the diagonal matrices with forward deviations for factors $\tilde{\mathbf{z}}^\alpha$, $\tilde{\mathbf{z}}_t^\rho$ and $\tilde{\mathbf{z}}_t^\mu$, respectively. Consider the following asymmetric uncertainty sets for the uncertain factors $\tilde{\mathbf{z}}^\alpha$, $\tilde{\mathbf{z}}_t^\rho$ and $\tilde{\mathbf{z}}_t^\mu$:

$$\mathcal{AU}^o = \Big\{\mathbf{z}^\alpha : \ \exists \mathbf{v^o}, \mathbf{w^o} \in \mathcal{R}_+^{G\alpha}, \ \mathbf{z}^\alpha = \mathbf{v^o} - \mathbf{w^o},$$

$$||(\mathbf{P^o})^{-1}\mathbf{v^o} + (\mathbf{Q^o})^{-1}\mathbf{w^o}|| \leq \Omega^o, \underline{\mathbf{z}}^\alpha \leq \tilde{\mathbf{z}}^\alpha \leq \overline{\mathbf{z}}^\alpha\Big\},$$

$$\mathcal{AU}_t^h = \Big\{\mathbf{z}_t^\rho : \ \exists \mathbf{v_t^h}, \mathbf{w_t^h} \in \mathcal{R}_+^{G_t^\rho}, \ \mathbf{z}_t^\rho = \mathbf{v_t^h} - \mathbf{w_t^h},$$

$$||(\mathbf{P_t^h})^{-1}\mathbf{v_t^h} + (\mathbf{Q_t^f})^{-1}\mathbf{w_T^h}|| \leq \Omega_t^h, \underline{\mathbf{z}}_t^\rho \leq \tilde{\mathbf{z}}^\rho \leq \overline{\mathbf{z}}_t^\rho \Big\}, \quad t = 1, \ldots, T, \text{and}$$

$$\mathcal{AU}_t^f = \Big\{\mathbf{z}_t^\mu : \exists \mathbf{v_t^f}, \mathbf{w_t^f} \in \mathcal{R}_+^{G_t^\mu}, \mathbf{z}_t^\mu = \mathbf{v_t^f} - \mathbf{w_t^f},$$

$$||(\mathbf{P_t^f})^{-1}\mathbf{v_t^f} + (\mathbf{Q_t^f})^{-1}\mathbf{w_t^f}|| \leq \Omega_t^f, \underline{\mathbf{z}}_t^\mu \leq \tilde{\mathbf{z}}^\alpha \leq \overline{\mathbf{z}}_t^\mu \Big\}, \quad t = 1, \ldots, T-1.$$

Theorem 2. *Given uncertainty sets $\mathcal{AU}^o$, $\mathcal{AU}_t^h$ and $\mathcal{AU}_t^f$ for the uncertain parameters, the robust counterpart of the ALM problem $(\mathcal{P}^R)$ can be written as follows:*

$$
\begin{aligned}
(\mathcal{P}_{asym}^R): \quad & \max\ \nu \\
& s.t.\ \hat{\boldsymbol{\alpha}}'\boldsymbol{\kappa} - \lambda W_T \gamma_t - \nu \geq \Omega^o||\mathbf{u}^\alpha|| + (\mathbf{r}^\alpha)'\overline{\mathbf{z}}^\alpha + (\mathbf{s}^\alpha)'\underline{\mathbf{z}}^\alpha \\
& u_j^\alpha \geq -p_j^\alpha\left(\mathbf{e}_j'(\boldsymbol{\Delta}^\alpha)'\boldsymbol{\kappa} + r_j^\alpha - s_j^\alpha\right), \quad j = 1, \ldots, M+T \\
& u_j^\alpha \geq q_j^\alpha\left(\mathbf{e}_j'(\boldsymbol{\Delta}^\alpha)'\boldsymbol{\kappa} + r_j^\alpha - s_j^\alpha\right), \quad j = 1, \ldots, M+T \\
& \mathbf{r}^\alpha, \mathbf{s}^\alpha \geq 0 \\
& \xi_t^m = \xi_{t-1}^m - \eta_t^m + \zeta_t^m, \quad t = 1, \ldots, T, m = 1, \ldots, M \\
& \xi_t^0 \leq \xi_{t-1}^0 + \hat{\boldsymbol{\rho}}_t'\boldsymbol{\pi}_t - \Omega_t^h||\mathbf{u}_t^h|| - (\mathbf{r}_t^\rho)'\overline{\mathbf{z}}_t^\rho - (\mathbf{s}_t^\rho)'\underline{\mathbf{z}}_t^\rho, \quad t = 1, \ldots, T \\
& u_{t,j}^\rho \geq -p_j^{\rho_t}\left(\mathbf{e}_j'(\boldsymbol{\Delta}_t^\rho)'\boldsymbol{\pi}_t + r_{t,j}^{\rho_t} - s_{t,j}^\rho\right), \ t = 1, \ldots, T, \quad j = 1, \ldots, 2M+2 \\
& u_{t,j}^\rho \geq q_j\left(\mathbf{e}_j'(\boldsymbol{\Delta}_t^\rho)'\boldsymbol{\pi}_t + r_{t,j}^\rho - s_{t,j}^\rho\right), \ t = 1, \ldots, T, j = 1, \ldots, 2M+2 \\
& \mathbf{r}_t^\rho, \mathbf{s}_t^\rho \geq 0, \quad t = 1, \ldots, T \\
& 0 \leq \hat{\boldsymbol{\mu}}_t'\boldsymbol{\tau}_t - \Omega_t^f||\mathbf{u}_t^\mu|| - (\mathbf{r}_t^\mu)'\overline{\mathbf{z}}_t^\mu - (\mathbf{s}_t^\mu)'\underline{\mathbf{z}}_t^\mu, \quad t = 1, \ldots, T-1 \\
& u_{t,j}^\mu \geq -p_j^{\mu_t}\left(\mathbf{e}_j'(\boldsymbol{\Delta}_t^\mu)'\boldsymbol{\pi}_t + r_{t,j}^\mu - s_{t,j}^\mu\right), \quad t = 1, \ldots, T, j = 1, \ldots, M+T-t \\
& u_{t,j}^\mu \geq q_j^{\mu_t}\left(\mathbf{e}_j'(\boldsymbol{\Delta}_t^\mu)'\boldsymbol{\pi}_t + r_{t,j}^\mu - s_{t,j}^\mu\right), \quad t = 1, \ldots, T, j = 1, \ldots, M+T-t \\
& \mathbf{r}_t^\mu, \mathbf{s}_t^\mu \geq 0, \ t = 1, \ldots, T \\
& \xi_t^m \geq 0, \quad t = 1, \ldots, T, \ m = 1, \ldots, M \\
& \eta_t^m \geq 0, \zeta_t^m \geq 0, \ t = 1, \ldots, T-1, \ m = 1, \ldots, M
\end{aligned}
$$

The proof of the theorem is provided in Appendix.

4.5.3 Selecting Inputs to the Robust Optimization Models

A very important piece of implementing the robust ALM models is determining uncertainty sets and inputs that make sense given available data. As a general approach, the required inputs—vectors of expected values, covariance matrices, and factor deviations—can be estimated if we have data on possible scenarios for the risk-free rate and for vectors of asset returns. Those scenarios can also be used to generate scenarios for the vectors $\tilde{\boldsymbol{\alpha}}$, $\tilde{\boldsymbol{\rho}}_t$ and $\tilde{\boldsymbol{\mu}}_t$, so that estimates of the expected values and covariance matrices of these vectors (which are inputs to the robust optimization formulations) can be calculated.

Even though we propose using scenarios for optimization model parameter estimation, note that there is a philosophical difference between how we use scenarios in the stochastic programming setting (Sect. 4.3) and in the robust optimization setting (this section). In the robust optimization setting, we would like to use data to generate estimates of summary probability distribution measures that we then use as inputs to the robust optimization models. In the stochastic programming approach, we use scenarios for the inputs directly in the formulation. This difference between approaches has a couple of different implications.

First, the representation of the uncertainties is done differently. In the stochastic programming approach, we come up with an exact strategy to be followed for each scenario. We expect that if nature behaves very similarly to the scenarios we have on hand, the optimal strategy would perform well on average. The robust optimization approach attempts to find a more general strategy (one not tied to specific scenarios) that works well in terms of worst-case performance.

Second, the size of the resulting optimization problems can be very different. Generally, robust optimization formulations based on summary measures of the probability distributions of the uncertainties have a much smaller size than stochastic programming formulations.

Symmetric Uncertainty: In the case of symmetric uncertainty set (Sect. 4.5.1), we need the expected values and the covariance matrices of the vectors $\tilde{\boldsymbol{\alpha}}$, $\tilde{\boldsymbol{\rho}}_t$, and $\tilde{\boldsymbol{\mu}}_t$. Suppose we have scenarios for the vectors of cumulative returns $\mathbf{R}_t$ at each time period t, as well as the cumulative risk-free rate R_t^0 for $t = 1, \ldots, T$. Those can be used to create scenarios for the uncertain vectors in each constraint of the optimization problem, $\tilde{\boldsymbol{\alpha}}$, $\tilde{\boldsymbol{\rho}}_t$ and $\tilde{\boldsymbol{\mu}}_t$. The scenarios for $\tilde{\boldsymbol{\alpha}}$, $\tilde{\boldsymbol{\rho}}_t$ and $\tilde{\boldsymbol{\mu}}_t$ in turn can be used to estimate the expected value vectors $\hat{\boldsymbol{\alpha}}$, $\hat{\boldsymbol{\rho}}_t$ and $\hat{\boldsymbol{\mu}}_t$, as well as the covariance matrices Ξ^α, Ξ_t^ρ and Ξ_t^μ. Finally, the input parameters are plugged into the optimization problem $(\mathcal{P}_{sym}^R)$, which is solved to determine the optimal strategy.

Asymmetric Uncertainty: In the case of asymmetric uncertainty set, we can use the following transformations. Let Ξ^α, Ξ_t^ρ, and Ξ_t^μ be the covariance matrices for the vectors $\tilde{\boldsymbol{\alpha}}$, $\tilde{\boldsymbol{\rho}}_t$, and $\tilde{\boldsymbol{\mu}}_t$, respectively, as defined in Sect. 4.5.1. Sets of uncorrelated factors $\tilde{\mathbf{z}}^\alpha$, $\tilde{\mathbf{z}}_t^\rho$ and $\tilde{\mathbf{z}}_t^\mu$ with zero means can be constructed such that

$$\tilde{\mathbf{z}}^\alpha = (\Xi^\alpha)^{-\frac{1}{2}} \cdot (\tilde{\boldsymbol{\alpha}} - \hat{\boldsymbol{\alpha}}), \tag{4.19}$$

$$\tilde{\mathbf{z}}_t^\rho = \left(\Xi_t^\rho\right)^{-\frac{1}{2}} \cdot (\tilde{\boldsymbol{\rho}}_t - \hat{\boldsymbol{\rho}}_t), \tag{4.20}$$

$$\tilde{\mathbf{z}}_t^\mu = \left(\Xi_t^\mu\right)^{-\frac{1}{2}} \cdot (\tilde{\boldsymbol{\mu}}_t - \hat{\boldsymbol{\mu}}_t). \tag{4.21}$$

Then,

$$\tilde{\boldsymbol{\alpha}} = \hat{\boldsymbol{\alpha}} + (\Xi^\alpha)^{\frac{1}{2}} \cdot \tilde{\mathbf{z}}^\alpha,$$

$$\tilde{\boldsymbol{\rho}}_t = \hat{\boldsymbol{\rho}}_t + \left(\Xi_t^\rho\right)^{\frac{1}{2}} \cdot \tilde{\mathbf{z}}_t^\rho,$$

$$\tilde{\boldsymbol{\mu}}_t = \hat{\boldsymbol{\mu}}_t + \left(\Xi_t^\mu\right)^{\frac{1}{2}} \cdot \tilde{\mathbf{z}}_t^\mu.$$

The diagonal matrices $\mathbf{P}^o$, $\mathbf{P}_t^h$, and $\mathbf{P}_t^f$ with backward deviations and the diagonal matrices $\mathbf{Q}^o$, $\mathbf{Q}_t^h$, and $\mathbf{Q}_t^f$ with forward deviations for the factors $\tilde{\mathbf{z}}^\alpha$, $\tilde{\mathbf{z}}_t^\rho$ and $\tilde{\mathbf{z}}_t^\mu$, respectively, can be calculated from data, using the following procedure. As explained in the case of symmetric uncertainty, suppose we have a set of scenarios for the vectors of cumulative returns $\mathbf{R}_t$ at each time period t, as well as the cumulative risk-free rate R_t^0 for $t = 1, \ldots, T$. Those can be used to create scenarios for the uncertain vectors in each constraint of the optimization problem, $\tilde{\boldsymbol{\alpha}}$, $\tilde{\boldsymbol{\rho}}_t$ and $\tilde{\boldsymbol{\mu}}_t$. These scenarios in turn can be used to estimate the expected value vectors $\hat{\boldsymbol{\alpha}}$, $\hat{\boldsymbol{\rho}}_t$ and $\hat{\boldsymbol{\mu}}_t$, as well as the covariance matrices Ξ^α, Ξ_t^ρ and Ξ_t^μ.

Scenarios for uncorrelated factors for each constraint in the optimization problems can then be calculated. For example, to calculate scenarios for uncorrelated factors $\tilde{\mathbf{z}}^\alpha$, one uses (4.19). Finally, the scenarios that are generated for the uncorrelated factors can be used to calculate estimates of the factors' backward and forward deviations, which are then plugged into the robust counterpart formulation $(\mathcal{P}_{asym}^R)$. For example, backward and forward deviations for the ith factor can be computed by solving the optimization problems

$$p_i(z) = \sup_{\varphi>0}\left\{\sqrt{2\frac{\ln(E(\exp(\varphi.\mathbf{z}))))}{\varphi^2}}\right\}$$

and

$$q_i(z) = \sup_{\varphi>0}\left\{\sqrt{2\frac{\ln(E(\exp(-\varphi.\mathbf{z})))}{\varphi^2}}\right\}$$

A proof of this relationship is given in Natarajan et al. (2008). Note that if the forward and backward deviation matrices are equal (i.e. $\mathbf{P} = \mathbf{Q}$), then the asymmetric uncertainty set produces the same portfolio composition as the ellipsoidal uncertainty set does.

Given uncertainty sets $\mathcal{AU}^o$, $\mathcal{AU}_t^h$ and $\mathcal{AU}_t^f$ for the uncertain parameters, one then solves the following robust counterpart of the ALM problem $(\mathcal{P}^R)$:

$$\begin{array}{lll}
(\mathcal{P}_{asym}^R): & \max & \nu \\
& \text{s.t.} & \hat{\boldsymbol{\alpha}}'\boldsymbol{\kappa} - \lambda W_T\gamma_t - \nu \geq \Omega^o\|\mathbf{u}^\alpha\| + (\mathbf{r}^\alpha)'\overline{\mathbf{z}}^\alpha + (\mathbf{s}^\alpha)'\underline{\mathbf{z}}^\alpha \\
& & u_j^\alpha \geq -p_j^\alpha\left(\mathbf{e}_j'\left((\Xi^\alpha)^{\frac{1}{2}}\right)'\boldsymbol{\kappa} + r_j^\alpha - s_j^\alpha\right), \quad j = 1, \ldots, M+T \\
& & u_j^\alpha \geq q_j^\alpha\left(\mathbf{e}_j'\left((\Xi^\alpha)^{\frac{1}{2}}\right)'\boldsymbol{\kappa} + r_j^\alpha - s_j^\alpha\right), \quad j = 1, \ldots, M+T \\
& & \mathbf{r}^\alpha, \mathbf{s}^\alpha \geq 0 \\
& & \xi_t^m = \xi_{t-1}^m - \eta_t^m + \zeta_t^m, \quad t = 1, \ldots, T, m = 1, \ldots, M \\
& & \xi_t^0 \leq \xi_{t-1}^0 + \hat{\boldsymbol{\rho}}_t'\boldsymbol{\pi}_t - \Omega_t^h\|\mathbf{u}_t^\rho\| - (\mathbf{r}_t^\rho)'\overline{\mathbf{z}}_t^\rho - (\mathbf{s}_t^\rho)'\underline{\mathbf{z}}_t^\rho, \quad t = 1, \ldots, T \\
& & u_{t,j}^\rho \geq -p_j^{\rho_t}\left(\mathbf{e}_j'\left((\Xi_t^\rho)^{\frac{1}{2}}\right)'\boldsymbol{\pi}_t + r_{t,j}^\rho - s_{t,j}^\rho\right), \quad t = 1, \ldots, T, j = 1, \ldots, 2M+2 \\
& & u_{t,j}^\rho \geq q_j^{\rho_t}\left(\mathbf{e}_j'\left((\Xi_t^\rho)^{\frac{1}{2}}\right)'\boldsymbol{\pi}_t + r_{t,j}^\rho - s_{t,j}^\rho\right), \quad t = 1, \ldots, T, j = 1, \ldots, 2M+2 \\
& & \mathbf{r}_t^\rho, \mathbf{s}_t^\rho \geq 0, \quad t = 1, \ldots, T
\end{array}$$

$$
\begin{aligned}
&0 \le \hat{\boldsymbol{\mu}}_t{}' \boldsymbol{\tau}_t - \Omega_t^f \|\mathbf{u}_t^\mu\| - (\mathbf{r}_t^\mu)' \overline{\mathbf{z}}_t^\mu - (\mathbf{s}_t^\mu)' \underline{\mathbf{z}}_t^\mu, \quad t = 1, \ldots, T-1 \\
&u_{t,j}^\mu \ge -p_j^{\mu_t} \left(\mathbf{e}_j' \left((\boldsymbol{\Xi}_t^\mu)^{\frac{1}{2}} \right)' \boldsymbol{\pi}_t + r_{t,j}^\mu - s_{t,j}^\mu \right), \quad t = 1, \ldots, T, j = 1, \ldots, M+T-t \\
&u_{t,j}^\mu \ge q_j^{\mu_t} \left(\mathbf{e}_j' \left((\boldsymbol{\Xi}_t^\mu)^{\frac{1}{2}} \right)' \boldsymbol{\pi}_t + r_{t,j}^\mu - s_{t,j}^\mu \right), \quad t = 1, \ldots, T, j = 1, \ldots, M+T-t \\
&\mathbf{r}_t^\mu, \mathbf{s}_t^\mu \ge 0 \quad t = 1, \ldots, T \\
&\xi_t^m \ge 0, \quad t = 1, \ldots, T, \quad m = 1, \ldots, M \\
&\eta_t^m \ge 0, \zeta_t^m \ge 0, \quad t = 1, \ldots, T-1, \quad m = 1, \ldots, M
\end{aligned}
$$

Generating Scenarios: How could the scenarios for $\mathbf{R}_t$ at each time period t, as well as the cumulative risk-free rate R_t^0 for $t = 1, \ldots, T$ be generated? One possibility is to use historical data, and consider sets of returns for each length of period $1, 2, \ldots, T$. Another is to assume an underlying model that drives returns.

Consider, for example, the factor model for returns used in Ben-Tal et al. [5]:

$$
\begin{aligned}
\ln(1 + r_t^m) &= \boldsymbol{\beta}_m' \left[\kappa \cdot \mathbf{e} + \sigma \cdot \mathbf{v}_t\right], \; t = 0, 1, \ldots, N-1, m = 1, \ldots, M \quad (4.22) \\
\ln(1 + r_t^0) &= \kappa, \; t = 0, 1, \ldots, N-1
\end{aligned}
$$

where $\{\mathbf{v}_0, \mathbf{v}_1, \ldots, \mathbf{v}_{N-1}\}$ are independent K-dimensional normal random vectors with zero mean and unit covariance matrix; $\mathbf{e} \in \mathcal{R}^K = (1, \ldots, 1)'$; $\boldsymbol{\beta}_m \in \mathcal{R}_+^K$ are fixed vectors; and $\kappa, \sigma > 0$ are fixed reals. (Single-period returns are therefore lognormal.)

These assumptions allow for computing scenarios for the vectors of uncertain coefficients in each constraint. Note that under these assumptions, the expected values $\hat{R}_t^m$ and the covariances $E\left[(\tilde{R}_t^m - \hat{R}_t^m)(\tilde{R}_t^l - \hat{R}_t^l)\right] = \{\boldsymbol{\Sigma}_t\}_{m,l}$ of the cumulative returns at time t can even be calculated in closed form:

$$
\begin{aligned}
\hat{R}_t^m &= e^{\kappa t \cdot \boldsymbol{\beta}_m' \mathbf{e} + \frac{\sigma^2 t}{2} \boldsymbol{\beta}_m' \boldsymbol{\beta}_m} \\
\{\boldsymbol{\Sigma}_t\}_{m,l} &= e^{\kappa t \cdot (\boldsymbol{\beta}_m + \boldsymbol{\beta}_l)' \mathbf{e} + \frac{\sigma^2 t}{2} (\boldsymbol{\beta}_m' \boldsymbol{\beta}_m + \boldsymbol{\beta}_l' \boldsymbol{\beta}_l)}.
\end{aligned}
$$

4.6 Computational Experiments

This section contains a computational study of the behavior of stochastic and robust strategies for ALM with generated data and real market data. The experiments we designed attempt to answer the following questions:

- How do the robust asset and liability management models perform when the underlying uncertainty is represented by symmetric and asymmetric uncertainty sets?
- How do the size and shape of the symmetric and asymmetric uncertainty sets affect the investment strategy?

- How does the performance of the stochastic programming strategy compare with the performance of the robust optimization strategy, specifically in terms of metrics such as expected value, variance, minimum, maximum and tail risk measures for final wealth?

4.6.1 Design of Experiments and Data

Our first data set is generated using the factor model from Ben-tal et al. [5] described in Sect. 4.5.3. We generate eight risky assets and one risk-free asset. All simulation parameters are selected as in Ben-Tal et al. [5].

We also report results from experiments with real market data from Goyal and Welch (2008). There are two assets available for investment: a risky asset (the market), and a riskless asset. A sample period of 24 years of quarterly data (between 1987 and 2010) for S&P 500 index returns and Treasury bill rates is used.

We consider four decision stages: $t = 0, 1, 2, 3$. Final fund holdings are recorded at $T = 4$ to keep track of performance. The only reason for selecting a small number of assets for investment and a small number of time periods is the stochastic programming formulation. It takes a very long time (h) to obtain an optimal solution to stochastic programming formulations with even a small number of scenarios. The robust optimization formulation, on the other hand, is capable of handling a large number of assets and time periods in a short period of time (s).

The real world data with quarterly prices over 24 years is divided into four time periods. The mean returns of each asset at each time period are estimated using the corresponding data set. The estimated expected values of the returns and the factors, as well as other descriptive statistics, such as standard deviations and backward and forward deviations, are presented in Table 4.2.

The descriptive statistics of the factors $\tilde{\mathbf{z}}^{\alpha}$, $\tilde{\mathbf{z}}_t^{\rho}$ and $\tilde{\mathbf{z}}_t^{\mu}$ derived from the real world data are summarized in Table 4.3.

The input parameters for the ALM models are selected as follows. The initial portfolio holdings (the amount with which the fund begins) is the same for all experiments, and equals \$1000. The wages W_t are \$2050, \$2100 and \$2150 for $t = 1, 2$ and 3, respectively. The liabilities l_t start at \$150, and increase at each time period: \$150, \$155, \$160, \$165. Transactions costs for buying (c_b) and selling (c_s) are both fixed as 2 %. The funding ratio is set at 0.9 for the computational results we present. Tests with different funding ratios ($\psi = 1$ and $\psi = 1.1$) lead to similar conclusions, and we do not show the results here in the interest of space. The experiments are run for a range of different degrees of robustness (robustness budgets), 0.001–1.

In all experiments, we follow a *rolling horizon optimization* procedure. A set of 1000 scenarios is simulated for each length of time period, and the input parameters for the multiperiod optimization problems are estimated. The optimization problems are solved, and the first step recommended by the optimal strategy is taken. Actual realizations of the returns are simulated again so that the realized performance of the strategy can be recorded:

- In Experiment 1, we use the originally estimated mean (μ) and variance (σ) for generating scenarios for evaluating performance.
- In Experiment 2, the scenarios for evaluating performance are generated with a mean value of $\mu - \frac{1}{2}\sigma$ and the same variance σ as in Experiment 1. This represents a market regime in which the investor invests optimally given a particular expected return, but in actuality asset returns follow a distribution that is worse than expected on average.
- Experiment 3 is similar to Experiment 2, but the scenarios for evaluating performance deviate substantially from expectations—the mean value is $\mu - \sigma$ and the variance is σ. This represents a market regime with the worst realized asset returns.

After updating the realized holdings for the first time period, scenarios for future returns are simulated again with the estimated mean (μ), and the multiperiod optimization problems with reduced time horizon are solved again. The recommended first step is taken, and scenarios for strategy evaluation are generated again as explained above to estimate the performance of the different strategies over the following time period and update the holdings. This is repeated until the last time period.

To estimate the input parameters for the robust ALM models with asymmetric uncertainty sets for returns, we follow the procedure described in Sect. 4.5.3. Uncorrelated factors of accumulated returns of each asset are estimated, and the backward and forward deviations for the ith factor are computed by using the procedure described in Natarajan et al. (2008). For both of our data sets, $\mathbf{P} \neq \mathbf{Q}$.

Table 4.2 Statistical summary of the historical data (returns for each period)

	t = 0	t = 1	t = 2	t = 3
Mean return				
Risky asset	0.029	0.051	−0.003	0.005
Risk-free asset	0.014	0.012	0.007	0.005
Standard deviation				
Risky asset	0.054	0.069	0.082	0.095
Risk-free asset	0.005	0.001	0.004	0.005
p: Forward deviation				
Risky asset	0.054	0.070	0.082	0.095
Risk-free asset	0.005	0.001	0.005	0.005
q: Backward deviation				
Risky asset	0.070	0.069	0.085	0.101
Risk-free asset	0.005	0.002	0.004	0.005

Table 4.3 Descriptive statistics for the entries of the vectors of factors $\tilde{\mathbf{z}}^{\alpha}$, $\tilde{\mathbf{z}}_t^{\rho}$ and $\tilde{\mathbf{z}}_t^{\mu}$ extracted from real data

		1	2	3	4	5
$\tilde{\mathbf{z}}^{\alpha}$						
	Std dev	0.999	0.975	0.848	0.826	0.976
	p: Forward dev	1.026	0.995	0.865	0.844	0.996
	q: Backward dev	0.999	0.975	0.848	0.826	0.976
$\tilde{\mathbf{z}}_t^{\rho}$						
t = 1	Std dev	0.712	0.702	0.986	0.166	
	p: Forward dev	0.712	0.703	0.986	0.166	
	q: Backward dev	0.713	0.702	0.986	0.166	
t = 2	Std dev	0.665	0.748	0.981	0.172	
	p: Forward dev	0.665	0.748	0.981	0.172	
	q: Backward dev	0.665	0.748	0.981	0.172	
t = 3	Std dev	0.646	0.765	0.980	0.163	
	p: Forward dev	0.646	0.765	0.980	0.163	
	q: Backward dev	0.646	0.766	0.980	0.163	
t = 4	Std dev	0.696	0.719	0.987	0.110	
	p: Forward dev	0.701	0.719	0.987	0.110	
	q: Backward dev	0.696	0.723	0.987	0.110	
$\tilde{\mathbf{z}}_t^{\mu}$						
t = 1	Std dev	0.999	0.969	0.749	0.958	0.994
	p: Forward dev	0.999	0.969	0.751	0.960	0.996
	q: Backward dev	1.000	0.969	0.749	0.958	0.994
t = 2	Std dev	0.999	0.986	0.952	0.983	
	p: Forward dev	0.999	0.987	0.953	0.985	
	q: Backward dev	0.999	0.986	0.952	0.983	
t = 3	Std dev	0.999	0.980	0.983		
	p: Forward dev	0.999	0.981	0.984		
	q: Backward dev	1.012	0.993	0.996		

Most of the observed factors (and, respectively, the cumulative returns) follow asymmetric distributions.

We present results on the performance of three strategies—nominal, robust, and traditional stochastic programming—for different values of the input parameters λ (the penalty multiplier for the amount of contributions) and the robustness budgets Ω associated with the constraints containing uncertain coefficients. As explained earlier, we also verified that the conclusions remain the same for different values of the funding ratio ψ in the range 0.9–1.1.

- The robust optimization strategy, abbreviated as R, is implemented for different values of the price of robustness (PoR), which is in fact the robustness budget parameter Ω in the formulations in Sects. 4.5.1 and 4.5.2.

- The nominal strategy, abbreviated as N, calculates the optimal investment strategy assuming that all uncertain coefficients are at their expected values. The optimization problem formulation is a deterministic problem solved by a risk-neutral investor, and is only used as a benchmark. Note that N is equivalent to the robust strategy when the price of robustness is fixed at zero.
- The stochastic programming strategy, abbreviated as SP, maximizes the expected value of the objective function over the generated scenarios. The formulation is described in Sect. 4.3.

All models are implemented in Matlab and solved with YALMIP [32]. The computational experiments are run on a laptop with Windows XP operating system with CPU 2.26 GHz Intel Core2Duo and 2 Gb of RAM.

4.6.2 Computational Results

As we mentioned in the description of the experiment design, we generate scenarios to evaluate the performance of the various strategies. In this section, we present statistical analysis of the simulated values for final wealth in terms of average, variance, minimum, and maximum out of the 1000 simulations. We also compute tail risk measures reminiscent of Value-at-Risk (VaR) and Conditional Value-at-Risk (CVaR). The VaR at 5 % is found by taking the 50th smallest realized portfolio value whereas the CVaR is calculated as the average of the 50 smallest portfolio values all simulations. The performance of the nominal, robust and stochastic ALM models for pension funds for simulated and real data is presented in Tables 4.4, 4.5 and 4.6, 4.7, respectively.

Some interesting observations can be made from the results in Tables 4.4, 4.5 and 4.6, 4.7:

- If the future asset returns follow the same distribution as the input distribution, then the expected terminal wealth as well as the variance of terminal wealth obtained by all robust ALM models decrease when the robustness budget increases (see the results of Experiment 1). In other words, as we expected based on our discussion on robust optimization, there is a tradeoff between the average performance and the amount of protection desired. The nominal and stochastic investment strategies provide higher average wealth than the robust strategy at any price of robustness.
- This pattern, however, completely reverses when future realized asset returns are worse than expected as in Experiments 2 and 3. In such cases, the expected terminal wealth obtained by all robust ALM models increases when the robustness budget increases, whereas the variance decreases. Both the nominal and stochastic investment strategies result in lower wealth than the robust strategy for symmetric and asymmetric uncertainty sets at any value for the price of robustness.

Table 4.4 Performance of ALM models using simulated data for different market regimes. $\lambda = 0.1$

PoR	Symmetric uncertainty set						Asymmetric uncertainty set					
	Mean	Variance	VaR 5 %	CVaR 5 %	Min	Max	Mean	Variance	VaR 5 %	CVaR 5 %	Min	Max
Experiment 1: Expected market structure (μ, σ)												
N = R(0)	1310.00	32.56	1297.45	1294.59	1292.17	1327.69	1310.00	32.56	1297.45	1294.59	1292.17	1327.69
R(0.01)	1308.98	27.91	1297.36	1294.72	1292.63	1325.82	1309.03	25.29	1297.72	1295.20	1293.32	1324.20
R(0.3)	1307.03	25.00	1296.03	1293.53	1291.35	1322.78	1307.06	15.83	1298.11	1296.12	1294.06	1319.30
R(0.5)	1305.95	18.82	1296.41	1294.24	1292.63	1319.92	1306.05	11.54	1298.41	1296.71	1294.89	1316.78
R(0.7)	1304.90	12.89	1297.00	1295.21	1294.00	1316.61	1304.93	6.79	1299.07	1297.76	1296.29	1312.99
R(1.0)	1303.15	8.80	1296.62	1295.14	1293.32	1312.11	1303.15	5.69	1297.79	1296.59	1295.02	1310.39
SP	1309.54	32.91	1296.5	1294.55	1291.23	1327.04	1309.54	32.91	1296.5	1294.55	1291.23	1327.04
Experiment 2: Worse than expected market structure $(\mu - \frac{1}{2}\sigma, \sigma)$												
N = R(0)	1183.64	31.29	1171.33	1168.53	1166.64	1201.32	1183.64	31.29	1171.33	1168.53	1166.64	1201.32
R(0.1)	1183.64	31.29	1171.33	1168.53	1166.64	1201.32	1186.25	24.65	1175.07	1172.59	1170.83	1201.84
R(0.3)	1192.44	12.51	1184.66	1182.89	1181.57	1203.40	1193.24	16.45	1184.11	1182.09	1180.64	1206.23
R(0.5)	1198.65	7.42	1192.66	1191.29	1190.02	1207.05	1199.63	10.89	1192.21	1190.56	1188.57	1210.17
R(0.7)	1195.88	5.43	1190.75	1189.58	1188.85	1203.70	1196.68	8.08	1190.28	1188.86	1187.86	1205.76
R(1.0)	1202.32	3.91	1197.97	1196.98	1196.36	1208.78	1203.01	5.08	1197.94	1196.81	1196.01	1210.09
SP	1183.57	31.48	1171.2	1168.21	1166.6	1200.37	1183.57	31.48	1171.2	1168.21	1166.6	1200.37
Experiment 3: The worst market structure $(\mu - \sigma, \sigma)$												
N = R(0)	1070.23	18.56	1060.75	1058.60	1057.19	1083.70	1070.23	18.56	1060.75	1058.60	1057.19	1083.70
R(0.1)	1072.48	10.80	1065.25	1063.61	1062.23	1082.96	1074.16	11.47	1066.54	1064.85	1063.12	1085.06
R(0.3)	1083.70	7.84	1077.55	1076.15	1074.73	1092.62	1084.75	9.19	1077.93	1076.41	1075.16	1093.99
R(0.5)	1095.89	6.70	1090.20	1088.90	1087.45	1104.03	1097.40	4.86	1092.44	1091.34	1090.11	1104.63
R(0.7)	1099.88	3.40	1095.82	1094.90	1093.55	1105.80	1100.42	3.33	1096.32	1095.41	1094.64	1106.09
R(1.0)	1112.09	2.81	1108.40	1107.56	1107.01	1117.83	1112.18	1.89	1109.09	1108.40	1107.91	1116.39
SP	1067.98	18.95	1059.99	1057.85	1056.51	1084.16	1067.98	18.95	1059.99	1057.85	1056.51	1084.16

Table 4.5 Performance of ALM models using simulated data for different market regimes. $\lambda = 0.5$

PoR	Symmetric uncertainty set						Asymmetric uncertainty set					
	Mean	Variance	VaR 5 %	CVaR 5 %	Min	Max	Mean	Variance	VaR 5 %	CVaR 5 %	Min	Max
Experiment 1: Expected market structure (μ, σ)												
N = R(0)	1008.83	29.55	996.87	994.15	991.98	1025.93	1008.83	29.55	996.87	994.15	991.98	1025.93
R(0.01)	1007.24	17.95	997.92	995.80	994.28	1020.54	1007.30	19.80	997.29	995.07	993.65	1020.73
R(0.3)	1005.51	9.58	998.70	997.16	996.01	1015.37	1005.55	16.63	996.38	994.34	992.11	1018.43
R(0.5)	1004.40	6.99	998.58	997.26	995.65	1012.62	1004.48	12.85	996.42	994.62	992.42	1016.10
R(0.7)	1002.70	5.25	997.65	996.51	995.04	1010.50	1002.71	6.47	996.98	995.71	994.32	1010.40
R(1.0)	1001.64	3.84	997.32	996.34	995.75	1007.98	1001.72	5.21	996.59	995.45	994.15	1008.65
SP	1008.21	29.8	996.65	993.84	990.99	1025.38	1008.21	29.8	996.65	993.84	990.99	1025.38
Experiment 2: Worse than expected market structure $(\mu - \frac{1}{2}\sigma, \sigma)$												
N = R(0)	922.17	30.17	910.08	907.34	905.15	938.77	922.17	30.17	910.08	907.34	905.15	938.77
R(0.1)	923.47	22.07	913.13	910.78	909.16	938.29	924.06	19.89	914.03	911.80	910.35	938.19
R(0.3)	927.65	16.71	918.65	916.61	915.28	940.08	928.57	15.99	919.57	917.57	915.36	940.90
R(0.5)	929.78	9.33	923.06	921.53	920.09	939.16	930.13	12.10	922.30	920.56	918.65	941.44
R(0.7)	929.08	8.09	922.82	921.40	920.14	937.74	929.57	7.89	923.25	921.84	920.84	938.60
R(1.0)	938.36	4.33	933.78	932.74	931.54	945.00	939.05	4.00	934.55	933.55	932.20	945.50
SP	921.19	30.22	909.71	907.29	904.59	938.32	921.19	30.22	909.71	907.29	904.59	938.32
Experiment 3: The worst market structure $(\mu - \sigma, \sigma)$												
N = R(0)	842.87	16.56	833.91	831.88	829.84	855.47	842.87	16.56	833.91	831.88	829.84	855.47
R(0.1)	845.24	13.84	837.06	835.20	833.64	856.98	846.94	8.20	840.50	839.07	837.45	855.98
R(0.3)	851.68	12.00	844.06	842.33	840.85	862.83	851.78	5.40	846.55	845.39	844.31	859.61
R(0.5)	859.18	8.17	852.89	851.46	850.47	867.80	860.45	4.27	855.80	854.77	853.62	867.59
R(0.7)	863.37	7.21	857.46	856.12	854.69	871.89	865.35	3.51	861.14	860.20	859.49	871.09
R(1.0)	877.51	6.47	871.91	870.64	869.76	885.78	878.85	2.30	875.44	874.68	873.57	883.96
SP	840.87	17.05	832.96	830.74	828.67	856.21	840.87	17.05	832.96	830.74	828.67	856.21

Table 4.6 Performance of ALM models using real data for different market regimes. $\lambda = 0.1$

	Symmetric uncertainty set						Asymmetric uncertainty set					
PoR	Mean	Variance	VaR 5 %	CVaR 5 %	Min	Max	Mean	Variance	VaR 5 %	CVaR 5 %	Min	Max
Experiment 1: Expected market structure (μ, σ)												
N=R(0)	1220.12	18.56	1210.64	1208.49	1206.29	1233.42	1220.12	18.56	1210.64	1208.49	1206.29	1233.42
R(0.1)	1219.40	13.29	1211.38	1209.56	1207.74	1231.02	1219.42	13.17	1211.26	1209.44	1207.87	1230.44
R(0.3)	1218.41	11.92	1210.82	1209.09	1207.76	1229.62	1218.44	9.96	1211.34	1209.77	1207.67	1228.35
R(0.5)	1217.76	7.86	1211.59	1210.19	1209.22	1226.24	1217.84	5.97	1212.34	1211.12	1210.22	1225.67
R(0.7)	1217.02	5.01	1212.09	1210.98	1210.25	1223.75	1217.10	4.44	1212.35	1211.30	1209.85	1223.82
R(1.0)	1216.59	2.52	1213.10	1212.31	1211.00	1221.56	1216.62	3.57	1212.37	1211.43	1210.01	1222.62
SP	1220.01	18.64	1210.57	1208.12	1205.67	1233.35	1220.01	18.64	1210.57	1208.12	1205.67	1233.35
Experiment 2: Worse than expected market structure $(\mu - \frac{1}{2}\sigma, \sigma)$												
N=R(0)	1137.76	18.98	1128.18	1126.00	1123.71	1150.86	1137.76	18.98	1128.18	1126.00	1123.71	1150.86
R(0.1)	1138.92	11.79	1131.36	1129.65	1127.99	1149.83	1139.20	12.07	1131.39	1129.65	1128.19	1150.54
R(0.3)	1141.65	8.90	1135.09	1133.60	1132.31	1151.17	1142.49	8.55	1135.91	1134.45	1133.07	1151.51
R(0.5)	1144.70	7.22	1138.79	1137.44	1135.71	1153.23	1144.76	5.66	1139.41	1138.22	1137.19	1152.65
R(0.7)	1145.21	5.76	1139.93	1138.74	1137.64	1152.73	1145.91	2.95	1142.04	1141.18	1139.64	1151.94
R(1.0)	1149.68	3.36	1145.64	1144.73	1143.42	1155.85	1150.18	2.20	1146.85	1146.11	1145.39	1154.87
SP	1137.58	19.97	1127.52	1125.43	1123.01	1150.69	1137.58	19.97	1127.52	1125.43	1123.01	1150.69
Experiment 3: Worse than expected market structure $(\mu - \sigma, \sigma)$												
N=R(0)	1060.96	18.56	1051.49	1049.33	1047.98	1074.57	1060.96	18.56	1051.49	1049.33	1047.98	1074.57
R(0.1)	1062.61	10.78	1055.39	1053.75	1052.14	1072.83	1063.32	9.65	1056.33	1054.77	1053.40	1073.48
R(0.3)	1067.45	9.35	1060.72	1059.19	1057.76	1077.34	1067.82	7.44	1061.68	1060.32	1058.88	1076.14
R(0.5)	1073.73	4.88	1068.87	1067.76	1066.46	1080.44	1074.39	4.50	1069.61	1068.55	1067.73	1081.54
R(0.7)	1079.94	3.36	1075.91	1074.99	1074.11	1085.92	1080.50	2.53	1076.92	1076.13	1074.61	1086.14
R(1.0)	1085.30	1.68	1082.44	1081.80	1080.77	1089.35	1087.28	2.11	1084.01	1083.29	1082.11	1091.83
SP	1060.28	19.21	1049.71	1048.72	1046.56	1074.04	1060.28	19.21	1049.71	1048.72	1046.56	1074.04

Table 4.7 Performance of ALM models using real data for different market regimes. $\lambda = 0.5$

	Symmetric uncertainty set						Asymmetric uncertainty set					
PoR	Mean	Variance	VaR 5 %	CVaR 5 %	Min	Max	Mean	Variance	VaR 5 %	CVaR 5 %	Min	Max
Experiment 1: Expected market structure (μ, σ)												
N=R(0)	928.83	16.56	919.88	917.84	916.33	941.35	928.83	16.56	919.88	917.84	916.33	941.35
R(0.1)	928.52	10.55	921.37	919.75	918.14	938.69	928.55	9.84	921.49	919.92	918.3	938.14
R(0.3)	928.36	7.79	922.22	920.82	919.41	937.55	928.37	8.13	921.96	920.53	918.86	937.11
R(0.5)	928.27	5.26	923.23	922.08	921.35	936.11	928.28	3.95	923.81	922.82	921.84	934.74
R(0.7)	927.42	4.51	922.75	921.68	920.54	934.53	927.44	2.49	923.89	923.11	922.25	933.08
R(1.0)	926.87	3.05	923.03	922.16	921.10	932.80	926.90	2.10	923.64	922.92	922.27	931.92
SP	928.50	16.42	919.69	917.71	916.09	941.18	928.50	16.42	919.69	917.71	916.09	941.18
Experiment 2: Worse than expected market structure $(\mu - \frac{1}{2}\sigma, \sigma)$												
N=R(0)	866.13	17.26	856.99	854.92	853.55	878.98	866.13	17.26	856.99	854.92	853.55	878.98
R(0.1)	867.24	14.85	858.76	856.83	855.17	879.07	867.84	12.08	860.02	858.29	856.87	878.53
R(0.3)	869.87	8.99	863.27	861.77	860.19	879.43	870.65	6.38	864.97	863.71	861.97	878.69
R(0.5)	872.57	4.92	867.7	866.59	865.09	879.35	873.43	4.3	868.76	867.73	866.86	880.23
R(0.7)	872.7	2.82	869.00	868.16	867.19	878.66	873.01	3.44	868.84	867.91	867.1	878.76
R(1.0)	875.89	2.43	872.46	871.68	870.47	880.78	875.91	1.81	872.89	872.21	871.5	880.09
SP	865.52	17.53	856.43	853.97	853.28	878.36	865.52	17.53	856.43	853.97	853.28	878.36
Experiment 3: Worse than expected market structure $(\mu - \sigma, \sigma)$												
N=R(0)	807.67	16.56	798.72	796.68	794.5	820.28	807.67	16.56	798.72	796.68	794.5	820.28
R(0.1)	809.13	8.8	802.61	801.12	799.73	818.2	811.02	11.41	803.42	801.73	800.09	822.1
R(0.3)	813.33	5.45	808.19	807.02	806.08	820.49	815.02	8.04	808.64	807.22	806.01	824.17
R(0.5)	818.47	2.8	814.79	813.95	813.13	824.33	819.08	5.36	813.87	812.71	811.55	826.33
R(0.7)	822.96	2.43	819.53	818.75	818.03	827.95	823.24	4.39	818.52	817.47	816.69	830.45
R(1.0)	826.84	1.80	823.89	823.22	821.93	831.72	827.77	3.63	823.48	822.53	821.58	833.98
SP	806.43	17.11	797.65	795.51	792.68	819.13	806.43	17.11	797.65	795.51	792.68	819.13

- The robust ALM formulation with symmetric uncertainty set results in lower expected wealth (as well as higher variance and worse tail performance) than the wealth obtained with the asymmetric uncertainty set formulation for any degree of robustness. This is notable, given that both the simulated and the real data used in our experiments were asymmetric.
- Robust optimization strategies appear to perform better than stochastic programming formulations in catastrophic situations like Experiments 2 and 3 when the realized returns are worse than expected. The robust strategy realizes higher expected profit and better tail risk (as reflected in the 5 % VaR and CVaR—like measures) than the expected value strategy obtained from the stochastic program for pension funds.

4.7 Concluding Remarks

This entry outlined a robust optimization approach to ALM for pension funds, and suggested formulations that can incorporate possible asymmetries in the uncertainties in the problem. We considered data-driven methods for generating inputs to the optimization problems, and evaluated the performance of the robust formulations in computational experiments with simulated and real market data. Robust ALM strategies are faster to implement and appear to have better performance, both in terms of average realized wealth and in terms of tail performance, when the uncertain parameters follow probability distributions with different means than expected. This is an important practical advantage because the probability distributions of asset returns are very difficult to estimate. When the uncertain parameters follow probability distributions with known parameters, the value of the price of robustness parameter plays an important role in determining the performance of robust ALM strategies, and should be calibrated to data.

Appendix

Proof of Theorem 1. Let us derive the robust counterpart of the first constraint in the optimization problem formulation ($\mathcal{P}^{R}_{sym}$). The derivation of the robust counterparts of the remaining constraints is similar. The first constraint can be rewritten as

$$\nu - \tilde{\boldsymbol{\alpha}}'\boldsymbol{\kappa} + \lambda\gamma_T W_T \leq 0.$$

The robust counterpart of the original constraint can be written as

$$\nu - \text{Min}_{\tilde{\boldsymbol{\alpha}} \in \mathcal{SU}^{\alpha}} \left\{ \tilde{\boldsymbol{\alpha}}'\boldsymbol{\kappa} \right\} + \lambda\gamma_T W_T \leq 0$$

Intuitively, we would like the constraint to be satisfied even if the uncertain coefficients take their worst-case values within the uncertainty set. In this case, the worst-case value of the uncertain expression is obtained when $\tilde{\boldsymbol{\alpha}}'\boldsymbol{\kappa}$ is at its minimum value for values of $\tilde{\boldsymbol{\alpha}}$ within the uncertainty set. (If it was at its maximum value, it would be easier, not harder, for the inequality to be satisfied because of the negative sign in front of the vector product.) We solve the inner minimization problem first. The inner problem is in the explicit form

$$\begin{array}{ll}\text{Min}_{\tilde{\boldsymbol{\alpha}}} & \tilde{\boldsymbol{\alpha}}'\boldsymbol{\kappa} \\ \text{s.t.} & \left\| (\Xi^o)^{-\frac{1}{2}} (\tilde{\boldsymbol{\alpha}} - \mathrm{E}[\tilde{\boldsymbol{\alpha}}]) \right\|_2 \leq \Omega^o \end{array}$$

Let $\pi \geq 0$ be a Lagrangian multiplier for the constraint in the optimization problem above. The Lagrangian function is

$$\mathcal{L}(\tilde{\boldsymbol{\alpha}}, \pi) = \tilde{\boldsymbol{\alpha}}'\boldsymbol{\kappa} + \pi \left(\left\| (\Xi^o)^{-\frac{1}{2}} (\tilde{\boldsymbol{\alpha}} - \mathrm{E}[\tilde{\boldsymbol{\alpha}}]) \right\|_2 - \Omega^o \right).$$

The first-order optimality condition is

$$\frac{\partial \mathcal{L}}{\partial \tilde{\boldsymbol{\alpha}}} = \boldsymbol{\kappa} + \frac{\pi}{\left\| (\Xi^o)^{-\frac{1}{2}} (\tilde{\boldsymbol{\alpha}} - \mathrm{E}[\tilde{\boldsymbol{\alpha}}]) \right\|_2} (\Xi^o)^{-1} (\tilde{\boldsymbol{\alpha}} - \mathrm{E}[\tilde{\boldsymbol{\alpha}}]) = 0 \tag{4.23}$$

and the complementarity condition is

$$\pi \left(\left\| (\Xi^o)^{-\frac{1}{2}} (\tilde{\boldsymbol{\alpha}} - \mathrm{E}[\tilde{\boldsymbol{\alpha}}]) \right\|_2 - \Omega^o \right) = 0 \tag{4.24}$$

Using the optimality conditions stated in (4.23) and (4.24), we can find the optimal value of the random parameter within the uncertainty set. Note that in (4.24), $\pi \neq 0$. In addition,

$$\left\| (\Xi^o)^{-\frac{1}{2}} (\tilde{\boldsymbol{\alpha}} - \mathrm{E}[\tilde{\boldsymbol{\alpha}}]) \right\|_2 = \Omega^o \tag{4.25}$$

From (4.23) and (4.25), we obtain

$$(\Xi^o)^{-1} (\tilde{\boldsymbol{\alpha}} - \mathrm{E}[\tilde{\boldsymbol{\alpha}}]) = -\frac{\Omega^o}{\pi}\boldsymbol{\kappa} \Rightarrow \tilde{\boldsymbol{\alpha}} - \mathrm{E}[\tilde{\boldsymbol{\alpha}}] = \left(-\frac{\Omega^o}{\pi}\right) \Xi^o \boldsymbol{\kappa}. \tag{4.26}$$

In addition, it can be easily shown that

$$\begin{aligned} \left\| (\Xi^o)^{-\frac{1}{2}} (\tilde{\boldsymbol{\alpha}} - \mathrm{E}[\tilde{\boldsymbol{\alpha}}]) \right\|_2 &= \sqrt{(\tilde{\boldsymbol{\alpha}} - \mathrm{E}[\tilde{\boldsymbol{\alpha}}]) (\Xi^o)^{-1} (\tilde{\boldsymbol{\alpha}} - \mathrm{E}[\tilde{\boldsymbol{\alpha}}])} \\ &= \sqrt{\left(-\frac{\Omega^o}{\pi}\Xi^o\boldsymbol{\kappa}\right)' (\Xi^o)^{-1} \left(-\frac{\Omega^o}{\pi}\Xi^o\boldsymbol{\kappa}\right)} \\ &= \frac{\Omega^o}{\pi}\sqrt{\boldsymbol{\kappa}'\Xi^o\boldsymbol{\kappa}}. \end{aligned} \tag{4.27}$$

Using (4.27) and (4.25), the Lagrangian multiplier is found as $\pi = \sqrt{\kappa' \Xi^o \kappa}$. Substituting this in (4.23) confirms that the optimal solution of the inner problem is

$$\tilde{\alpha}' \kappa = \mathrm{E}[\tilde{\alpha}]' \kappa - \Omega^o \sqrt{\kappa' \Xi^o \kappa}.$$

The robust counterparts of the other constraints that contain uncertain coefficients, namely the liability and the cash balance constraints, can be derived in a similar manner. As a result, the robust model for $(\mathcal{P}^R)$ can be obtained as stated in Theorem 1.

Proof of Theorem 2. The theorem follows from (4.17) and the fact that, given the representation of the uncertain coefficients as linear combinations of factors, the constraints can be written in the form (4.15). To see how the constraint $\tilde{\alpha}'\kappa - \lambda W_T \gamma_T - \nu \geq 0$, for example, can be written in the form (4.15), note that it can be written in terms of the uncertain factors $\tilde{\mathbf{z}}^\alpha$ as

$$\tilde{\alpha}'\kappa - \lambda W_T \gamma_T - \nu = \underbrace{\hat{\alpha}'\kappa - \lambda W_T \gamma_T - \nu}_{f_0(\kappa, \gamma_T)} + \sum_{j=1}^{M+T} \underbrace{\mathbf{e}_j' (\boldsymbol{\Delta}^\alpha)' \kappa}_{f_j(\kappa)} \cdot \tilde{z}_j^\alpha \tag{4.28}$$

Given (4.17), the robust counterpart of the constraint

$$\tilde{\alpha}'\kappa - \lambda W_T \gamma_t - \nu \geq 0$$

when the uncertain factors $\tilde{\mathbf{z}}^\alpha$ vary in uncertainty set $\mathcal{AU}^o$ is

$$\begin{aligned}
\hat{\alpha}'\kappa - \lambda W_T \gamma_t - \nu &\geq \Omega^o ||\mathbf{u}^\alpha|| + (\mathbf{r}^\alpha)' (\overline{\mathbf{z}}^\alpha) + (\mathbf{s}^\alpha)' (\underline{\mathbf{z}}^\alpha) \\
u_j^\alpha &\geq -p_j \left(\mathbf{e}_j' (\boldsymbol{\Delta}^\alpha)' \kappa + r_j^\alpha - s_j^\alpha \right), \quad j = 1, \ldots, M+T \\
u_j^\alpha &\geq q_j \left(\mathbf{e}_j' (\boldsymbol{\Delta}^\alpha)' \kappa + r_j^\alpha - s_j^\alpha \right), \quad j = 1, \ldots, M+T \\
\mathbf{r}^\alpha, \mathbf{s}^\alpha &\geq 0
\end{aligned}$$

The robust counterparts of the remaining constraints can be derived in a similar manner.

References

1. A. Ben-Tal, A. Nemirovski, Robust convex optimization. Math. Oper. Res. **23**(4), 769–805 (1998)
2. A. Ben-Tal, A. Nemirovski, Robust solutions of uncertain linear programs. Oper. Res. Lett. **25**(1), 1–13 (1999)

3. A. Ben-Tal, A. Nemirovski, Robust solutions of linear programming problems contaminated with uncertain data. Math. Program. **88**, 411–424 (2000)
4. A. Ben-Tal, A. Nemirovski, *Lectures on Modern Convex Optimization: Analysis, Algorithms, and Engineering Applications*. MPS-SIAM Series on Optimization (Society for Industrial and Applied Mathematics, Philadelphia, 2001)
5. A. Ben-Tal, T. Margalit, A. Nemirovski, Robust modeling of multi-stage portfolio problems, in *High-Performance Optimization*, ed. by H. Frenk, K. Roos, T. Terlaky, S. Zhang (Kluwer Academic Publishers, Dordrecht, 2000), pp. 303–328
6. A. Ben-Tal, L. El Ghaoui, A. Nemirovski, *Robust Optimization* (Princeton University Press, Princeton, 2009)
7. D. Bertsimas, D. Pachamanova, Robust multiperiod portfolio management in the presence of transaction costs. Comput. Oper. Res. **35**(1), 3–17 (2008)
8. D. Bertsimas, M. Sim, The price of robustness. Oper. Res. **52**(1), 35–53 (2004)
9. D. Bertsimas, D. Pachamanova, M. Sim, Robust linear optimization under general norms. Oper. Res. Lett. **32**, 510–516 (2004)
10. G.C.E. Boender A hybrid simulation/optimization scenario model for asset/liability management. Eur. J. Oper. Res. **99**(1), 126–135 (1997)
11. E. Bogentoft, H.E. Romeijn, S. Uryasev, Asset/liability management for pension funds using CVaR constraints. J. Risk Financ. **3**(1), 57–71 (2001)
12. J. Campbell, A. Lo, A. MacKinlay, *The Econometrics of Financial Markets* (Princeton University Press, Princeton, NJ, 1997)
13. S. Ceria, R. Stubbs, Incorporating estimation errors into portfolio selection: robust portfolio construction. J. Asset Manage. **7**(2), 109–127 (2006)
14. X. Chen, M. Sim, P. Sun, A robust optimization perspective on stochastic programming. Oper. Res. **55**(6), 1058–1071 (2007)
15. G. Consigli, M.A.H. Dempster, Dynamic stochastic programming for asset-liability management. Ann. Oper. Res. **81**, 131–162 (1998)
16. A. Consiglio, F. Cocco, S.A. Zenios, Asset and liability modeling for participating policies with guarantees. Eur. J. Oper. Res. **186**(1), 380–404 (2008)
17. C.L. Dert, A dynamic model for asset liability management for defined benefit pension funds, in *Worldwide Asset and Liability Modeling*, ed. by W.T. Ziemba, J.M. Mulvey (reprinted) (Cambridge University Press, Cambridge, 1999)
18. G. Duffee, The long-run behavior of firms' stock returns: evidence and interpretations, working paper, Haas School of Business, University of California at Berkeley, Berkeley, CA (2002)
19. L. El Ghaoui, H. Lebret, Robust solutions to least-squares problems with uncertain data. SIAM J. Matrix Anal. Appl. **18**(4), 1035–1064 (2007)
20. L. El Ghaoui, M. Oks, F. Oustry, Worst-case value-at-risk and robust portfolio optimization: a conic programming approach. Oper. Res. **51**(4), 543–556 (2003)
21. L.F. Escudero, A. Garín, M. Merino, G. Pérez, On multistage stochastic integer programming for incorporating logical constraints in asset and liability management under uncertainty. Comput. Manage. Sci. **6**, 307–327 (2009)
22. E. Fama, *Foundations of Finance* (Basic Books, New York , 1976)
23. D. Goldfard, G. Iyengar, Robust portfolio selection problem. Math. Oper. Res. **28**(1), 1–37 (2003)
24. A. Goyal, I. Welch, A comprehensive look at the empirical performance of equity premium prediction. Rev. Financ. Stud. 21:1455–1508 (2008)
25. N. Gulpinar, B. Rustem, Optimal decisions and robust methods for forecast errors. Comput. Stat. Data Anal. **51**, 3595–3611 (2007a)
26. N. Gulpinar, B. Rustem, Worst-case optimal robust decisions for multi-period portfolio optimization. Eur. J. Oper. Res. **183**(3), 981–1000 (2007b)
27. N. Gulpinar, B. Rustem, R. Settergren, Simulation and optimization approaches to scenario generation. J. Econ. Dyn. Control **28** (1–2), 1291–1315 (2004)
28. P. Hilli, M. Koivu, T. Pennanen, Optimal construction of a fund of funds. Eur. Actuar. J. **1**, 345–359 (2011)

29. R. Kouwenberg, Scenario generation and stochastic programming models for asset liability management. Eur. J. Oper. Res. **134**(2), 279–292 (2001)
30. K. Natarajan, D. Pachamanova, M. Sim, Incorporating asymmetric distributional information in robust value-at-risk optimization. Manage. Sci. **54**(3):573–585 (2008)
31. K. Natarajan, D. Pachamanova, M. Sim, Constructing risk measures from uncertainty sets. Oper. Res. **57**(5):1129–1141 (2009)
32. J. Löfberg, YALMIP: a toolbox for modeling and optimization in MATLAB, in *Proceedings of the CACSD Conference*, Taipei (2004)
33. W.T. Ziemba, J.M. Mulvey (eds.), *Worldwide Asset and Liability Modeling* (Cambridge University Press, Cambridge, 1998)

Chapter 5
Liability-Driven Investment in Longevity Risk Management

Helena Aro and Teemu Pennanen

Abstract This paper studies optimal investment from the point of view of an investor with longevity-linked liabilities. The relevant optimization problems rarely are analytically tractable, but we are able to show numerically that liability driven investment can significantly outperform common strategies that do not take the liabilities into account. In problems without liabilities the advantage disappears, which suggests that the superiority of the proposed strategies is indeed based on connections between liabilities and asset returns.

Keywords Longevity risk • Mortality risk • Stochastic mortality • Stochastic optimization • Hedging

5.1 Introduction

Longevity risk, the uncertainty in future mortality developments, affects pension providers, life insurers, and governments. The population structure of developed countries is increasingly leaning towards the old, and the effects of medical advances and lifestyle choices on mortality remain unpredictable, which creates an increasingly acute need for life insurance and pension plans to hedge themselves against longevity risk.

Various longevity-linked instruments have been proposed for the management of longevity risk; see e.g. [6, 7, 11, 21, 27]. It has been shown how such instruments, once in existence, can be used to hedge mortality risk exposures in pensions or life insurance liabilities [7, 12, 13, 15, 18, 26]. Indeed, demand for longevity-linked instruments appears to exist, and some longevity transactions have already taken place. However, a major challenge facing the development of longevity markets

H. Aro
Finnish Financial Supervisory Authority, Helsinki, Finland
e-mail: helena.aro@BOF.fi

T. Pennanen (✉)
Department of Mathematics, King's College London, London, UK
e-mail: teemu.pennanen@kcl.ac.uk

© Springer International Publishing Switzerland 2017
G. Consigli et al. (eds.), *Optimal Financial Decision Making under Uncertainty*, International Series in Operations Research & Management Science 245,
DOI 10.1007/978-3-319-41613-7_5

is the hedging of the risk that stems from issuing longevity-linked securities. The supply for mortality-linked instruments might increase if their cash-flows could be (partially) hedged by appropriately trading in assets for which liquid markets already exist. Such a development has been seen e.g. in options markets, which flourished after the publication of the seminal Black–Scholes–Merton model. Even though the cash flows of mortality-linked instruments cannot be perfectly replicated, it may be possible to diminish the residual risk by an appropriate choice of an investment strategy.

This paper addresses the above issues by studying optimal investment from the point of view of an insurer with longevity-linked liabilities. As such problems are rarely analytically tractable, we employ a numerical procedure that adjusts the investment strategy according to the statistical properties of assets and liability as well as a given risk measure. The approach can be applied in pricing and hedging of longevity-linked instruments, as well as in asset-liability management of pension plans and life insurers. Rather than aiming at general investment principles, we illustrate the technique in the specific example of hedging a survivor bond whose payments are tied to a given cohort over a fixed time interval.

The most straightforward approach to hedging of longevity-linked instruments is *natural hedging*, where an insurer hedges longevity risk by taking positions with opposite exposures to longevity developments [17, 29]. Such an approach is obviously limited by the demand on the relevant insurance products. Another popular approach builds on *risk neutral valuation* which is based on the no-arbitrage principle from financial economics; see e.g. [3, 12, 16, 20] and Sect. 10 of [30]. In analogy with the Black–Scholes–Merton theory, it has been suggested that longevity-linked instruments could then be hedged using *delta hedging* by determining price sensitivities with respect to traded securities. This approach is, however, invalidated by the fact that the payouts of longevity-linked instruments cannot be replicated by liquidly traded assets as assumed by the risk neutral valuation theory; see the discussion in [4, 5, 30].

This paper employs a computational technique that constructs diversified strategies from a family of simpler *basis strategies*. We find that the risk associated with the diversified strategy diminishes significantly when one includes basis strategies suggested by the statistical connections between mortality and financial markets observed in [2]. To assess to which extent the reduction of risk is due to the asset-liability connections, we performed the same computations also without liabilities. In this case, the inclusion of the liability-driven investment strategies had negligible effect on risk, which suggests that the reduction of risk in the asset-liability management problem was indeed due to the connections between assets and liabilities.

The rest of this paper is organized as follows. Section 5.2 formulates the asset-liability management problem of a longevity-linked cash flow. Section 5.3 introduces investment strategies that serve as basis strategies for the computational procedure described in Sect. 5.4. Section 5.5 presents results from a simulation study, and Sect. 5.6 concludes.

5.2 The Asset-Liability Management Problem

Consider an insurer with given initial capital w_0 and longevity-linked liabilities with claims c_t over time $t = 1, 2, \dots, T$. After paying out c_t at time t, the insurer invests the remaining wealth in financial markets. We look for investment strategies whose proceeds fit the liabilities as well as possible in the sense of a given risk measure ρ on the residual wealth at time T.

We assume that a finite set J of liquid assets (bonds, equities, ...) can be traded at $t = 0, \dots, T$. The total return on asset j over period $[t-1, t]$ will be denoted $R_{t,j}$. The amount of cash invested in asset j over period $(t, t+1]$ will be denoted by $h_{t,j}$. The asset-liability management problem of the insurer can then be written as

$$\begin{aligned} &\text{minimize} && \rho\Big(\sum_{j\in J} h_{T,j}\Big) \quad \text{over} \quad h \in \mathcal{N} \\ &\text{subject to} && \sum_{j\in J} h_{0,j} \le w_0 \\ & && \sum_{j\in J} h_{t,j} \le \sum_{j\in J} R_{t,j} h_{t-1,j} - c_t \quad t = 1, \dots, T \\ & && h_t \in D_t, \quad t = 1, \dots, T \end{aligned} \tag{ALM}$$

The liabilities $(c_t)_{t=0}^T$ and the investment returns $(R_{t,j})_{t=0}^T$ will be modeled as stochastic processes on a filtered probability space $(\Omega, \mathcal{F}, (\mathcal{F})_{t=1}^T, P)$. The set $\mathcal{N}$ denotes the $\mathbb{R}^J$-valued $(\mathcal{F}_t)_{t=1}^T$-adapted investment strategies $(h_t)_{t=0}^T$. Being adapted means that the portfolio h_t chosen at time t may only depend on information observed by time t. The last constraint describes portfolio constraints. The set D_t of feasible strategies is allowed to be random but known at time t.[1] Short-selling constraints, for instance, correspond to the deterministic constraint $D_t(\omega) = \mathbb{R}^J_+$. Describing limitations on investment strategies in case of negative wealth require a random D_t.

The risk measure ρ is a convex function on the space of real-valued random variables. It describes the insurer's preferences over random terminal wealth distributions. We refer the reader to [22, Chap. 4] for a general treatment of risk measures. In addition to the terminal wealth, one may also wish to take into consideration the overall trajectory of wealth either in the objective or the constraints. For simplicity, we will concentrate on the above formulation, which is more in line with established models of mathematical finance; see e.g. [22, Chap. 8] or [19, Chap. 3].

Liability-driven investment refers to the general principle that optimal investment strategies depend on the liabilities. The same idea is behind the famous

[1]More precisely, D_t is assumed $\mathcal{F}_t$-measurable, i.e. $\{\omega \in \Omega \mid D_t(\omega) \cap U \neq \emptyset\} \in \mathcal{F}_t$ for every open U.

Black–Scholes–Merton option pricing model where the price of an option is defined as the least amount of initial capital that allows for the implementation of an investment strategy whose proceeds match the option payout exactly. In the case of longevity-linked liabilities, exact replication is not possible so one has to evoke the risk preferences as is done in problem (ALM) in terms of the risk measure ρ.

Problems of the form (ALM) arise naturally in reserving for existing insurance liabilities as well as in underwriting new insurance contracts. Optimal risk adjusted reserves are obtained as the least initial wealth w_0 that allow for acceptable levels of risk in (ALM). In underwriting, one looks for a premium that would allow the insurer to take on the additional liabilities without worsening the optimal value of (ALM); see [28] for a general study of risk management-based valuation of uncertain cash-flows.

5.3 Investment Strategies

In this section we present investment strategies that are used in subsequent numerical illustrations. We recall some well-known trading strategies recommended for long-term investment, and also introduce new strategies that try to employ the connections between the longevity-linked liabilities and asset returns.

We will describe the proportions of wealth invested in each asset $j \in J$ at time t by vector $\pi_t = [\pi_t^1, \pi_t^2, \dots, \pi_t^J]$ whose components sum up to one. The amount of wealth invested in each asset can be expressed as $h_t = \pi_t w_t$, where for $t = 1, \dots, T$

$$w_t = \sum_{j \in J} R_{t,j} h_{t-1,j} - c_t$$

is the *net wealth* of the investor at time t.

5.3.1 Non-liability-Driven Investment Strategies

In *buy and hold* (B and H) strategies the initial asset allocation π_0 is held over the subsequent time periods. To cover a nonzero claim process c_t, each asset is liquidated in proportion to the initial investment. In other words, one invests

$$h_{t,j} = \begin{cases} \pi_0^j w_0 & t = 0, \\ R_{t,j} h_{t-1,j} - \pi_0^j c_t & t = 1, \dots, T, \end{cases}$$

units of cash in asset $j \in J$ at the beginning of the holding period starting at time t.

Fixed proportions (FP) is a strategy where the allocation is rebalanced at the beginning of each holding period into fixed proportions given by a constant vector $\pi \in \mathbb{R}^J$ as

$$h_t = \pi w_t.$$

A *target date fund* (TDF) is a well-known strategy in the pension industry [14]. In a TDF, the proportion invested in risky assets diminishes as a pre-determined target date approaches. We implement TDFs by adjusting the allocation between two complementary subsets J^r and J^s of the set of all assets J. Here J^s consists of "safe" assets and J^r consists of the remaining "risky" assets. In the simulations presented in Sect. 5.5 safe assets consist of government bonds, and risky assets comprise equities and corporate bonds. The *proportional exposure* to J^r at time t is given by

$$e_t = a - bt.$$

The parameter a determines the initial proportion invested in J^r and b defines how fast the proportion decreases in time. Choices of such a and b that

$$a \geq 0 \quad \text{and} \quad a - bT \geq 0.$$

guarantees that the exposure e_t in the risky assets remains nonnegative. A TDF can be written as

$$h_t = \pi_t w_t$$

where the vector π_t is adjusted to give the specified proportional exposure:

$$\sum_{j \in J^r} \pi_t^j = e_t.$$

Within J^r and J^s the wealth is allocated using FP rules.

5.3.2 Liability-Driven Investment Strategies

This subsection presents strategies in which the proportions invested in different assets are connected to the development of the longevity-linked liabilities. Some of the strategies utilize the connections between mortality and financial markets observed in [2], while others employ, directly or indirectly, the current and forecast future cash flows of longevity-linked liabilities to determine the asset allocation.

A well-known liability-driven strategy is the *constant proportion portfolio insurance* (CPPI) strategy; see e.g. [8–10]. The proportion of wealth invested in risky assets is given by

$$\begin{aligned} e_t &= \frac{m}{w_t} \max\{w_t - F_t, 0\} \\ &= m \max\{1 - \frac{F_t}{w_t}, 0\}, \end{aligned}$$

where the *floor* F_t represents the value of outstanding claims at time t and the parameter $m \geq 0$ determines the fraction of the cushion (wealth over the floor) invested in risky assets. Risky and safe assets are here the same as in TDF strategies. Within J^r and J^s the wealth is allocated using FP rules. We define the floor through

$$F_T = 0,$$
$$F_t = (1+r)F_{t-1} - \bar{c}_t \quad t = 0, \ldots, T,$$

where r is a deterministic discount factor and $\bar{c}_t$ is the median of claim amount at time t. In this type of strategies, the liabilities are taken into account not only in the projected claim amounts $\bar{c}_t$ but also in the remaining wealth w_t, which for a given w_0 depends on the realized values of the claims c_t.

The idea behind *spread strategies* is to capture the connections between mortality and asset returns described in [2]. Statistical analysis suggests that long-term increases in GDP have a positive effect on old-age mortality. The rationale behind this is that increases in the national income are reflected in the wellbeing of the old. On the other hand, high levels of GDP growth are connected to high term spreads. The interpretation of this connection is that interest rates reflect the future changes in the level of economic activity.

Further, when term spread is high, the yields of long-term treasury bonds are then relatively high, compared to short-term treasury bonds. Hence, relatively high yields on long-term bonds are connected with high survival probabilities of the old. *Term spread strategies* aim to utilize this notion. The proportion of wealth invested in long-term treasury bonds is determined as a function of the term spread s_t^T by

$$\pi_t^L = \sigma^{a,b}(s_t^T),$$

where

$$\sigma^{a,b}(s) = \frac{1}{1 + e^{-b(s+a)}},$$

and $b > 0$ and $a \in \mathbb{R}$ are user-defined parameters. The remaining wealth is invested in short-term government bonds. LISAA linkkien selitykset, ks. morfin.

Correspondingly, the analysis in [2] corroborate that high levels of GDP are connected to low credit spreads, which suggests that the low credit spreads are connected to high survival probabilities of over 50-year-olds. Simultaneously, the returns on riskier corporate bonds are relatively low compared to less risky bonds. Hence, in the case of *credit spread strategies* the proportional exposure to riskier corporate bonds is

$$\pi_t^C = \sigma(s_t^C),$$

The remaining wealth is invested in less risky bonds.

In *survival index strategies* the wealth allocated amongst assets depends on the *survival index* S of a given population. The value of the survival index S_t time t is defined as the fraction of the population that survives until time t. The value of S_t gives indication on future liabilities: the smaller the remaining number of survivors, the smaller the future cash flows are likely to be. Hence, the proportional exposure in asset j at time t is given by

$$\pi_t^j = g^a(S_t),$$

where

$$g^a(s) = \min\{as, 1\},$$

and $a \in \mathbb{R}$. The rest is invested in other assets using fixed-mix strategies.

In *wealth strategies* the proportion invested in asset j depends on the proportion of initial wealth w_t/w_0 remaining at time t. The proportional exposure at time t is given by

$$\pi_t^j = g^a(w_t/w_0),$$

where $a \in \mathbb{R}$. The rest is invested in other assets. Wealth-dependent strategies resemble CPPI strategies in the sense that both define the proportions of wealth invested in various assets in terms of the present wealth. However, the wealth strategies do not depend on median liabilities like CPPI, but the liabilities are reflected only in the present level of wealth.

5.4 Diversification Procedure

We now briefly recall the numerical procedure presented in [23, 25]. It is a computational method for diversifying the initial wealth w_0 amongst a set of simple parametric strategies called *basis strategies*. The convex combination of feasible basis strategies is always feasible, since the optimization problem is convex. The investment strategies presented in the previous section serve as basis strategies in the numerical illustrations in Sect. 5.5.

Consider a finite set $\{h^i \mid i \in I\}$ of basis strategies that invest the amount $h^i_{t,j}$ in asset j at time t. The problem of finding an optimal diversification amongst the basis strategies can be written as

$$\underset{\alpha \in X}{\text{minimize}} \quad \rho\left(\sum_{i \in I} \alpha^i w_T^i\right),$$

where $w^i_T = \sum_{j\in J} h^i_{Tj}$ is the terminal wealth obtained by following strategy h^i when starting with initial capital w_0, and

$$X = \{\alpha \in \mathbb{R}^I_+ \mid \sum_{i\in I} \alpha^i = 1\}.$$

are the weights in the convex combination. In this work we employ the entropic risk measure

$$\rho(X) = \frac{1}{\gamma} \log E[e^{-\gamma X}],$$

in which case the minimization problem becomes

$$\underset{\alpha\in X}{\text{minimize}} \quad \frac{1}{\gamma} \log E[e^{-\gamma(\sum_{i\in I}\alpha^i w^i_T)}].$$

It is to be noted that our choice of the entropic risk measure is rather arbitrary and it was mainly chosen for computational convenience. Other possibilities include the *Conditional Value at Risk*, which is employed in an analogous setting in [23].

Because of the convexity of D_t, $\sum_{i\in I} \alpha^i h^i_t \in D_t$ for $t = 0, \dots, T$. In addition, the budget constraint of the aggregate strategy $\sum_{j\in J} h_{t,j} \le \sum_{j\in J} R_{t,j} h_{t-1,j} - c_t$ holds, if it holds for individual strategies. This is a finite-dimensional convex optimization problem, but the objective involves high-dimensional integration.

In order to solve (5.4), we form the following quadrature approximation of the objective. A finite number N of return and claim scenarios (R^k, c^k), $k = 1, \dots, N$ is generated over time $t = 0, \dots, T$. Here R^k denotes a realization of the $|J|$-dimensional process $(R_t)^T_{t=1}$ where $R_t = (R_{t,j})_{j\in J}$. The expectation is then approximated by

$$\frac{1}{N} \sum_{k=1}^{N} e^{-\gamma\left(\sum_{i\in I}\alpha^i w^{i,k}_T\right)},$$

where $w^{i,k}_T$ is the terminal wealth in scenario k, obtained by following strategy h^i. For a more detailed description of the method, see e.g. [23, 25]. Given a realization (R^k, c^k) and a strategy h^i, the corresponding wealth process $w^{i,k} = (w^{i,k}_t)^T_{t=0}$ is given recursively by

$$w^{i,k}_t = \begin{cases} w_0 & \text{for t} = 0, \\ \sum_{j\in J} R^k_{t,j} h^{i,k}_{t-1,j} - c^k_t & \text{for t} > 0. \end{cases}$$

The resulting minimization problem is of a form that is, in principle, straightforward to solve using numerical optimization algorithms. In the numerical study below, we employ the standard SQP solver of Matlab.

5.5 Numerical Results

In the following numerical illustrations, the termination date was set to $T = 30$, and the cash flows c_t were defined as the survival index S_t of a cohort of US females aged 65 at time $t = 0$. The structure of this instrument is the same as in the first longevity bond issued in 2004 by the European Investment Bank (for a more detailed description see e.g. [6]). The asset returns R_t and liabilities c_t were modelled as a multivariate stochastic process as described in Appendix A. Using Latin hypercube sampling, we constructed $N = 10^6$ scenarios for the numerical procedure described above. Each problem instance was generated and solved in no more than five minutes using Matlab's parallel computing on an Intel Xeon X5650 @ 2.67 GHz processor.

Our aim was to investigate if liability-driven investment strategies can lead to reductions in the risk associated with a cash flow of longevity-linked liabilities. To this end, we used two sets of basis strategies. The first set consisted of non-liability-driven basis strategies, namely 30 FP strategies, 24 TDF strategies, and four buy and hold strategies. The second set encompassed both the above non-liability-driven and additional liability-driven basis strategies, including 15 term spread strategies, 15 credit spread strategies, 50 survival index strategies and 50 wealth strategies. We computed the optimal aggregate investment strategy and the corresponding value of the risk measure function ρ for each set, using the numerical procedure of the previous section. We then proceeded to compare the optimal values of the objective ρ associated with each set. In order to discern to which extent a possible reduction in risk can be attributed to considering the liabilities, as opposed to merely having a larger number of strategies, we also considered a portfolio optimization problem without liabilities for both sets of basis strategies. The optimal allocations were computed for different values of risk aversion parameters γ. The larger the parameter, the more risk averse the investor.

Table 5.1 summarizes the resulting values of the objective function. We observe that as the risk aversion grows, so does the reduction in risk of the ALM problem with liabilities as the liability-driven strategies are included. This is plausible since the higher the risk aversion, the more the risk measure places importance to the fact that the asset returns conform to the liabilities. As for the optimization problem with zero liabilities, the effect of adding the liability-driven strategies was negligible and independent of the level of risk aversion.

Table 5.1 Values of objective function ρ

	$\gamma = 0.05$		$\gamma = 0.1$		$\gamma = 0.3$		$\gamma = 0.5$	
	$c_t = S_t$	$c_t = 0$	$c_t = S_t$	$c_t = 0$	$c_t = S_t$	$c_t = 0$	$c_t = S_t$	$c_t = 0$
Basis strategies								
Non-LDI	−27.46	−75.14	−18.64	−60.82	−11.16	−46.73	−9.17	−41.81
All	−27.90	−75.14	−19.84	−60.84	−12.40	−46.87	−10.16	−42.14
Reduction (%)	1.6	0.006	6.47	0.04	11.14	0.3	10.71	0.8

Tables 5.2 and 5.3 show the optimal allocations to each set of the basic investment strategies and both problems for risk aversion parameter $\gamma = 0.3$. Asset indexes are as indicated in Appendix A. After the liability-driven strategies were included in the optimization procedure, none of the non-liability driven strategies were included in the optimal allocation of the problem with $c_t = S_t$, whereas in the optimal allocation of the portfolio optimization problem a non-liability driven fixed proportions strategy still had the highest weight.

Table 5.2 Diversified strategy, non-liability-driven strategies, $w_0 = 15$, $\gamma = 0.3$

$c_t = S_t$	Weight (%)	Type	π
	97.7	FP	$\pi^2 = 1 - 0.25$ $\pi^4 = 0.25$
	2.3	FP	$\pi^2 = 1 - 0.35$ $\pi^4 = 0.35$
$c_t = 0$	Weight	Type	π
	59.8	FP	$\pi^2 = 1 - 0.25$ $\pi^4 = 0.25$
	40.2	FP	$\pi^2 = 1 - -0.15$ $\pi^4 = 0.15$

Table 5.3 Diversified strategy, all strategies, $w_0 = 15$, $\gamma = 0.3$

$c_t = S_t$	Weight (%)	Type	π	
	52.7	Survival index	$\pi^2 = g^a(S_t)$ $\pi^4 = 1 - g^a(S_t)$	$a = 1$
	19.0	Wealth	$\pi^2 = g^a(w_t/w_0)$ $\pi^4 = 1 - g^a(w_t/w_0)$	$a = 0.5$
	13.8	Survival index	$\pi^2 = g^a(S_t)$ $\pi^4 = 1 - g^a(S_t)$	$a = 0.75$
	7.4	Wealth	$\pi^2 = g^a(w_t/w_0)$ $\pi^4 = 1 - g^a(w_t/w_0)$	$a = 0.75$
	7.1	Term spread	$\pi^1 = 1 - \sigma(s_t^T)^{a,b}$ $\pi^2 = \sigma(s_t^T)^{a,b}$	$a = -0.5, b = 5$
$c_t = 0$	Weight	Type	π	
	44.3	FP	$\pi^2 = 1 - 0.35$ $\pi^4 = 0.35$	–
	37.6	Term spread	$\pi^1 = 1 - \sigma(s_t^T)^{a,b}$ $\pi^2 = \sigma(s_t^T)^{a,b}$	$a = -0.5, b = 5$
	9.6	Survival index	$\pi^2 = g^a(S_t)$, $\pi^4 = 1 - g^a(S_t)$,	$a = 1$
	8.4	Wealth	$\pi^2 = g^a(w_t/w_0)$ $\pi^4 = 1 - g^a(w_t/w_0)$	$a = 0.5$

Table 5.4 Five best basis strategies, with liabilities, $w_0 = 15$, $\gamma = 0.3$

Type	Parameters	π	ρ
Survival index	$a = 0.75$	$\pi^2 = g^a(S_t)$ $\pi^4 = 1 - g^a(S_t)$	−11.80
Survival index	$a = 1$	$\pi^2 = g^a(S_t)$ $\pi^4 = 1 - g^a(S_t)$	−11.23
Wealth	$a = 1$	$\pi^2 = g^a(w_t/w_0)$ $\pi^4 = 1 - g^a(w_t/w_0)$	−11.22
FP	–	$\pi^2 = 1 - 0.25$ $\pi^4 = 0.25$	−11.13
CPPI	$m = 0.2, r = 0.04$	$\pi^2 = 1 - e_t$ $\pi^4 = e_t$	−10.89

Table 5.5 Five best basis strategies, without liabilities, $w_0 = 15$, $\gamma = 0.3$

Type	Parameters	π	ρ
CPPI/FP	$m = 0.2, r = 0.03$	$\pi^2 = 1 - e_t$ $\pi^4 = e_t$	−46.62
FP	–	$\pi^2 = 1 - 0.25$ $\pi^4 = 0.25$	−46.46
FP	–	$\pi^2 = 1 - 0.15$ $\pi^4 = 0.15$	−46.13
TDF	$a = 0.2, b = 0.003$	$\pi^2 = 1 - e_t$ $\pi^4 = e_t$	−45.98
TDF	$a = 0.25, b = 0.005$	$\pi^2 = 1 - e_t$ $\pi^4 = e_t$	−45.08

Tables 5.4 and 5.5 show the best five individual strategies with the smallest risks for both problems. While all the best strategies of the problem with liabilities were liability-driven, all the best ones for the problem without liabilities were non-liability driven strategies. Note that when $c_t = 0$, CPPI reduces to a fixed proportions strategy.

Figure 5.1 illustrates the effect of the liability link by plotting the proportions π^2 of wealth invested in 5-year government bonds as a function of remaining wealth w_t at $t = 15$ in different scenarios. In the case of the ALM problem of the longevity-linked cash flow, π^2 is higher when w_t is higher. For the portfolio optimization problem, however, the connection is much less clear.

5.6 Conclusions

This paper presented several liability-driven investment strategies for longevity-linked liabilities. We were able to show numerically that liability-driven investment can significantly outperform common strategies that do not take into account the

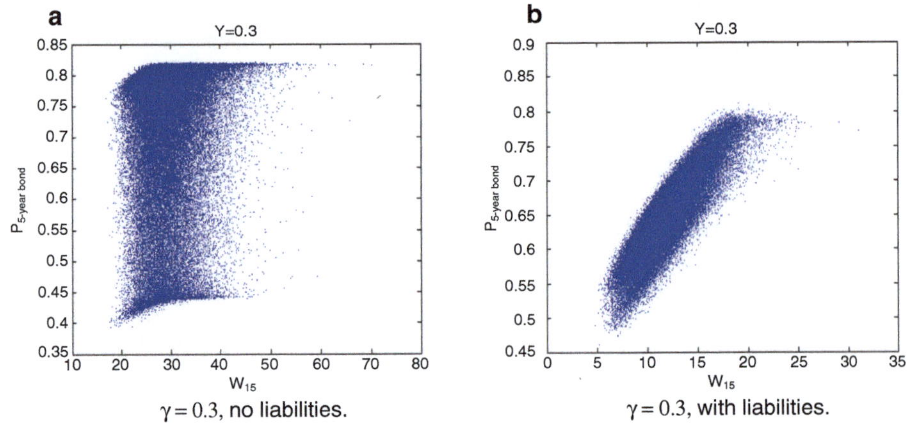

Fig. 5.1 Proportion of wealth invested at time $t = 15$ in 5-year bonds as a function of W_{15}. All strategies. (**a**) $\gamma = 0.3$, no liabilities. (**b**) $\gamma = 0.3$, with liabilities

liabilities. These strategies may help pension insurers and issuers of longevity-linked instruments in asset-liability management, reserving, and in underwriting new insurance contracts.

While encouraging, the results still leave substantial room for improvement. The basis strategies employed in the simulations are only an example of liability-driven strategies. Discovering and utilizing new connections between longevity-linked cash flows and asset returns would further improve the overall hedging strategy.

5.7 Assets and Liabilities

We consider a set J of assets consisting of

1. US Treasury bills (1-year rate)
2. US Treasury bonds (5-year rate)
3. US Corporate bonds
4. US equity (S and P total return index).

These are the asset classes between which the investment strategies distribute the existing wealth, and the above numbers are the indices with which the strategies are referred to. Returns on government bonds are given by the formula

$$R_t^i = \exp(Y_{t-1}^i \Delta t - D \Delta Y_t^i),$$

where Y_t^i is the yield to maturity of each bond $i \in \{1, 2\}$ at time t, and D is the duration. Following [24], corporate bond returns are computed by

$$R_t^3 = \exp(c + (Y_t^3 - \alpha S_t^C)\Delta t - D\Delta Y_t^3),$$

where Y_t^3 is the yield to maturity of the bond, S_t^C is here the credit spread between the yields of corporate bonds and longer-term government bonds Y_t^2, and D is again the duration. Setting $c = 1$ and $\alpha = 1$ yields

$$R_t^3 = \exp(Y_t^2 \Delta t - D\Delta Y_t^3).$$

The total return of the equity is calculated in terms of its total return index S_t^E,

$$R_t^4 = \frac{S_t^E}{S_{t-1}^E}.$$

The value of the liabilities depends on the survival index of cohort of US females aged 65 at the beginning of the observation period. The population dynamics are governed by three mortality risk factors of the mortality model presented in [1].

We briefly recall the stochastic model proposed in [1]. Let $E_{x,t}$ be the number of individuals aged $[x, x+1)$ years at the beginning of year t in a given population. The number of survivors $E_{x+1,t+1}$ among the $E_{x,t}$ individuals during year $[t, t+1)$ can be described by the binomial distribution:

$$E_{x+1,t+1} \sim \text{Bin}(E_{x,t}, p_{x,t}), \tag{5.1}$$

where $p_{x,t}$ is the probability that an x year-old individual randomly selected at the beginning of year t survives until $t+1$.

The future values of E_{t+1} are obtained by sampling from $\text{Bin}(E_t, p_{x,t})$. However, as the population grows, the fraction $E_{t+1}/[E_t p_{x+t,t}]$ converges in distribution to constant 1. For large populations, the population dynamics are well described by $E(x+1, t+1) = E_{x,t} p_{x,t}$, when the main uncertainty comes from unpredictable variations in the future values of $p_{x,t}$. In this work, we employ the latter approach.

As in [1], we model the *survival probabilities* $p_{x,t}$ with the formula

$$p_{x,t} = \frac{\exp\left(\sum_{i=1}^n v_t^i \phi^i(x)\right)}{1 + \exp(\sum_{i=1}^n v_t^i \phi^i(x))}, \tag{5.2}$$

where ϕ^i are user-defined *basis functions* and v_t^i are stochastic *risk factors* that may vary over time.

As in [1], we will use the three piecewise linear basis functions given by

$$\phi^1(x) = \begin{cases} 1 - \frac{x-18}{32} & \text{for x} \le 50 \\ 0 & \text{for x} \ge 50, \end{cases}$$

$$\phi^2(x) = \begin{cases} \frac{1}{32}(x-18) & \text{for x} \le 50 \\ 2 - \frac{x}{50} & \text{for x} \ge 50, \end{cases}$$

$$\phi^3(x) = \begin{cases} 0 & \text{forx} \le 50 \\ \frac{x}{50} - 1 & \text{for x} \ge 50. \end{cases}$$

The linear combination $\sum_{i=1}^{3} v_t^i \phi^i(x)$ will then be piecewise linear and continuous as a function of the age x. The risk factors v_t^i now represent points on logistic survival probability curve:

$$v_t^1 = \operatorname{logit} p_{18,t},\ v_t^2 = \operatorname{logit} p_{50,t},\ v_t^3 = \operatorname{logit} p_{100,t}.$$

Once the basis functions ϕ^i are fixed, the realized values of the corresponding risk factors v_t^i can be easily calculated from historical data using standard max-likelihood estimation.

As in [2], we model the future development of and connections between mortality risk factors and spreads with the following system of equations

$$\begin{aligned}
\Delta v_t^1 &= a^{11} v_{t-1}^1 + b^1 + \varepsilon_t^1 \\
\Delta v_t^2 &= b^2 + \varepsilon_t^2 \\
\Delta v_t^3 &= a^{33} v_{t-1}^3 + a^{34} g_{t-1} + b^3 + \varepsilon_t^3 \\
\Delta g_t &= a^{45} s_{t-1}^T + a^{46} s_{t-1}^C + b^4 + \varepsilon_t^4 \\
\Delta s_t^T &= a^{55} s_{t-1}^T + b_5 + \varepsilon_t^5 \\
\Delta s_t^C &= a^{66} s_{t-1}^C + b_6 + \varepsilon_t^6 \\
\Delta y_t^1 &= a^{77} y_{t-1}^1 + b_7 + \varepsilon_t^7 \\
\Delta s_t^E &= b_8 + \varepsilon_t^8.
\end{aligned}$$

where v_t^i are mortality risk factors, g_t is the natural logarithm of per capita real GDP, s_t^T is the term spread between the logarithms of yields to maturity for 5-year and 1-year government bonds, and s_t^C is the logarithm of the credit spread between the logarithmic yields to maturity of BAA and AAA rated corporate bonds. In addition to the risk factors included in the original model, the 1-year government bond yield $y_t^1 = \log(Y_t^1)$ was added to enable computation of bond returns, as well as the S and P total return index $s_t = \log(S_t^E)$ as pension plans typically invest in the stock market. The terms ε_t^i are random variables describing the unexpected development in the risk factors.

Once the 1-year government bond yield is known, the 5-year government bond yield can be computed by means of the term spread. Due to lack of data, we approximate the credit spread between government bonds and corporate bonds with the spread s_t^C between corporate bonds of varying riskiness, obtaining the corporate bond yield.

Final year of available mortality data was 2007. Parameters of the time series model were estimated as in [2], with the exception that the mean reversion yields of 1-year and 5-year government bonds and corporate bonds were set to 2.5, 3.5, and 4.5 %, respectively, and expected return on equity was set to 8 %. Durations D for the 1-year and 5-year Treasury bonds were 1 and 5 years, respectively, and 5 years for the corporate bonds. In the case of negative wealth, required funds were borrowed from the money market at the 1-year rate adjusted by a loan margin of 1 %.

References

1. H. Aro, T. Pennanen, A user-friendly approach to stochastic mortality modelling. Eur. Actuar. J. **1**, 151–167 (2011)
2. H. Aro, T. Pennanen, Stochastic modelling of mortality and financial markets. Scand. Actuar. J. **2014**(6), 483–509 (2014)
3. L. Ballotta, S. Haberman, The fair valuation problem of guaranteed annuity options: the stochastic mortality environment case. Insur. Math. Econ. **38**(1), 195–214 (2006)
4. P. Barrieu, H. Bensusan, N. El Karoui, C. Hillairet, S. Loisel, C. Ravanelli, Y. Salhi, Understanding, modelling and managing longevity risk: key issues and main challenges. Scand. Actuar. J. **2012**(3), 203–231 (2012)
5. D. Bauer, M. Borger, J. Russ, On the pricing of longevity-linked securities. Insur. Math. Econ. **46**(1), 139–149 (2010)
6. E. Biffis, D. Blake, Mortality-linked securities and derivatives. Pensions Institute Discussion Paper, PI 0901 (2009)
7. D. Blake, W. Burrows, Survivor bonds: helping to hedge mortality risk. J. Risk Insur. **68**(2), 339–348 (2001)
8. F. Black, R. Jones, Simplifying portfolio insurance. J. Portf. Manag. **14**(1), 48–51 (1987)
9. F. Black, A.F. Perold, Theory of constant proportion portfolio insurance. J. Econ. Dyn. Control **16**, 403–426 (1992)
10. F. Black, A.F. Perold, Dynamic strategies for asset allocation. Financ. Anal. J. **51**, 149–160 (1995)
11. D. Blake, A. Cairns, K. Dowd, Living with mortality: longevity bonds and other mortality-linked securities. Br. Actuar. J. **12**(45), 153–197 (2006)
12. D. Blake, A. Cairns, K. Dowd, R. MacMinn, Longevity bonds: financial engineering, valuation, and hedging. J. Risk Insur. **73**(4), 647–672 (2006)
13. D. Blake, A. Cairns, K. Dowd, G. Coughlan, Longevity hedge effectiveness: a decomposition. Quant. Finan. **14**(2), 2017–235 (2014)
14. Z. Bodie, J. Treussard, Making investment choices as simple as possible, but not simpler. Financ. Anal. J. **63**(3), 42–47 (2007)
15. A. Cairns, Modelling and management of longevity risk: approximations to survival functions and dynamic hedging. Insur. Math. Econ. **49**, 438–453 (2011)
16. A. Cairns, D. Blake, K. Dowd, Pricing death: frameworks for the valuation and securitization of mortality risk. ASTIN Bull. **36**(1), 79–120 (2006)
17. S. Cox, Y. Lin, Natural hedging of life and annuity mortality risks. North Am. Actuar. J. **11**(3), 1–15 (2007)

18. M. Dahl, M. Melchior, T. Müller, On systematic mortality risk and risk-minimization with survivor swaps. Scand. Actuar. J. **2008**(2–3), 114–146 (2008)
19. F. Delbaen, W. Schachermayer, *The Mathematics of Arbitrage*. Springer Finance (Springer, Berlin/Heidelberg, 2006)
20. K. Dowd, A. Cairns, D. Blake, Mortality-dependent financial risk measures. Insur. Math. Econ. **38**(3), 427–440 (2006)
21. K. Dowd, D. Blake, A. Cairns, P. Dawson, Survivor swaps. J. Risk Insur. **73**(1), 1–17 (2006)
22. H. Föllmer, A. Schied, *Stochastic Finance* (Walter de Gruyter, Berlin, 2004)
23. P. Hilli, M. Koivu, T. Pennanen, Optimal construction of a fund of funds. Eur. Actuar. J. **1**, 345–359 (2011)
24. M. Koivu, T. Pennanen, Two-factor models for index linked bond portfolios. The Journal of Portfolio Management, **41**(1) 78–84 (2014)
25. M. Koivu, T. Pennanen, Galerkin methods in dynamic stochastic programming. Optimization **59**, 339–354 (2010)
26. J.S.-H. Li, M.R. Hardy, Measuring basis risk in longevity hedges. North Am. Actuar. J. **15**, 177–200 (2011)
27. Y. Lin, S.H. Cox, Securitization of mortality risks in life annuities. J. Risk Insur. **72**(2), 227–252 (2005)
28. T. Pennanen, Risk management and contingent claim valuation in illiquid markets. Optimal investment and contingent claim valuation in illiquid markets; Finance and Stochastics. **18**(4) 733–754 (2014)
29. J.L. Wang, H.C. Huang, S.S. Yang, J.T. Tsai, An optimal product mix for hedging longevity risk in life insurance companies: the immunization theory approach. J. Risk Insur. **77**(2), 473–497 (2010)
30. A. Wilkie, H. Waters, S. Yang, Reserving, pricing and hedging for policies with guaranteed annuity options. Br. Actuar. J. **9**(02), 263–391 (2003)

Chapter 6
Pricing Multiple Exercise American Options by Linear Programming

Monia Giandomenico and Mustafa Ç. Pınar

Abstract We consider the problem of computing the lower hedging price of American options of the call and put type written on a non-dividend paying stock in a non-recombinant tree model with multiple exercise rights. We prove using a simple argument that an optimal exercise policy for an option with h exercise rights is to delay exercise until the last h periods. The result implies that the mixed-integer programming model for computing the lower hedging price and the optimal exercise and hedging policy has a linear programming relaxation that is exact, i.e., the relaxation admits an optimal solution where all variables required to be integral have integer values.

Keywords American options • Swing options • Multiple exercise rights • Linear programming • Mixed-integer programming • Lower hedging price

6.1 Introduction

Pricing and hedging American options has been an important subject of mathematical finance. Starting with the work of Harrison and Kreps [22], Bensoussan [6] and Karatzas [25], finding a no-arbitrage price for American options has been studied in various settings ranging from discrete-time, discrete probability space to continuous time infinite state space settings in complete and incomplete markets; see e.g., [8, 9, 13, 15, 26, 29, 32, 38]. For a text-book treatment of American options in discrete time the book by Föllmer and Schied [20] is an authoritative source while the monograph by Detemple [16] concentrates on models in continuous time.

For options with early exercise possibility (thus, of American type) but with multiple exercise rights such as the swing options of energy markets [24], one can

M. Giandomenico
Department of Computer Science, University of L'Aquila, L'Aquila, Italy
e-mail: monia.giandomenico@univaq.it

M.Ç. Pınar (✉)
Department of Industrial Engineering, Bilkent University, Ankara, Turkey
e-mail: mustafap@bilkent.edu.tr

© Springer International Publishing Switzerland 2017
G. Consigli et al. (eds.), *Optimal Financial Decision Making under Uncertainty*,
International Series in Operations Research & Management Science 245,
DOI 10.1007/978-3-319-41613-7_6

consult the following literature [1–5, 11, 12, 18, 21, 23, 27, 31, 36, 37, 39, 40]. Thompson [37] uses lattice based claim evaluation techniques for commodity options with multiple exercise rights. Keppo [27] gives an elementary introduction to electricity swing options. Bardou et al. [1, 2] and Barrera-Esteve et al. [3] consider swing options in complete markets within a stochastic control framework. In particular, in [2] the bang-bang nature of the optimal exercise policy is studied. In [12], Carmona and Touzi study American options with multiple exercise rights in a Black-Scholes [7] framework, whereas in [11] a more general case using linear diffusion models is treated. Bender [5] studies multiple exercise options in continuous time with a finite maturity and proves the existence of the Snell envelope, a reduction principle as nested single stopping problems, and a Doob-Meyer type decomposition for the Snell envelope. He also derives a dual representation that generalizes that of Schoenmakers [36] and gives a primal-dual Monte-Carlo algorithm. In [31] a dual representation in discrete time is given, and its extension to volume constraints is studied in [4]. Haarbrücker and Kuhn [21] use multi-stage stochastic programming to price electricity swing options while Winter and Wilhelm [40] use the finite element method to evaluate swing options. Vayanos et al. [39] consider electricity swing options in incomplete markets as in the present paper using forward contracts for hedging, and compute buyer and seller prices using robust control and constraint sampling techniques. Longstaff and Schwartz [29], Ibáñez [23] and Figueroa [18] use Monte-Carlo simulation techniques to price single and multiple exercise claims. Chalasani and Jha [13], and Pınar and Camcı [34] study American options in the discrete time finite state probability setting as in the present paper, but allow for proportional transaction costs. Camcı and Pınar [10] and Flåm [19] and Pennanen and King [33] treat similar problems from a finite-dimensional optimization point of view.

In the present paper, we concentrate on the problem of finding an optimal exercise and hedging policy, and hence a fair buyer's price for an American option with multiple exercise rights, written on a stock evolving in a non-recombinant tree in the presence of a risk free asset paying no interest, a problem on which little (if anything at all) has been written. We formulate the problem as a mixed-integer programming problem. It is well-known that in discrete-time complete (and arbitrage free) markets the price of a single exercise American call option on a non-dividend paying asset behaves as a sub-martingale, and hence, it is optimal to delay exercise until maturity; see e.g., [20]. The assertion remains true also for an American single exercise put option on a non-dividend paying asset in a zero-interest rate environment [20]. Our main result provides an extension of this well-known fact (delaying exercise until maturity is optimal) for American options with multiple exercise rights. The result not only shows the optimal exercise policy, but also proves the exact nature of the LP relaxation of the mixed-integer model. Therefore, one can obtain the lower hedging price by solving a linear programming problem, a problem that can be solved in polynomial time. To the best of our knowledge, this simple result was not previously available in the mathematical finance literature. We also obtain a min–max expression for the price of an American claim with multiple (two) exercise rights as follows:

$$\max_{\tau \in \mathscr{T}^2(T)} \min_{Q \in \tilde{\mathscr{Q}}} \mathbb{E}^Q[F_\tau] = \min_{Q \in \tilde{\mathscr{Q}}} \max_{\tau \in \mathscr{T}^2(T)} \mathbb{E}^Q[F_\tau]$$

where T is the maturity date of the claim, $\mathscr{T}^2(T)$ is the collection of all vectors of stopping times $\tau = (\tau_1, \tau_2) \in [0, 1, \ldots, T] \cup \{+\infty\}$ satisfying some conditions (c.f., end of Sect. 6.4) and $\tilde{\mathscr{Q}}$ represents (the closure of) all equivalent martingale measures. This is reminiscent of the representation

$$\max_{\tau \in \mathscr{T}} \min_{Q \in \tilde{\mathscr{Q}}} \mathbb{E}^Q[F_\tau] = \min_{Q \in \tilde{\mathscr{Q}}} \max_{\tau \in \mathscr{T}} \mathbb{E}^q[F_\tau]$$

for American claims with $\mathscr{T}$ denoting the set of all stopping times. The above representation can easily be generalized to h exercise rights.

A word of caution is in order here. One should bear in mind that for the swing options traded in energy markets, the underlying (e.g., electricity) is not traded in the spot market whereas our analysis in the present paper is based on the assumption that the underlying can be traded.

The rest of the paper is organized as follows. In Sect. 6.2 we review the basics of the stochastic scenario tree and American claims. In Sect. 6.3 we present an optimization model to compute a fair price for an American claim with multiple exercise rights. We prove the main result in Sect. 6.4. We conclude in Sect. 6.5.

6.2 The Stochastic Scenario Tree and American Contingent Claims

An American contingent claim (abbreviated ACC) F is a financial instrument generating a real-valued stochastic (cash-flow) process $(F_t)_{t=0,\ldots,T}$ with $h \geq 1$ exercise rights to the holder. At any stage $t = 0, \ldots, T$, the holder of a single-exercise ACC may decide to take F_t in cash and terminate the process. In the case of $h > 1$ exercise rights, the holder may decide to make up to and including h exercises (at h different time points). The process terminates when the h-th exercise is performed. Of course, the holder may choose to exercise less than h times during the lifetime of the claim. An American call option on a stock S with strike price K has a payoff equal to $F = S - K$. American put is obtained by reversing the sign of F. In our finite probability space setting an American option F with h exercise rights generates payoff opportunities F_n ($F_n = \max\{S_n^1 - K, 0\}$ or $F_n = \max\{K - S_n^1, 0\}$ for some strike price K), ($n \geq 0$) and h exercise possibilities to its holder depending on the states n of the market that we define below.

To lay down a pricing framework based on no-arbitrage arguments for contingent claims, we assume that security prices and other payments are discrete random variables supported on a finite probability space $(\Omega, \mathscr{F}, P)$ whose atoms are sequences of real-valued vectors (asset values) over discrete time periods $t = 0, 1, \ldots, T$. We further assume the market evolves as a discrete, non-recombinant scenario tree. A non-recombinant tree structure is suitable for incomplete markets as discussed in [17] since it allows to work with path-dependent portfolio strategies whereas in recombinant trees one optimizes over path-independent strategies which may be suboptimal. In the scenario tree, the partition of probability atoms $\omega \in \Omega$ generated by matching path histories up to time t corresponds one-to-one with nodes $n \in \mathscr{N}_t$ at level t in the tree. The set $\mathscr{N}_0$ consists of the root node $n = 0$, and the leaf nodes $n \in \mathscr{N}_T$ correspond one-to-one with the probability atoms $\omega \in \Omega$. The σ-algebras are such that, $\mathscr{F}_0 = \{\emptyset, \Omega\}$, $\mathscr{F}_t \subset \mathscr{F}_{t+1}$ for all $0 \leq t \leq T-1$ and $\mathscr{F}_T = \mathscr{F}$. A stochastic process is said to be $(\mathscr{F}_t)_{t=0}^T$-adapted if for each $t = 0, \ldots, T$, the outcome of the process only depends on the element of $\mathscr{F}_t$ that has been realized at stage t. Similarly, a decision process is said to be $(\mathscr{F}_t)_{t=0}^T$-adapted if for each $t = 0, \ldots, T$, the decision depends on the element of $\mathscr{F}_t$ that has been realized at stage t. In the scenario tree, every node $n \in \mathscr{N}_t$ for $t = 1, \ldots, T$ has a unique parent denoted $\pi(n) \in \mathscr{N}_{t-1}$, and every node $n \in \mathscr{N}_t$, $t = 0, 1, \ldots, T-1$ has a non-empty set of child nodes $\mathscr{C}(n) \subset \mathscr{N}_{t+1}$. We denote the set of all nodes in the tree by $\mathscr{N}$. For a given node n, the inverse mapping $t(n)$ gives the time period to which the node n belongs to. The set $\mathscr{A}(n)$ denotes the collection of ascendant nodes or path history of node n including itself. The probability distribution P is obtained by attaching positive weights p_n to each leaf node $n \in \mathscr{N}_T$ so that $\sum_{n \in \mathscr{N}_T} p_n = 1$. For each non-leaf (intermediate level) node in the tree we have, recursively,

$$p_n = \sum_{m \in \mathscr{C}(n)} p_m, \quad \forall\, n \in \mathscr{N}_t, \; t = T-1, \ldots, 0.$$

Hence, each non-leaf node has a probability mass equal to the combined mass of its child nodes.

A random variable X is a real valued function defined on Ω. It can be *lifted* to the nodes of a partition $\mathscr{N}_t$ of Ω if each level set $\{X^{-1}(a) : a \in \mathbb{R}\}$ is either the empty set or is a finite union of elements of the partition. In other words, X can be lifted to $\mathscr{N}_t$ if it can be assigned a value on each node of $\mathscr{N}_t$ that is consistent with its definition on Ω [28]. This kind of random variable is said to be measurable with respect to the information contained in the nodes of $\mathscr{N}_t$. A stochastic process $\{X_t\}$ is a time-indexed collection of random variables such that each X_t is measurable with respect $\mathscr{N}_t$. The expected value of X_t is uniquely defined by the sum

$$\mathbb{E}^P[X_t] := \sum_{n \in \mathscr{N}_t} p_n X_n.$$

The conditional expectation of X_{t+1} on $\mathscr{N}_t$ is given by the expression

$$\mathbb{E}^P[X_{t+1} | \mathscr{N}_t] := \sum_{m \in \mathscr{C}(n)} \frac{p_m}{p_n} X_m.$$

The market consists of two traded securities with prices at node n given by the vector $S_n = (S_n^0, S_n^1)$. We assume that the security indexed by 0 has strictly positive prices at each node of the scenario tree. Our blanket assumption throughout the paper is that $S_n^0 = 1$ for all n i.e., a zero interest rate for the risk-free asset.

This assumption is crucial e.g., for the put option case where non-zero interest rate may lead to strict optimality of exercise earlier than the last h periods. We give a counterexample supporting this claim after the proof of Proposition 1.

The number of shares of security j held by the investor in state (node) $n \in \mathscr{N}_t$ is denoted θ_n^j. Therefore, to each state $n \in \mathscr{N}_t$ is associated a vector $\theta_n \in \mathbb{R}^2$. The value of the portfolio at state n is

$$S_n \cdot \theta_n = \sum_{j=0}^{1} S_n^j \theta_n^j.$$

We need the following definition.

Definition 1. If there exists a probability measure $Q = \{q_n\}_{n \in \mathscr{N}_T}$ such that

$$S_t = \mathbb{E}^Q[S_{t+1}|\mathscr{N}_t] \; (t \leq T-1)$$

then the vector process $\{S_t\}$ is called a vector-valued martingale under Q, and Q is called a martingale probability measure for the process.

It is well-known that a market is free of arbitrage opportunities if and only if the price process S is a martingale; see [28] for a discussion of arbitrage and martingales in finite-state markets. We shall assume this situation to be the case throughout the present paper.

6.3 The Formulation

The buyer's problem can be formulated as the following problem that we will refer to as AP1:

$$\begin{aligned}
\max \quad & V \\
\text{s.t.} \quad & S_0 \cdot \theta_0 = F_0 e_0 - V \\
& S_n \cdot (\theta_n - \theta_{\pi(n)}) = F_n e_n, \; \forall\, n \in \mathscr{N} \setminus \mathscr{N}_0 \\
& S_n \cdot \theta_n \geq 0, \; \forall\, n \in \mathscr{N}_T \\
& \sum_{m \in \mathscr{A}(n)} e_m \leq h, \; \forall\, n \in \mathscr{N}_T \\
& e_n \in \{0, 1\}, \; \forall\, n \in \mathscr{N}
\end{aligned}$$

where $h \geq 2$ is a fixed integer. In mathematical finance, the theory of incomplete markets involves the price of the seller and the price of the buyer for a contingent claim. These two values can be quite different, leading to an interval in which no arbitrage opportunities for the buyer and seller exist [28, 33]. The fact that these two prices may differ is a matter of active research and discussion in the financial mathematics community (see e.g. [28]) since it brings about the following question: if the maximum the buyer can pay is strictly less than the minimum a seller can settle for, then how are the claims traded in markets? It appears that the present theory—at least in its present form—is not fully capable to explain the prices of contingent claims actually traded in the market. King [28] addresses this problem using existing liabilities of buyers and sellers.

Setting this question aside, for the seller the problem is to form the least costly initial portfolio of traded assets that will cover the potential payments to the holder of the claim (if and when exercised) such that no losses are incurred at the end. By contrast, from the buyer's perspective the problem is to build the most valuable portfolio that can be formed against the ownership rights of the claim. In other words, the buyer initiates a portfolio process (by shorting some instrument(s)), and closes the short positions later by self-financing transactions and the proceeds from the claim in such a way that no losses are incurred at the end of the horizon.

In model AP1, the optimal value of V represents the largest amount that a potential buyer is willing to disburse for acquiring a given American contingent claim F with h exercise rights. The computation of this quantity via the above integer programming problem is performed by construction of the most valuable (today) adapted portfolio process using the proceeds from the exercise of the contingent claim and self-financing transactions using the market-traded securities to avoid any terminal losses. More precisely, the proceeds obtained from the exercise of the claim are used to finance (cover short positions) portfolio transactions initiated by the buyer at time $t = 0$ to acquire the claim. This is expressed in the first and second sets of constraints above in AP1. They represent the requirement that the proceeds from the claim, if exercised, are used in revising the portfolio positions without injection or withdrawal of funds. If there is no exercise at a node, the equation represents self-financing portfolio rebalancing. The third set of constraints makes sure that all terminal portfolio values are non-negative. The integer variables and related constraints represent the h-times exercise of the American contingent claim. The linear programming relaxation of AP1 is the following problem AP2:

$$
\begin{array}{lrcl}
\max & V & & \\
\text{s.t.} & S_0 \cdot \theta_0 & = & F_0 e_0 - V \\
& S_n \cdot (\theta_n - \theta_{\pi(n)}) & = & F_n e_n, \ \forall\, n \in \mathcal{N} \setminus \mathcal{N}_0 \\
& S_n \cdot \theta_n & \geq & 0, \ \forall\, n \in \mathcal{N}_T \\
& \sum_{m \in \mathcal{A}(n)} e_m & \leq & h, \ \forall\, n \in \mathcal{N}_T \\
& e_n & \leq & 1, \ \forall\, n \in \mathcal{N} \\
& e_n & \geq & 0, \ \forall\, n \in \mathcal{N}.
\end{array}
$$

6.4 The Main Result

The main result of this paper is the following.

Proposition 1. *Assuming that the underlying is a traded instrument, in a financial market described as a non-recombinant tree with two traded instruments (one risky*

Table 6.1 Cash flows of the two strategies 1 and 2 for a call option

Strategy	t_i	t_j	$T-1$	T
Strategy 1	$S_{t_i}-K$	$S_{t_j}-K$	0	0
Strategy 2	$S_{t_i}-K$	$S_{t_j}-K$	$K-S_{T-1}+(S_{T-1}-K)_+$	$K-S_T+(S_T-K)_+$

Table 6.2 Cash flows of the two strategies 1 and 2 for a put option

Strategy	t_i	t_j	$T-1$	T
Strategy 1	$K-S_{t_i}$	$K-S_{t_j}$	0	0
Strategy 2	$K-S_{t_i}$	$K-S_{t_j}$	$S_{T-1}-K+(K-S_{T-1})_+$	$S_T-K+(K-S_T)_+$

asset which is the underlying, and one riskless asset), T time periods to maturity, and zero interest rate, the following holds for an American contingent claim with $h \geq 2$ exercise rights :

1. *It is optimal to delay exercise until the periods $T-h+1, T-h+2, \ldots, T-1$ and T,*
2. *AP2 has an optimal solution with all e variables binary.*

Proof. [1] For the sake of simplicity we shall give the proof of part 1 for the case of $h = 2$. The proof is based on a simple argument of no-arbitrage adapted from the book by Cox and Rubinstein [14], pp. 139–140 for the case $h = 1$.

Assume an exercise strategy that exercises the two rights of a call at times t_i, t_j with $t_i < t_j \leq T, S_{t_i} \geq K, S_{t_j} \geq K$. Now, we can see that exercising at times $T-1$ and T does no worse, in a path-wise sense, than exercising at times t_i and t_j. To see this, compare the cash flows generated by the exercise strategy of times t_i, t_j (referred to as strategy 1), and strategy that exercises the option at times $T-1$ and T, together with shorting a unit of the stock and lending K dollars at times t_i and t_j while closing the positions at times $T-1$ and T (referred to as strategy 2) for a call option. In the case of a put, simply reverse strategy 2 in the following sense: borrow K dollars and go long one unit of stock to close positions at times $T-1$ and T. The following two tables show the cash flows of the two respective strategies in the case of call and put options (Tables 6.1 and 6.2).

It is immediate to see from the cash flows of the two strategies that either strategy 2 has a cash flow identical to strategy 1 or it dominates strategy 1. To see this, note that if $K - S_{T-1} < 0$ then $(S_{T-1} - K)_+ = -(K - S_{T-1})$. On the other hand if $K - S_{T-1} > 0$ then $(S_{T-1} - K)_+ = 0 < (K - S_{T-1})$. A similar observation holds for period T. Therefore, using strategy 2, the holder has a non-negative surplus which is immediately translated into a portfolio process with an objective function value at least as large as that of strategy 1. The reason is that the potential surplus at the

[1] An earlier version of the paper had quite a long proof for the case $h = 2$ and restricted to binomial and trinomial trees. It was based on an elaborate primal-dual construction. The present proof was offered by an anonymous reviewer of the earlier version, to whom we are thankful.

last two periods can be placed in the riskless asset, which (carried backward at no interest) corresponds to a larger initial short position (borrowing) in one of the two assets at period 0, thus a larger value for V. Hence, exercising at the last two periods is at least as good a strategy as any other exercise strategy.

Based on part 1, we can fix the binary variables e to one in the nodes of the last two periods where the payoff is positive, and solve the resulting linear program. The result is an optimal hedging strategy. Therefore, AP1 is equivalent to a linear programming problem.

For the general case of $h > 2$ it suffices to extend the above construction using h exercise rights. □

Note that the above result would be valid also in recombining trees. However, we stated the result for the more general non-recombinant tree structure where one can optimize over path-dependent policies.

A result similar in spirit to Proposition 1 is given in Bardou et al. [2] where the bang-bang nature of the optimal exercise quantities for swing options is proved in complete markets using a stochastic control framework. The bang-bang property corresponds in our case to the fact that the LP relaxation allows an optimal solution with 0–1 valued exercise variables, i.e., either no exercise or full exercise at each time point.

We proved that it is always optimal to use the exercise rights at the final h periods. This statement does not mean that earlier exercise is sub-optimal, though. There exist examples where exercise at node 0 may also be part of another optimal exercise policy as the following example demonstrates.

Example (Put Option in a Four-Period Market). Consider a financial market with four trading points, i.e., $T = 3$ evolving as a trinomial tree up to $t = 2$, and from each node of the tree at $t = 2$ two nodes emerge, i.e., the tree behaves binomially at the last period. Hence, the tree has 31 nodes. The risky asset price evolves as follows: at time $t = 0$, we have $S_0 = 8$. At time $t = 1$ the price evolves to either $S_1 = 20$, or $S_2 = 15$ or $S_3 = 7.5$ with equal probability. At time $t = 2$, if the price were equal to 20 at $t = 1$, it becomes either $S_4 = 22$ or $S_5 = 21$ or $S_6 = 19$ with equal probability. If the price were equal to 15 at $t = 1$, it becomes either $S_7 = 17$ or $S_8 = 14$ or $S_9 = 13$ with equal probability. Finally, given that the price were equal to 15 at $t = 1$, it evolves into either $S_{10} = 9$ or $S_{11} = 8$ or $S_{12} = 7$ with equal probability. The remaining nodes, numbered 13–30, have the following price values respectively, $(24, 20, 23, 18, 21, 16, 19, 16.5, 17, 12, 15, 11, 10, 8, 9.5, 7.5, 8.5, 6)$. A partial representation of the tree is given in Fig. 6.1 for the convenience of the reader. The number next to each node is the stock price at that node. An option of the put type with two exercise rights and strike $K = 15$ is introduced into this financial market. Solving the optimization problem (AP1) we observe that it is equally optimal to use one exercise right at the node 3 or suppressing exercise at node 3 and delay exercise to periods $t = 2$ and $t = 3$. Both strategies lead to equal objective function value, hence there exist two different optimal hedging strategies resulting in identical price for the option.

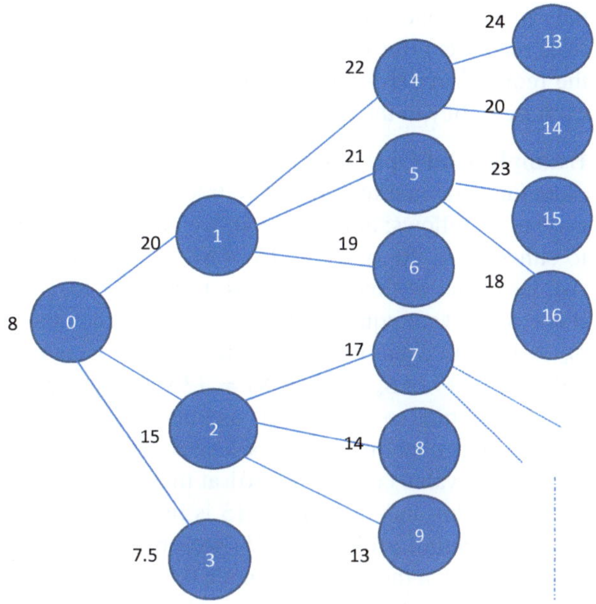

Fig. 6.1 The non-recombinant tree of example for put option in four periods with 31 nodes (partially depicted)

6.4.1 The Case of Non-zero Interest Rate

Corollary 1. *The statement of Proposition 1 is valid for a call option in a market where the risk-less asset has positive per period growth equal to* $R > 1$.

Proof. The proof is similar to the proof of Proposition 1 with a slight modification. For the sake of simplicity, let us consider again the case $h = 2$. Due to non-zero interest rate, the cash flows at the last two periods change as shown in the table below (Table 6.3). It is immediate to see that the cash flows of strategy 2 are at least as good as those of strategy 1. □

However, a similar statement cannot be made in the case of an American put in the presence of a non-zero interest rate even in complete markets and single exercise. According to Luenberger [30] which has an elementary discussion and numerical example for American (single exercise) put options in complete markets, "intuitively, early exercise of a put may be optimal because the upside profit is bounded (unlike the case of call options). Clearly, for example, if the stock price

Table 6.3 Cash flows of the two strategies 1 and 2 for a call option under non-zero interest rate

Strategy	t_i	t_j	$T-1$	T
Strategy 1	$S_{t_i} - K$	$S_{t_j} - K$	0	0
Strategy 2	$S_{t_i} - K$	$S_{t_j} - K$	$KR^{T-1-t_i} - S_{T-1} + (S_{T-1} - K)_+$	$KR^{T-t_j} - S_T + (S_T - K)_+$

tails to zero, one should exercise there, since no greater profit can be achieved." (The reader is referred to pp. 334–335 of [30].) The following example shows that the removal of zero-interest rate assumption may lead to a change in the optimal exercise policy in the case of multiple exercise and incomplete markets as well.

Example (Put Option with Non-zero Interest Rate). Consider a trinomial incomplete financial market with three trading points, i.e., $T = 2$. The risky asset price evolves as follows: at time $t = 0$, we have $S_0 = 8$. At time $t = 1$ the price evolves to either $S_1 = 20$, or $S_2 = 15$ or $S_3 = 7.5$ with equal probability while 1 unit of risk-less asset at time $t = 0$ has a value of 1.01 at time $t = 1$. At time $t = 2$, if the price were equal to 20 at $t = 1$, it becomes either $S_4 = 22$ or $S_5 = 21$ or $S_6 = 19$ with equal probability. If the price were equal to 15 at $t = 1$, it becomes either $S_7 = 17$ or $S_8 = 14$ or $S_9 = 13$ with equal probability. Finally, given that the price were equal to 15 at $t = 1$, it evolves into either $S_{10} = 9$ or $S_{11} = 8$ or $S_{12} = 7$ with equal probability. The risk-less account again appreciates by a factor of 1.01, i.e., it has a value equal to 1.0201 at time $t = 2$. An option of the put type with two exercise rights and strike $K = 15$ is introduced into this financial market. Solving the optimization problem (AP1) we observe that it is strictly optimal to use one exercise right at the root node, node 0, i.e., suppressing exercise at node 0 leads to a strictly smaller objective function value, hence a sub-optimal price for the option.

In our computational experience the exactness property of the LP relaxation appears to continue to hold also in that case.

Conjecture 1. The LP relaxation AP2 is tight in the case of a put option with h exercise rights in a market where the risk-less asset has positive per period growth equal to $R > 1$.

If the conjecture is true, then one can obtain the buyer's price for an American put with multiple exercise rights in a non-zero interest rate market by solving a linear programming problem.

6.4.2 A Min–Max Representation

The usual method to describe exercise strategies of American contingent claims involves stopping times. These are functions $\tau : \Omega \to \{0, \dots, T\} \cup \{+\infty\}$ such that $\{\omega \in \Omega \mid \tau(\omega) = t\} \in \mathscr{F}_t$, for each $t = 0, \dots, T$. The relation $e_t = 1 \Leftrightarrow \tau = t$ defines a one-to-one correspondence between stopping times and decision processes $e \in E$ where

$$E = \{e \mid e \text{ is } (\mathscr{F}_t)_{t=0}^{T}\text{-adapted}, \sum_{t=0}^{T} e_t \leq 1 \text{ and } e_t \in \{0, 1\}\ P\text{-a.s.}\}.$$

The set of stopping times will be denoted by $\mathscr{T}$. Let $\tilde{\mathscr{Q}}$ denote the closure of the set of all martingale measures equivalent to P, i.e., the set

$$\tilde{\mathscr{Q}} = \{q \mid q_0 = 1, q_n S_n = \sum_{m \in \mathscr{C}(n)} q_m S_m, \forall n \in \mathscr{N} \setminus \mathscr{N}_T; 0 \leq q_n, \forall n \in \mathscr{N}_T\}.$$

The following expression for American contingent claims is well-known:

$$\max_{\tau \in \mathscr{T}} \min_{Q \in \tilde{\mathscr{Q}}} \mathbb{E}^Q[F_\tau] = \min_{Q \in \tilde{\mathscr{Q}}} \max_{\tau \in \mathscr{T}} \mathbb{E}^q[F_\tau].$$

In the case of multiple rights we can also obtain a similar expression as a result of Proposition 1. For $h = 2$ we shall denote by $\mathscr{T}^2(T)$ the collection of all vectors of stopping times $\tau = (\tau_1, \tau_2)$ such that

$$\tau_1 \leq T \text{ and } \tau_2 - \tau_1 \geq 1 \text{ on } \{\tau_2 \leq T\} \text{ a.s.},$$

where we implicitly assumed that the minimum allowed elapsed time (a.k.a. latency) between two consecutive exercise dates is smaller than (or equal to) the discrete time step used in constructing the scenario tree (e.g., using an appropriate discretization of a continuous stochastic process). If this is not the case, then the constraint $\tau_2 - \tau_1 \geq 1$ should be modified accordingly.

Define the sets

$$E_2 = \{e \mid e \text{ is } (\mathscr{F}_t)_{t=0}^T\text{-adapted}, \sum_{t=0}^T e_t \leq 2 \text{ and } e_t \in \{0,1\}\ P\text{-a.s.}\},$$

$$\tilde{E}_2 = \{e \mid e \text{ is } (\mathscr{F}_t)_{t=0}^T\text{-adapted}, \sum_{t=0}^T e_t \leq 2 \text{ and } 0 \leq e_t \leq 1\ P\text{-a.s.}\}.$$

The following result follows the ideas of Theorem 4 in [33].

Proposition 2. *If there is no arbitrage in a financial market represented by a non-recombinant tree with two traded instruments (one risky asset which is the underlying, and one riskless asset), T time periods to maturity, the buyer's price for American contingent claim F (call option under zero or positive interest rate, put option with zero interest rate) with two exercise rights can be expressed as*

$$\max_{\tau \in \mathscr{T}^2(T)} \min_{Q \in \tilde{\mathscr{Q}}} \mathbb{E}^Q[F_\tau] = \min_{Q \in \tilde{\mathscr{Q}}} \max_{\tau \in \mathscr{T}^2(T)} \mathbb{E}^q[F_\tau]. \quad (6.1)$$

Proof. If we set e fixed in AP1 and maximize with respect to θ, we have a contingent claim with payoffs $F_t e_t$ for $t = 0, 1, \ldots, T$. Then, for the buyer's price of this claim, we have

$$\min_{Q\in\tilde{\mathscr{Q}}} \mathbb{E}^Q[\sum_{t=0}^{T} F_t e_t].$$

Then, maximizing with respect to e, for the buyer's price of the American claim with two exercise rights we have

$$\max_{e\in E_2} \min_{Q\in\tilde{\mathscr{Q}}} \mathbb{E}^Q[\sum_{t=0}^{T} F_t e_t].$$

The correspondence between multiple stopping times in $\mathscr{T}^2(T)$ and the vectors $e \in E_2$ implies that the buyer's price for the American claim with two exercise rights can be expressed as the left hand side of Eq. (6.1) since maximization over $\mathscr{T}^2(T)$ is equivalent to maximization over E_2 after making the appropriate change in the objective function. By Proposition 1, instead of the last expression we can use

$$\max_{e\in \tilde{E}_2} \min_{Q\in\tilde{\mathscr{Q}}} \mathbb{E}^Q[\sum_{t=0}^{T} F_t e_t]. \tag{6.2}$$

Since $\tilde{E}_2$ and $\tilde{\mathscr{Q}}$ are bounded convex sets, by Corollary 37.6.1 of [35] we can change the order of max and min without changing the value. Then, for each fixed $Q \in \tilde{\mathscr{Q}}$, the objective in (6.2) is linear in e. So the maximum over $\tilde{E}_2$ is attained at an extreme point of $\tilde{E}_2$. We know that the extreme points of $\tilde{E}_2$ are the elements of the set E_2 since $\tilde{E}_2$ is an integral polytope. Thus, we reach the expression on the right hand side in Eq. (6.1). □

6.5 Conclusions

In this paper we have dealt with the pricing of American options with multiple exercise rights in a financial market composed of a risky stock following a non-recombinant tree process and a risk free asset. We established that it is optimal to delay exercise until the last h periods. The result also implies that the LP relaxation of the associated mixed-integer programming formulation to find a no-arbitrage price and hedging policy has an integral solution. Hence, the lower hedging price can be obtained by solving a linear programming problem.

An open problem remains to confirm or refute the claim (made after numerical experimentation) that the LP relaxation continues to be exact in the case of a put option in the presence of a non-zero interest rate.

Acknowledgements Supported in part by a grant from the University of L'Aquila, as part of Research Grant *Reti per la conoscenza e l'orientamento tecnico-scientifico per lo sviluppo della competitività* (RE.C.O.TE.S.S.C.) POR 2007–2013-Action 4, funded by Regione Abruzzo and the European Social Fund 2007–2013. Revised twice, November and December 2013. The paper benefited from the comments of two anonymous reviewers.

References

1. O. Bardou, S. Bouthemy, G. Pagès, Optimal quantization for the pricing of swing options. Appl. Math. Financ. **16**(2), 187–217 (2009)
2. O. Bardou, S. Bouthemy, G. Pagès, When are swing options bang-bang and how to use it? Int. J. Theor. Appl. Finance **13**(6), 867–899 (2010)
3. C. Barrera-Esteve, F. Bergeret, C. Dossal, E. Gobet, A. Meziou, R. Munos, D. Reboul-Salze, Numerical methods for the pricing of swing options: a stochastic control approach. Methodol. Comput. Appl. Probab. **8**(4), 517–540 (2006)
4. C. Bender, Dual pricing of multi-exercise options under volume constraints. Finance Stochast. **15**(1), 1–26 (2010)
5. C. Bender, Primal and dual pricing of multiple exercise options in continuous time. SIAM J. Finan. Math. **2**, 562–586 (2011)
6. A. Bensoussan, On the theory of option pricing. Acta Appl. Math. **2**, 139–158 (1984)
7. F. Black, M. Scholes, The pricing of options and corporate liabilities. J. Polit. Econ. **81**, 637–654 (1973)
8. B. Bouchard, E. Temam, On the hedging of American options in discrete time markets with proportional transaction costs. Electron. J. Probab. **10**(22), 746–760 (2005)
9. R. Buckdahn, Y. Hu, Pricing of American contingent claims with jump stock price and constrained portfolios. Math. Oper. Res. **23**, 177–203 (1998)
10. A. Camcı, M.Ç. Pınar, Pricing American contingent claims by stochastic linear programming. Optimization **58**(6), 627–640 (2009)
11. R. Carmona, S. Dayanık, Optimal multiple stopping of linear diffusions. Math. Oper. Res. **33**(2), 446–460 (2008)
12. R. Carmona, N. Touzi, Optimal multiple stopping and valuation of swing options. Math. Financ. **18**(2), 239–268 (2008)
13. P. Chalasani, S. Jha, Randomized stopping times and American option pricing with transaction costs. Math. Financ. **11**, 33–77 (2001)
14. J.C. Cox, M. Rubinstein, *Options Markets* (Prentice-Hall, Englewood Cliffs, NJ, 1985)
15. M. Davis, T. Zariphopoulou, American options and transaction fees, in *The IMA Volumes in Mathematics and Its Applications*, vol. 65, ed. by M. Davis (Springer, Berlin, 1995), pp. 47–62
16. J. Detemple, *American Style Derivatives: Valuation and Computation*. Financial Mathematics Series (Chapman & Hall/CRC, Boca Raton, 2005)
17. C. Edirisinghe, V. Naik, R. Uppal, Optimal replication of options with transaction costs and trading restrictions. J. Financ. Quant. Anal. **28**, 117–138 (1993)
18. M. Figueroa, Pricing Multiple Interruptible-Swing Contracts. Technical report, Birkbeck, School of Economics, Mathematics & Statistics, Birkbeck Working Papers in Economics and Finance: 0606 (2006)
19. S.D. Flåm, Option pricing by mathematical programming. Optimization **57**(1), 165–182 (2008)
20. H. Föllmer, A. Schied, *Stochastic Finance: An Introduction in Discrete Time*. De Gruyter Studies in Mathematics, vol. 27, 2nd edn. (Walter de Gruyter, Berlin, 2004)
21. G. Haarbrücker, D. Kuhn, Valuation of electricity swing options by multistage stochastic programming. Automatica **45**(4), 889–899 (2009)
22. J.M. Harrison, D.M. Kreps, Martingales and arbitrage in multiperiod securities markets. J. Econ. Theor. **20**, 381–408
23. A. Ibáñez, Valuation by simulation of contingent claims with multiple exercise opportunities. Math. Financ. **14**(2), 223–248 (2004)
24. P. Jaillet, E.I. Ronn, S. Tompaidis, Valuation of commodity-based swing options. Manage. Sci. **50**(7), 909–921 (2004)
25. I. Karatzas, On the pricing of American options. Appl. Math. Optim. **17**, 37–60 (1988)
26. I. Karatzas, S.G. Kou, Hedging American contingent claims with constrained portfolios. Finance Stochast. **2**, 215–258 (1998)
27. J. Keppo, Pricing of electricity swing options. J. Derivatives **11**, 26–43 (2004)

28. A.J. King, Duality and martingales: a stochastic programming perspective on contingent claims. Math. Programm. Ser. B **91**, 543–562 (2002)
29. F. Longstaff, E. Schwartz, Valuing American options by simulation: a simple least squares approach. Rev. Financ. Stud. **14**(1), 113–147 (2001)
30. D.G. Luenberger, *Investment Science* (Oxford University Press, Oxford, 1998)
31. N. Meinshausen, B.M. Hambly, Monte Carlo methods for the valuation of multiple-exercise options. Math. Financ. **14**(4), 557–583 (2004)
32. R. Myeni, The pricing of the American option. Ann. Appl. Probab. **2**, 1–23 (1992)
33. T. Pennanen, A. King, Arbitrage Pricing of American Contingent Claims in Incomplete Markets - A Convex Optimization Approach, Working paper, June (2006)
34. M.Ç. Pınar, A. Camcı, An integer programming model for pricing American contingent claims under transaction costs. Comput. Econ. **39**, 1–12 (2012)
35. T.R. Rockafellar, *Convex Analysis* (Princeton University Press, Princeton, NJ, 1970)
36. J. Schoenmakers, A pure martingale dual for multiple stopping. Finance Stochast. **16**, 319–334 (2012)
37. A. Thompson, Valuation of path-dependent contingent claims with multiple exercise decisions over time: the case of take-or-pay. J. Financ. Quant. Anal. **30**(2), 271–293 (1995)
38. K. Tokarz, T. Zastawniak, American contingent claims under small proportional transaction costs. J. Math. Econ. **43**, 65–85 (2006)
39. P. Vayanos, W. Wiesemann, D. Kuhn, Hedging electricity swing options in incomplete markets, in *Proceedings of the 18th IFAC World Congress* (2011), pp. 846–853
40. C. Winter, M. Wilhelm, Finite element valuation of swing options. J. Comput. Finance **11**(3), 107–132 (2008)

Chapter 7
Optimizing a Portfolio of Liquid and Illiquid Assets

John M. Mulvey, Woo Chang Kim, and Changle Lin

Abstract Current market conditions pose new challenges for institutional investors. Traditional asset and liability models are struggling to meet investors' needs due to poor performance of equity and bond markets. The move of portfolio allocation to alternative assets is evident. As a result, illiquidity issues and rebalancing difficulty arise. We propose some new tactics of commodity futures to enhance the performance of portfolio return as well as solving illiquidity issues. Hidden Markov Model and multistage stochastic optimization are used to systematically optimize portfolio over a set of assets.

Keywords Asset liability model • Illiquidity • Commodity futures • Hidden Markov Model • Multi-stage optimization

7.1 Introduction

During the good times from 1982 to 1999, institutional investors could solely rely on the two traditional investment pillars—equities and fixed income instruments—to construct portfolios that meet their goals. During that period, stocks outperform bonds in the long run. And there existed a relative stable relationship between asset return and volatility. Bonds provided a cushion against stock retrenchment. As a

J.M. Mulvey
Department of Operations Research and Financial Engineering, Bendheim Center for Finance, Princeton University, Princeton, NJ, USA
e-mail: mulvey@princeton.edu

W.C. Kim
Department of Industrial and System Engineering, KAIST, Daejeon, South Korea
e-mail: wkim@kaist.ac.kr

C. Lin (✉)
Department of Operations Research and Financial Engineering, Princeton University, Princeton, NJ, USA

Merrill Lynch Wealth Management, Chief Investment Office, New York, NY, USA
e-mail: changlel@princeton.edu

© Springer International Publishing Switzerland 2017

G. Consigli et al. (eds.), *Optimal Financial Decision Making under Uncertainty*, International Series in Operations Research & Management Science 245,
DOI 10.1007/978-3-319-41613-7_7

result, one can construct a well-diversified and stable portfolio using stocks and bonds. Asset liability models worked well for institution investors. Pension plans in general had positive funding ratios due to the high returns. The typical 60/40, 50/50, 70/30 stock/bond allocations[1] were widely employed and served as ideal performance benchmarks.

Since the advent of twenty-first century, the market conditions pose big challenges for institutional investors of traditional asset allocation styles. During the first decade of twenty-first century, we observed equity markets with nearly zero return, and bond return is even below. The investment environment is even worse than in 1930s. Until recently, global equities had lower return than global bonds since 1990. Pension surpluses in 1999 were replaced by substantial deficits as a result. The situation is even worse in Japan, because the country has been experiencing 20 years' decline. European insurance companies have been suffering as well. Due to the poor performance in both bond and equity market, the performance of their investment methodology has been struggling to catch up with their financial goals.

Moreover, the aging population in a number of countries poses a bigger challenge for pensions and insurance companies, like Japan and several European countries. Projected deficit will be deeper than before due to longevity risk. And since expected economic growth declines in these countries, the longevity risk will be even harder to mitigate by investing solely in equity and fixed income markets. In addition, correlations between equity markets in different regions are much higher than before due to globalization (Fig. 7.1. Data is collected through Bloomberg terminal of Princeton Firestone Library: S&P 500 index for U.S. stocks, MSCI EAFE index for equity market performance of developed markets outside of the U.S. and Canada, MSCI EM index for equity market performance across 23 emerging markets). It is increasingly harder to diversify a portfolio using traditional assets. Bonds are not a safe haven either due to the current ultra low rates. Bonds can experience as great drawdown as stocks when interest rate rises.

To meet the investment objectives in such a dire environment, institution investors are adopting different approaches. Some investors aim to immunize their liabilities by investing in bonds to hedge duration, convexity exposures to interest rate. Others lean their allocation to alternatives with high returns to catch up with their investment goals. Another important factor for institution investors is liquidity. The recent 2008 financial crisis made a lot of institutions suffer from liquidity crisis. Investors need better and more systematic system of liquidity management.

In this paper, we show how to construct a quantitative index strategy to meet the goals of investors: good return, low correlation with traditional assets and high liquidity. Then we show how to use Hidden Markov Model and stochastic programming techniques to integrate this quantitative index in investor's portfolio in an optimal fashion. The quantitative index, combined with Hidden Markov Model and stochastic programming, will greatly enhance the performance of investor's portfolio.

---

[1]The "golden rule" of 60/40 stock bond mix is explained in [1] by Malkiel.

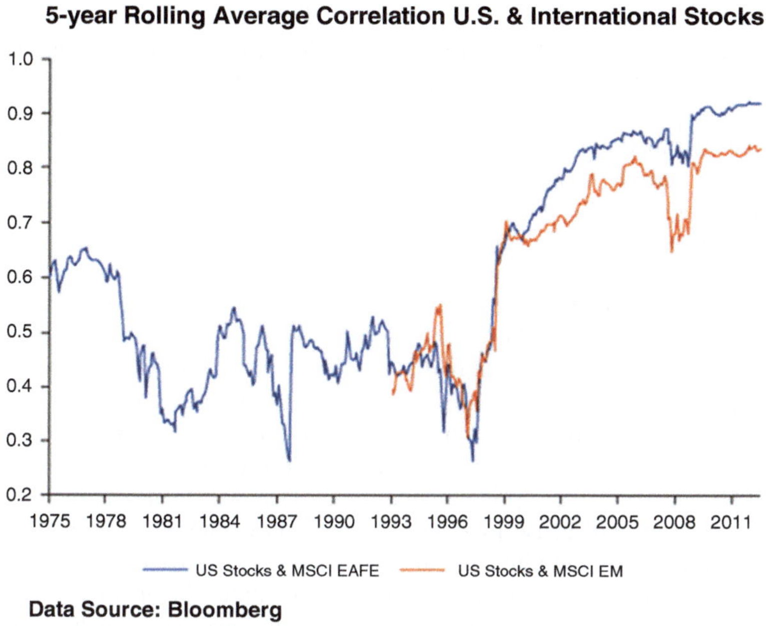

Fig. 7.1 Increasing correlation between the U.S. and international stocks

The following sections are organized as below: In Sect. 7.2, we will discuss in detail the current approaches used by investors: All Bonds v.s. All Alternatives and illustrate the issues of these approaches. In Sect. 7.3, we will show a quantitative index strategy to address with good features to solve these issues. In Sect. 7.4, we will show how to use this index strategy in a portfolio of tactics for performance enhancement. We will show the steps to optimize the index strategy with Hidden Markov Model and stochastic programming techniques. In Sect. 7.5, we will summarize our findings and list the future directions of potential research.

7.2 All Bonds Strategy vs. All Alternative Strategy

To address the aforementioned recent challenges, an enormous range of solutions have been experimented by institutional investors. Some investors adopt the Liability Driven Investment (LDI) or All Bonds (AB) methodology. These investors, including German life insurers and Japanese pension plans, allocate most of their assets into fixed income instruments. The philosophy behind this investment style can be seen in its name—liability driven. These investors try to immunize their liabilities using fixed income instruments. This is a very conservative approach with low expected return. For example, many German life insurers have 80 % + of their assets in fixed income, less than 10 % in real estate and equity and only few alternatives.[2] Japanese pension plans have 70–90 % in fixed income, 10–30 % in equities, and a dose of alternatives.[3] For an All -Bonds portfolio, the expected return is low so that high contribution (e.g. contribution to a pension plan from the pension sponsor) is expected. This is not a desirable situation for the sponsor. Under current market conditions, there are several issues with all-bonds approach. First of all, current interest rates are ultra low. Quantitative easing measures in many countries have been pushing down the interests rates. The interest rates can hardly go anywhere but up. As soon as the interest rates bounce back, the bond markets will drop a lot. This situation poses serious risks for the all-bonds investors. Also, since pension plans and insurance companies normally have long durations in their liabilities, they tend to buy large amount of long-duration bonds. For example, defined-benefit pensions have around $30 trillion or 33 %[4] of assets in bonds. This increased demand in long-duration bonds push up the prices of these bonds and results in even lower returns from fixed income. Moreover, if most of the asset is invested in fixed income, less asset there are left to invest in long-term growth areas, like venture capital, private equity, etc. This will have macro impacts on the economy and will cause long-term decrease in economic growth. This chain effect will make the future investment environment worse off. Plus, the unforeseen risks like longevity risks make the situation more complex. With longevity risks, the liabilities are harder to immunize. And if the institutional investors try to mitigate this risk by increasing the duration of portfolios, the demand in long-duration bonds will be increased further and returns on these bonds even lower. Ang et al. (2013) built a framework for liability driven investment with downside risk. Amenc et al. (2010) showed how to construct optimal liability-hedging portfolios. These papers built new strategies of constructing liability driven portfolios. However, like Ang et al. (2013) states, the asset classes should be extended to broader classes instead of limiting to cash, equity and fixed income that they are currently using in the models.

Contrary to the All-Bonds approach, another common approach adopted by institutional investors follows a totally different philosophy. All-alternatives or AA approach is adopted by the institutional investors seeking for higher returns by allocating more assets to hedge funds, private equity, real assets, etc. These alternatives have higher expected return due to illiquidity premium. Also, since there are more opportunities for mispricing, sophisticated investors can take advantage of the mispricing opportunities to generate excess return. Numerous institutional investors are gradually adopting this approach. For example, alternatives have grown from 5 % in 1995 to 19 % in 2012 in pension plans' asset allocation. In 2012, 19 % of $30 trillion in 13 largest countries' pension plans are alternatives. CalSTRS is the California State Teachers' Retirement System, and is the largest teachers' retirement fund in the United States. According to its official website, the current

[2]Please refer to [2].

[3]Please refer to [3].

[4]Please refer to [4].

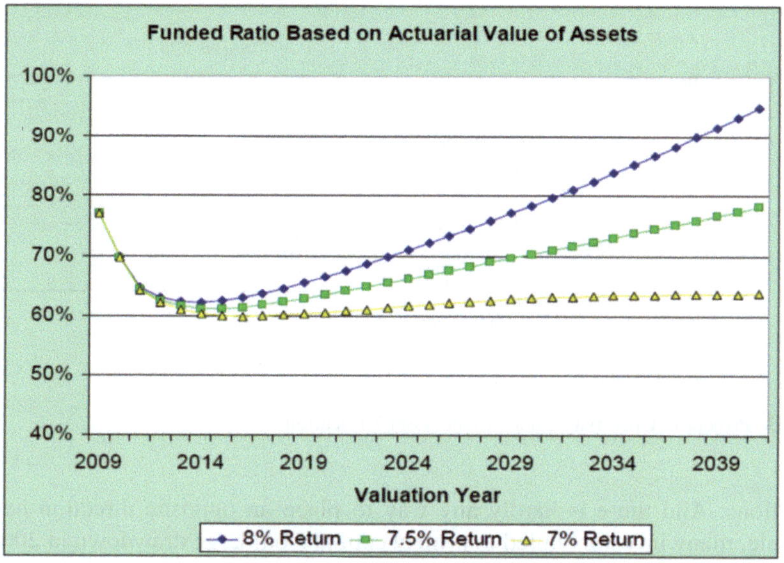

Fig. 7.2 Simulated funding ratio path of different portfolio returns

asset allocation of CalSTRS consists of 12.9 % in real estates and 12.7 % in private equity. The motivation of this All-Alternatives approach is that small increase in return can make significant difference in a long run. See Fig. 7.2 of different funded ratio path with different expected return of portfolio.

This approach do not have the issues associated with All-Bonds approach. Good examples of this approach include leading university endowments like Yale University endowment led by David Swensen. Almost 80 % of Princeton University endowment is in alternatives too. However, it also has its own disadvantages. This approach requires great expertise of the investors. And the illiquidity of alternatives poses a lot of challenges. See Fig. 7.3 the asset allocation table of Princeton University Endowment.[5]

The policy target and actual allocation in the private equity class are very different. Due to the illiquidity issue, it can be hard to maintain portfolio at policy targets for illiquid alternatives. As a result, the portfolio can't be rebalanced as much as wanted. The capital gains from rebalancing are low. Rebalancing gains are additional gains from rebalancing portfolio from time to time. Luenberger gives a good mathematical derivation to illustrate rebalancing gains in his book: Investment Science (Chapter 15, Example 15.2 Volatility Pumping). Preliminary study indicates that the loss is 150–200 basis points gain per year in geometric returns.

Plus, the illiquidity issue makes it difficult to develop a dynamic asset allocation policy. Illiquid portfolios are extremely hard to cash out under turbulent market

[5]The figure is from course lecture notes of Golden [5].

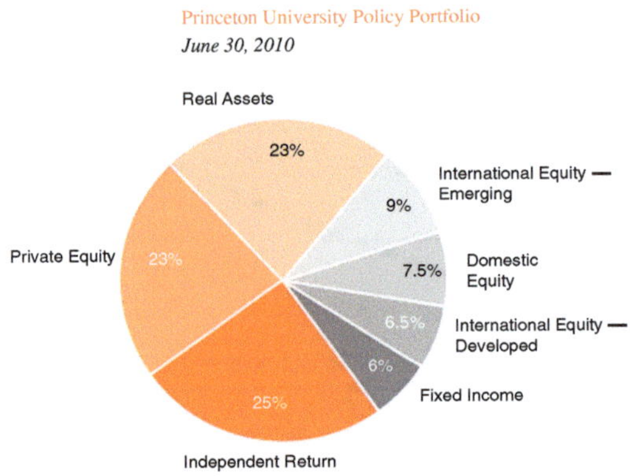

Asset Allocation
June 30, 2010

Allocation	*Policy Target*	*Actual*
Domestic Equity	7.5%	5.2%
International Equity		
Developed Markets	6.5%	3.4%
Emerging Markets	9.0%	5.7%
Independent Return	25.0%	20.2%
Private Equity	23.0%	37.2%
Real Assets	23.0%	21.6%
Fixed Income	6.0%	2.5%
Cash	0%	4.2%

Fig. 7.3 Princeton University Endowment Asset Allocation

conditions. And there is hardly any way to place an opposite direction bet. For example, many investors failed to protect capital from large drawdown in 2008 and their asset levels are still below previous high water mark, even after exceptional performance over past years. Also, for pension plans and related investors with contribution requirements, surplus protection is easier with liquid assets. In addition, with illiquid assets, it is harder to make opportunistic deals following Warren Buffet's philosophy.[6]

Cash requirement is another issue. For example, paying for operating expenses is vital to keeping a university running. During crash periods, cash requirement has to be met. Illiquid assets cannot readily be sold during fire sale. And borrowing may be very expensive. For example, Harvard University Endowment tried to sell off a $1.5 billion chunk of its private equity portfolio in the fall of 2008. But no one was willing to pay near the asking price for those assets. Desperate for cash, the university sold $2.5 billion worth of bonds, increasing its total debt to over $6 billion. Servicing that debt alone will cost Harvard an average of $517 million a year through 2038, according to Standard & Poor's news.

The traditional asset liability model, All-Bonds and All-Alternatives approaches are all struggling to meet investors' goals. Long-term investors need new perspectives. Long-term investors can take advantage of having a long horizon. They can endure short-term market choppiness. And institutions have access to more advanced technologies. However, significant drawdown should be avoided as much as possible. Capitals need to be carefully protected and contributions should be minimized while meeting cash requirements. Risk management of long-term investors needs new approaches. Risk management via hedging, namely paying premium should not be the first option. Paying premium is not desirable under already dire market conditions. An old but correct trick should be used. That is

[6]Please refer to [6] by Buffet for detailed discussion of the legendary investor's investment philosophy.

diversification among assets or strategies with positive expected returns and low correlations between them, especially during contagion periods. Diversification in fact enhances the expected return under the multi-period setting for rebalancing gains or volatility pumping. Thus, in order to rebalance the portfolio, liquidity matters a lot. Also, liquidity requirements are necessary for various reasons we discussed above. So our goal is clear, to obtain a set of independent assets or strategies with positive expected returns with liquidity. Because of the increased correlation among traditional asset classes, we look aside to alternatives. In the following sections, we will propose a new strategy that is independent of traditional asset classes and highly liquid. This new strategy will generate high return for investors and meet the liquidity requirements, and thence will be very desirable for institutional investors.

7.3 Tracking Indexes for Alternative Asset Categories

There are increasing needs of institutional investors for constructing fundamental tracking indexes of alternative classes. Take the example of Princeton University Endowment: the university endowment had unacceptably high proportions in high tech venture capital during high tech bubble at the beginning of twenty-first century. However, they could not cut the venture capital position due to its illiquidity. So they shorted the Nasdaq 100 index instead. This move turned out to be a wise rebalancing decision as the high tech bubble exploded and venture capitals lost a lot of value. If there had not been an index like Nasdaq 100 well correlated with its venture capital asset, Princeton University Endowment would have suffered great losses after bubble explosion. Thus, fundamental tracking indexes for alternative classes are important. To construct a tracking index for an illiquid alternative is to "match" the performance of it with a liquid security. We want to distinguish "fundamental" tracking from purely "statistical replication" tactics. Statistical replication is to track the target performance by pure technical measures. For example, one can use PCA, bootstrapping, all the econometric models GARCH, factor models, etc. to get a fantastic quantitative model of the target without any understanding about the underlying dynamics of the target. However, there is no guarantee that the statistical model will always track the underlying performance well. Fundamental tracking, on the other hand, relies on understandings of fundamental dynamics of the target. We believe this approach to be robust. Our goal is to develop strategies that "mimic" as close as possible. One may ask whether fundamental tracking can be achieved. Below are some examples of fundamental tracking indexes.

Mulvey and Kim (2007) showed that the performance of top equity managers can be replicated by long-only industry-level momentum strategy, especially after 1992. See Fig. 7.4. The recipe for the industry-level momentum strategy is described as follows:

Excess Return Correlation: Lagre-Cap Growth

Sample Period	Funds Group Based on Excess Returns 1st	2nd	3rd	4th
1987~1988	0.069	0.139	0.054	0.236
1989~1990	0.258	-0.057	0.086	-0.203
1991~1992	0.122	0.122	-0.093	-0.225
1993~1994	0.457**	0.532***	0.484**	0.338
1995~1996	0.601***	0.344*	0.412**	0.325
1997~1998	0.510**	0.499**	0.165**	0.224
1999~2000	0.730***	0.235	-0.022	-0.258
2001~2002	0.879***	0.832***	0.803***	0.429**
2003~2004	0.371*	0.405**	0.195	0.23
2005~2006	0.765***	0.690***	0.521**	0.301

*(**, ***): Correlations are significant at 90% (95%, 99%) Level

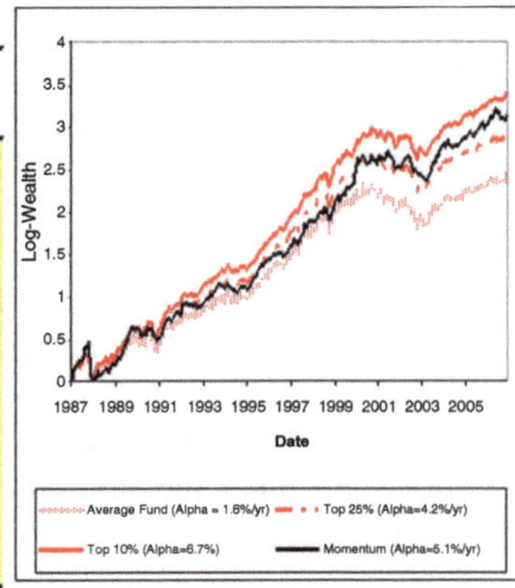

Fig. 7.4 Top equity managers performance can be replicated by long-only industry-level momentum strategy. *Left table* shows significant correlation between top equity managers performance and industry-level momentum strategy. *Right graph* shows the paths of equity managers' performance and momentum strategy

1. Choose winner industries with equal weights based on their past 3-, 6-, 9- and 12-month return to get up to 16 industries.
2. If an industry is seen more than once, put more weight on it accordingly.
3. Hold the chosen industries for predetermined holding period.
4. To reduce the timing bias, form the portfolio using overlapping time windows.
5. Holding periods are 3, 6, 9, 12 months.

Mulvey and Ling (2010) showed that the performance of private equity can be replicated by investing to designated domains via ETFs with moderate leverage.

Jurek and Stafford (2013) showed that the risks and pre-fee returns of broad hedge fund indices can be accurately matched with simple equity index put writing strategies. See Fig. 7.5 for the total return and risks tracking.

The examples above showed fundamental tracking for various underlying processes. We will propose a fundamental tracking tactic for CTAs' performance. CTA stands for Commodity Trading Advisor. They are registered managers who exercise or advise to exercise trades of futures contracts, options on futures, retail off-exchange forward contracts or swaps. The tracking index will provide an access to investors seeking for a transparent, efficient means to gain long-short exposure to commodity markets. Our goal is to obtain a set of independent assets or strategies with positive expected returns with liquidity. And the tactics in commodity markets to mimic CTA funds should be constructed to achieve this goal. Why should we turn our look to commodities? We have covered basically the reasons for us to find new tactics in alternatives. We will now discuss them in detail.

As we discussed before, the impact of quantitative easing on financial markets is huge. The purchase of debt securities by the Fed and other Central Banks has driven

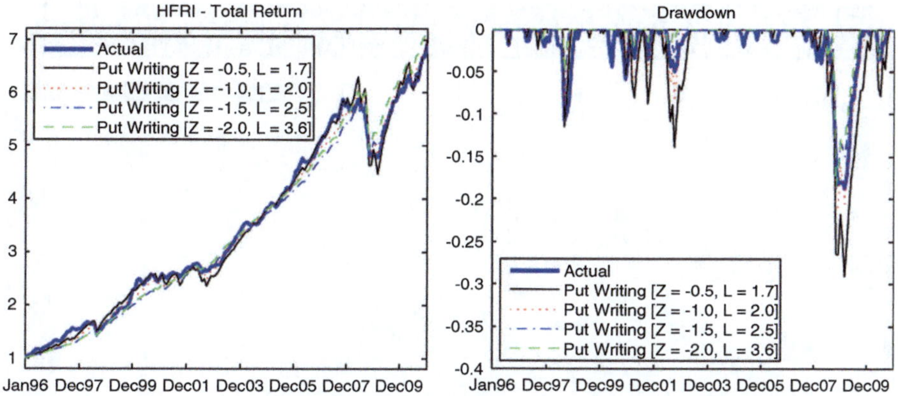

Fig. 7.5 Hedge fund index and put writing strategy. *Left graph* shows hedge fund index performance in comparison to different put writing strategies. *Right graph* shows hedge fund index drawdown in comparison to put writing strategies drawdown. Z, L are parameters in the put-writing strategy. L is the leverage of the portfolio. Z is a parameter defined by the authors: $K(Z) = S_t \exp(\sigma_{t+1} Z)$. S_t is the prevailing level of the S&P 500 index and σ_{t+1} is the 1-month stock index implied volatility, observed at time t. Basically, if one specifies Z, one specifies K, the strike price level

global government bond yields to record-low levels. Rates of The U.S. bonds have been hit particularly hard. For example, the U.S. 10-year yield is traded as low as 1.45 % in June 2012.

We can see that bonds are no longer a safe heaven. The existing potential inflationary pressures are severely bearish for bonds prices. Interest rates cannot stay at the current level forever and the only direction to go is up. With yields so low, an inflationary shock of any sort would be devastating, as rates would spike in response. If yields on the U.S. Treasury bonds rose 2 percentage points to levels that prevailed as recently as 2007, the resulting bond price decline would be approximately 20 % for the 10-year Treasury bonds and a 30 % drop for the 30-year Treasury bonds.

The equity markets are shaky. Increased nervousness is observed in equity markets. A small shock could affect equity markets significantly. See Fig. 7.6 of S&P 500 index in May 2013. On May 22, 2013, the U.S. equity markets reacted drastically to the release of Fed minutes and the index plunged rapidly.

Inflation is another issue. The Feds is trying to generate inflation since it is afraid of deflation. Many economists and Central Banks maintain that “money printing” creates significant inflation. Given the unprecedented measures, investors should be prepared for inflation, especially because the inflation has been benign so far. The gap between yields on 10-year Treasuries and same-maturity inflation-protected notes is a gauge of consumer-price expectations. And the gap jumped following the Feds’ announcement of QE3 in September 2012. This means expected inflation shock can be sudden.

Due to the issues mentioned above, commodities as an asset class have unparallel long-term advantages. It has low correlation with the traditional asset classes of

Fig. 7.6 Shaky markets: a small shock could affect equity markets significantly

	S&P GSCI TR Index	FTSE All World Index	JP Morgan Global Aggregate Bond Index
S&P GSCI TR Index	1.00	**0.33**	**0.08**
FTSE All World Index		1.00	0.04
JP Morgan Global Aggregate Bond Index			1.00

Fig. 7.7 Correlation matrix of S&P GSCI TR Index, FTSE All World Index and J.P. Morgan Global Aggregate Bond Index

equity and bond. It can be used to hedge inflation. Figure 7.7 shows the correlation matrix of monthly returns between S&P GSCI TR Index, FTSE All World Index and J.P. Morgan Global Aggregate Bond Index. S&P GSCI Index serves as a benchmark for investment in the commodity markets and a measure of commodity performance over time. FTSE All World Index is the Large/Mid Cap aggregate of 2800 stocks from the FTSE Global Equity Index Series. It covers 90–95 % of the investable

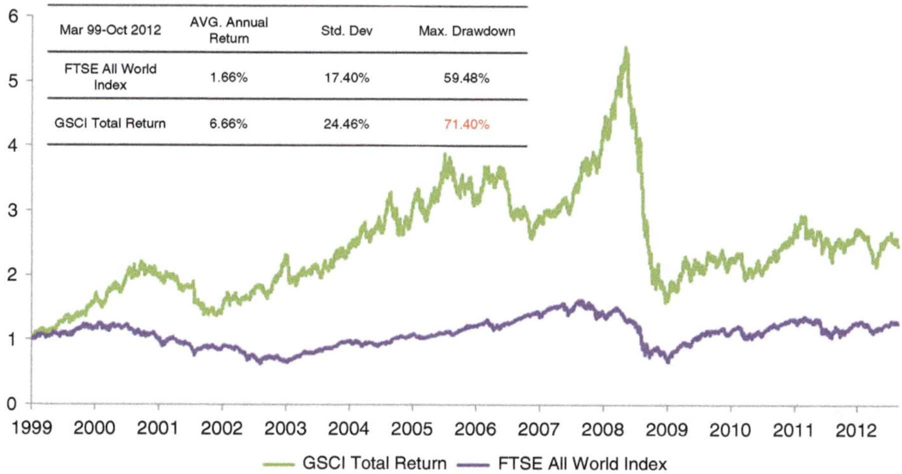

Fig. 7.8 Large drawdown happens in commodity market

market capitalization. J.P. Morgan Global Aggregate Bond Index is a measure of global bonds performance. It is shown that the correlation of commodity markets and traditional markets is low.

Constructing an index strategy in commodities is a promising approach. Though, to construct a good-performing index is not so simple and takes an understanding of the performance of CTA funds. We will show that the simple approaches do not work.

Although the commodities as an asset class on average have low correlation with stock markets, the increasing correlation between them during contagion leads to large drawdown during equity market crash. As a result, the simple long strategy is not a good one (Fig. 7.8).

The shape of futures curve is another issue against long-only approach. The futures curve can be in backwardation, wherein the price of futures contract is trading below the expected spot price at contract maturity; or contango, wherein the price of futures contract is trading above the expected spot price at contract maturity. Backwardation happens when the predominant hedgers are producers and contango happens when the predominant hedgers are consumers. Backwardation is no longer the predominant futures curve shape. See Fig. 7.9. So when an investor rolls the contract, there is a rolling loss due to contango futures curve shape. Performance differential between GSCI Spot price and GSCI Excess (the investible version) reflects the losses due to rolling of futures in contango markets. In a contango market, the roll yield is negative. In a contango market, the price of futures contract is trading above the expected spot price at contract maturity. So the price will roll down to the spot price. Thus, investor will lose money in rolling the contract.

The downsides common to long-only approaches include large and protracted drawdown, exposure to price bubbles, losses resulting from contango futures curve shape, over-concentration in particular commodity types (e.g. energy, which

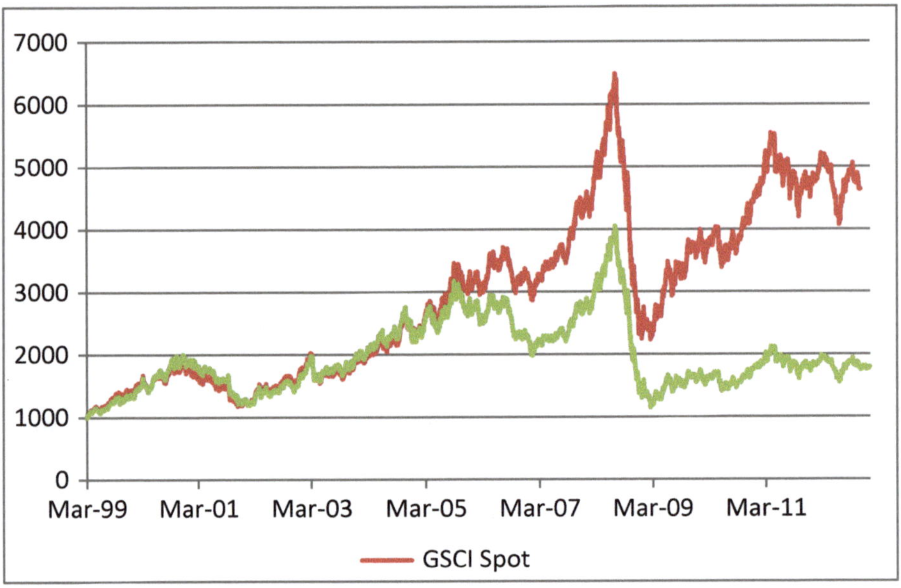

Fig. 7.9 Performance differential between GSCI Spot and GSCI Excess reflects the losses due to rolling of futures in contango markets

accounts for a big proportion in commodity indices), and high degree of volatility. To avoid these disadvantages, we need to do some thinking and construct the Target commodity index wisely. CTA-type funds have a lot of benefits we can borrow. Managed futures funds, in particular, give us inspiration. They have low correlation with other types of fund strategies, even in contagion periods. And it can even generate positive returns in contagion periods.

Based on the underlying performance of CTA funds, like managed futures, we go on to describe the construction of the Target Commodity Index.[7] The components of the index involve two long-only strategies and two long-short dynamic tilting tactics. The long-only strategies are momentum strategy and futures curve strategy. The dynamic tilting strategies are breakout strategy and trend following strategy. Mulvey [7] proposes two different levels of exposure: 100 % Exposure Commodity Index and 80 % Exposure Commodity Index. These indexes are now under the names: FTSE Target 100 % Exposure Commodity Index and FTSE Target 80 % Exposure Commodity Index respectively. See Table 7.1 for their compositions.

The long-only strategies include momentum strategy and futures curve strategy. The momentum strategy assigns long or flat positions to constituents according to their recent performance relative to other commodities in the index. Basically, it selects top performing commodities in the commodity pool based on their returns of

[7]DPT Management constructed this index for FTSE, called the FTSE Target Commodity Family. All rights reserved. See Mulvey [7] for details.

Table 7.1 Compositions of target 80 % and target 100 % Exposure Commodity Index

Index	Trend				
	Momentum (%)	Futures curve (%)	Following (%)	Breakout (%)	Buffer (%)
FTSE target 80 % Exposure Commodity Index	20	20	20	20	20
FTSE target 100% Exposure Commodity Index	25	25	25	25	0

In FTSE Target 80 % Exposure index, 20 % of the portfolio is allocated to Momentum strategy, 20 % to Futures curve strategy, 20 % to Trend following strategy, 20 % to breakout strategy and 20 % are used as buffer; In FTSE Target 100 % Exposure index, 25 % of the portfolio is allocated to each of the four strategies

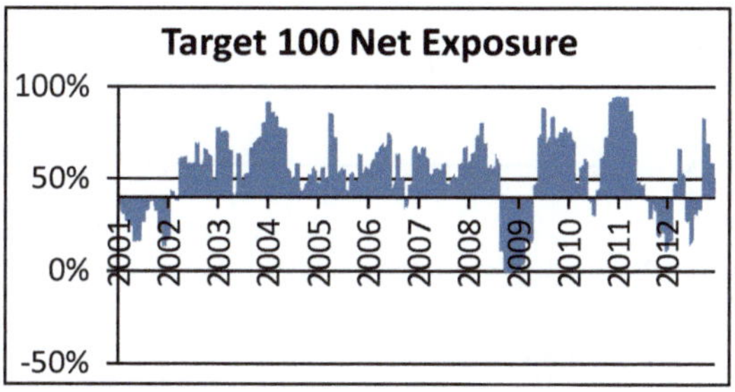

Fig. 7.10 Target 100 index net exposure

a certain past period. The strategy capitalizes on an empirically observed tendency for rising commodity prices to rise even further, and for falling prices to keep falling. The futures curve strategy assigns long or flat positions to constituents according to the relative degree of contango or backwardation in their futures curve. By primarily being long commodities in backwardation, this strategy potentially reduces the rolling losses that affect long-only passive indices.

The dynamic tilting strategies include trend following strategy and breakout strategy. Trend following strategy assigns long or short positions based on mid-term price performance. If the current price is above long-term mean, we will assign long position to it. Otherwise, we will assign short position to it. Breakout strategy assigns long or short positions to constituents that are trading above or below a set price range. If one constituent is trading above a set price, a long position will be set to it. Otherwise, a short position will be set to it.

The Target Commodity Index is an equal mix of above four tactics (25 % each for 100 % exposure and 20 % each for 80 % exposure), with monthly rebalancing. Below is the graph of the net exposure (long minus short) of Target 100 Index (Fig. 7.10).

Next, we examine the performance characteristics of this commodity index. The index series aims to track the performance of a typical commodity CTA. The local patterns on a short-term scale of the Target 100 index are very similar to

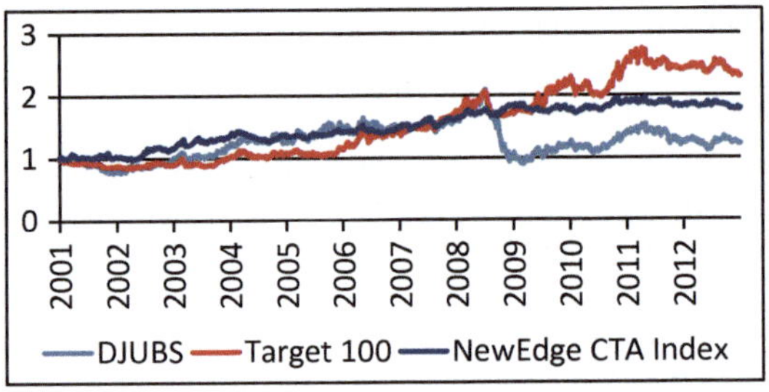

Fig. 7.11 Performance of Target 100 index in comparison to DJUBS index and NewEdge CTA index

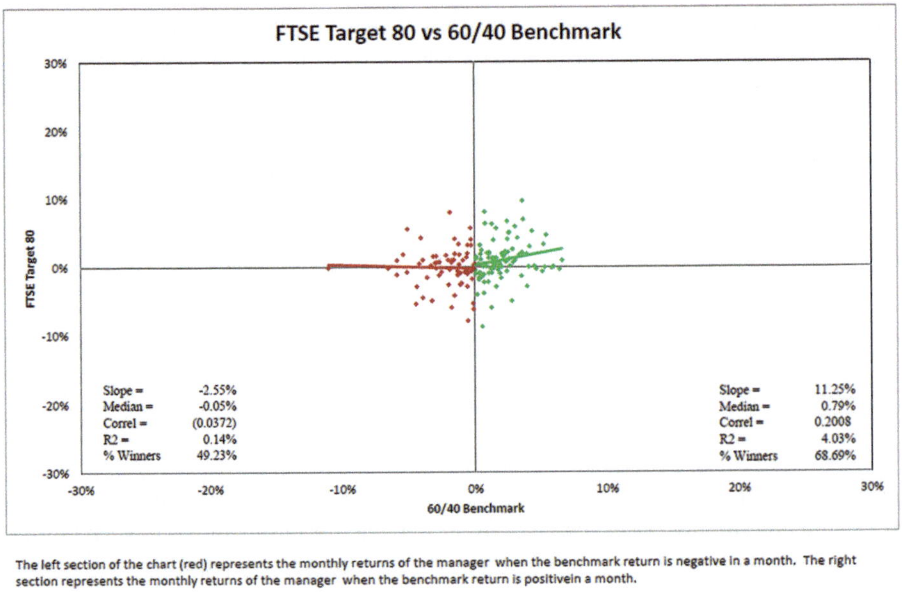

Fig. 7.12 Characteristics of correlation between Target 80 index and 60/40 stock–bond mix benchmark

that of NewEdge CTA index (The NewEdge CTA index provides a reliable daily performance benchmark of major CTAs). Moreover, the index series also has the potential to outperform the benchmark. One can see Fig. 7.11 and compare their performances.

If we compare the performance of the Target Index and the traditional 60 % stock, 40 % bond mix, we can see that the Target Index tends to move up when the 60–40 % mix moves up. However, the Target Index tends to stay flat when the 60–40 % mix goes down. This pattern of the Target Index provides important portfolio diversification. It is good evidence that our Target Index is a good diversifier to traditional portfolios. Figure 7.12 shows this pattern.

Another desirable feature of the Target Index is its low correlations with other asset classes, which is one of the properties we aim for. See Table 7.2 for the correlations between our index and other benchmarks.

Table 7.2 Performance characteristics (5-year correlation)

Based on monthly data	FTSE target 80 % Exposure Commodity Index	FTSE target 100 % Exposure Commodity Index	FTSE All World Index	FTSE Global Government Bond Index	FTSE EPRA/NAREIT Global Index	FTSE StableRisk Commodity Trend Index
FTSE target 80% Exposure IMO Commodity Index	**1.0000**	1.0000	0.2657	0.0951	0.0747	0.5680
FTSE target 100% Exposure Commodity Index		**1.0000**	0.2657	0.0954	0.0744	0.5684
FTSE All World Index			**1.0000**	−0.0180	0.6935	−0.0533
FTSE Global Government Bond Index				**1.0000**	−0.3151	0.0464
FTSE EPRA/NAREIT Global Index					**1.0000**	−0.0641
FTSE StableRisk Commodity Trend Index						**1.0000**

Now that we have described the Target Index, we can use it in portfolio construction. The good features of it—low correlation with other asset classes, liquidity, and good returns are exactly what we aimed for in the first place.

7.4 A Portfolio of Tactics

We will show how the FTSE Target Commodity Index can be used in various ways in portfolio construction. First of all, it serves as a performance benchmarks for investing in CTA funds. The commodity investors can compare their portfolio performance with the index performance. Moreover, since the commodity markets are highly liquid, the index can be used to rebalance portfolio, for example, to achieve rebalancing gains. Investors in CTA funds cannot rebalance their portfolios easily for lock-up policies. But investors can short or long this index to rebalance their portfolios. It can be used as a performance enhancer as well. Since it is highly liquid, it is very easy to take opportunistic investments by investing in it. During crash periods, it can be used to protect capital by either longing or shorting it. During crash periods, it may be difficult to withdraw funds from CTA funds due to lock-up policies. And also, investors may have concerns as to keeping relationship with top CTA managers. Good fund managers will generally recover quickly from a crash period. But if one withdraw funds from the managers during crash period due to cash flow constraints, it will probably jeopardize the relationship with the top managers and hard to put money back in their good performing funds after crash period. However, it is very easy to balance their portfolio by longing or shorting the Target Index. We will discuss various uses of the index in details.

7.4.1 Overlay Approach

Long-term investors can improve their investment performance by incorporating specialized "overlay" strategies. Overlay strategies are investing in futures market using core assets to fulfill margin requirements. The overlays require no dedicated capital beyond the core portfolio. Traditional assets in the core portfolio are employed as margin capital for targeted positions in the futures markets. Mulvey et al. [8, 9] show the benefits of general overlay strategies within asset allocation and asset and liability models. These strategies seek to widen diversification of the portfolio, while generating higher growth rates. Trading volumes in the futures market is large enough so that investors can quickly rebalance the portfolio mix as conditions warrant. Mulvey and Kim [10] show how to enhance portfolio duration and generate higher growth rates for pension plans. The added duration reduces contribution risks while increasing expected portfolio performance.

The overlays can provide higher risk-adjusted portfolio returns than approaches based on traditional leverage. One can see that from Fig. 7.13.

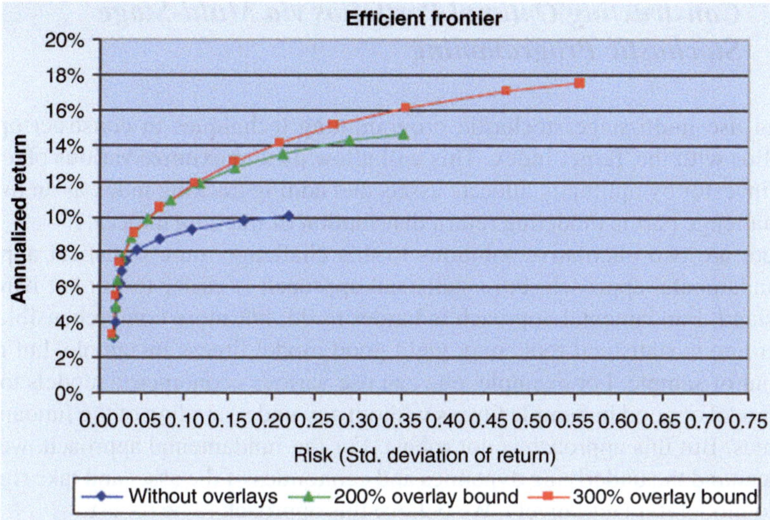

Fig. 7.13 Efficient frontier with different overlay bound

Table 7.3 Performance characteristics of overlay approach

	50 % SP500+50 % JPM Agg. Bond Index	Add 50 % target 100
Start Date	1/31/92	1/31/92
End Date	12/31/12	12/31/12
Geo. Return	7.75 %	11.49 %
Volatility	8.26 %	10.78 %
Sharpe ratio	0.57	0.79
Max-Drawdown	28.04 %	29.51 %
Ret/Drawdown	0.28	0.39
Ulcer Index	5.77 %	6.55 %
Return/Ulcer	0.82	1.30
Corr w/SP500	0.938	0.786

Table 7.3 gives another example of overlay strategies using the target 100 Index. The left column is a combination of 50 % S&P 500 index and 50 % J.P. Morgan Aggregate Bond index. The right column is 50 % S&P 500 index, 50 % J.P. Morgan Aggregate Bond index and 50 % overlay of the Target 100 index. Geo.Return is geometric average return. Max-drawdown is maximal drawdown during the period. Ret/Drawdown is average return per maximal drawdown. Ulcer index is a risk measure to compute the drawdown variance. Return/Ulcer is ratio of average return over Ulcer index. The Sharpe ratio is higher with negligible increase in Max-Drawdown. The return to Ulcer index is higher and correlation with S&P 500 index is lower. One can see from this numerical example that the index enhance to portfolio performance a lot.

7.4.2 *Constructing Optimal Portfolios via Multi-Stage Stochastic Programming*

We can use multi-stage stochastic programming techniques to construct optimal portfolios with the Target Index. This will allow us to maximize various objectives of the investor by optimally allocate assets and adding tracking index as an overlay. The challenge part is modeling return distribution of tracking indices.

There are two alternative solutions to this challenge: pure statistical approach and fundamental approach. Pure statistical approach is easier to do, but harder to understand. Fundamental approach is harder to do, but more comprehensible. The bottom line is statistical tools may yield good model fitness in-sample, but not so good out of sample. For example, one can use various econometric models to get a good model of good in-sample fitness without any understanding of the fundamental dynamics. But this approach is not robust. For the fundamental approach, we need to understand the underlying dynamics and economics of the asset and take rigorous steps to model it quantitatively. We believe this approach to be robust.

One way to identify probabilistic laws is through scenario generation. The basic idea for this approach is as follows: The behaviors of constituent assets are better understood than the whole index. So firstly, we employ widely accepted models for constituents. Then, we identify probabilistic laws of tracking indexes conditioned on specific constituents. Mulvey et al. [11] built an asset and liability management system for Towers Perrins-Tillinghast. They develop a cascade scenario generation structure called CAP:Link, where variables at the top of the structure influence those below, but not vice-versa. This approach eases the task of calibrating parameters. The ordering does not reflect causality between economic variables, but rather captures significant co-movements. Mulvey et al. [8, 12, 13] show several examples of using multi-stage stochastic programming to construct optimal portfolios.

To construct the scenario generation framework, we need to delve into the factors affecting commodity prices. First of all, global demand pattern is one factor. For example, strong growth in emerging markets will push commodity prices up. Supply disruption is another one. The Iran oil embargo makes oil prices go up every time. Moreover, movements in value of dollar clearly have an effect on commodities. The more valuable dollar is, the lower prices of commodities will be. Interest rates are yet one more important factor.

There are numerous models in this field explaining how different factors affect commodity prices. Frankel and Andrew [14] identifies the channel of effect of interest rate on commodity prices. If interest rate lowers, inventory costs will be lower. And inventory demand will be higher. Thus, commodities prices will be higher. Gruber and Vigfusson [15] identify the effect of interest rate on commodities correlations. If interest rate lowers, inventory will be higher. And individual shock tolerance will be higher. Thus, macro effect is relatively higher, resulting in high correlations among commodities. Tang and Xiong [16] show the effect of commodity index investment on commodities correlations. When commodity index investment goes up, volatility from other markets will spill over. And then, co-movement among commodities will go up, as do correlations between commodities. Futures prices of non-energy commodities became increasingly correlated with oil after 2004. This trend was significantly more pronounced for indexed commodities than for off-index commodities. And this trend intensified after the financial crisis of 2008, which is a well-noted fact. There is also evidence of an increasing return correlation between commodities and the MSCI Emerging Markets Index in recent years. Index investors typically focus on strategic portfolio allocation between the commodity class and other asset classes, such as stocks and bonds, and hence tend to trade in and out of all commodities in a given index at the same time. Thus, their portfolio rebalancing can spill price volatility from outside markets on and across commodity markets.

From knowledge and evidence in various factors affecting commodities, one can then construct scenario generator using CAP:Link cascade structure. Based on the structure of scenarios, one can then apply multi-stage stochastic programming techniques to it.

Another way to build the probabilistic laws is using Hidden Markov Models (HMM). The Hidden Markov Model is a popular method for regime identification, which has been widely used in engineering and science. Hamilton [17] uses HMM to predict business cycles of the U.S. economy by analyzing the U.S. Gross National Product (GNP). Turner et al. [18], Hansen [19], Hamilton and Susmel [20], and Garcia [21] have further discussion of HMM in finance. These researches commonly describe the regimes of equity markets in terms of return and volatility. Guidolin and Timmerman identify four regimes in the joint return series of stock and bond market with HMM. Bae et al. [22] show how to employ Hidden Markov Model to identify states in varied financial markets and construct a stochastic program to optimize portfolios under the regime switching framework.

First we use Hidden Markov Model to estimate the joint probabilistic laws of the index and other asset classes in the portfolio. Then we can create optimal portfolios via multi-stage stochastic programming. The index is an overlay strategy added to the whole portfolio to enhance its performance. One can again refer to Bae et al. [22] for detailed procedures. The formulation of stochastic programming requires seven steps. (1) Decide the number of children for the current node of the scenario tree. (2) Assign child node based on filtered probabilities. (3) For each assigned child node, apply estimated distribution from Hidden Markov Model. (4) Generate sample returns. (5) Calculate filtered probability by forward algorithm. (6) Repeat for each child node and update time. (7) Repeat till the time hits final time. After formulation of multi-stage structure, one can optimize the objective function incorporating regulation and other constraints.

Following Bae et al. [22], we will give a formulation of combining Hidden Markov Model and stochastic programming to optimize the portfolio performance with the index strategy as overlay. Fraser [23] documents the basic framework of

Hidden Markov Model. First we show the framework of Hidden Markov Model estimation (following [23]):

- S(t): A random variable of (unobservable) state at time t
- Y(t): A random variable of observation (in this case, three-dimensional daily return series) of different asset classes (like stock, bond, etc.) and tactics (like our commodity index)
- S_{t_1,t_2}: A sequence of random variables of states from time t_1 to time t_2
- Y_{t_1,t_2}: A sequence of random variables of states from time t_1 to time t_2
- s(t): A realized(unobservable) state at time t
- y(t): A realized observation at time t
- s_{t_1,t_2}: A sequence of realized states from time t_1 to time t_2
- y_{t_1,t_2}: A sequence of realized observations from time t_1 to time t_2
- $\boldsymbol{\Theta}$: A set of variables of HMM parameters to be estimated
- θ: A set of estimated HMM parameters
- N: The dimension of observations (depends on the number of asset classes or tactics in the portfolio)

$$t \in \{1, \ldots, T\},\ \forall\, s(t) \in S = \{1, \ldots, K\}$$

We assume the daily return series has normal distribution under each state, or regime:

$$Y(t)\,|S(t) = y(t)|\, s(t) \sim N\left(\mu_{s(t)},\ \Sigma_{s(t)}\right)$$

$\mu_{s(t)}$ is the mean of the daily return series, and $\Sigma_{s(t)}$ is the covariance matrix under state s(t). Baum-Welch algorithm can be used to estimate the parameters μ_s, Σ_s under each state or regime, and the initial probability, the transition matrix of the states. We denote the initial probability of different states by $\pi = (\pi_1, \pi_2, \ldots, \pi_K)$ and transition matrix between different states by $A = \left(A_{ij}\right) \cdot A_{ij} = P\left(s(t+1) = j \middle| s(t) = i\right)$.

Baum-Welch algorithm is essentially a two-step EM algorithm. Denote all parameters in Hidden Markov Model μ_s, Σ_s, π, as θ. Denote all observations $\{y(1), y(2), \ldots, y(T)\}$ by $\mathcal{Y}$ and all state throughout time by Bishop [24] has detailed derivation of the algorithm. Baum-Welch algorithm can be simply described as repeating the following steps until convergence:

1. Compute $Q(\theta, \theta^m) = \sum_{s\in S} \log\left[P(\mathcal{Y}, s; \theta)\right] P\left(s \middle| \mathcal{Y}; \theta^m\right)$.
2. Set $\theta^{m+1} = \text{argmax}_\theta\, Q(\theta, \theta^m)$.

The algorithm always converges with the initial setting of θ^0. The parameters can be solved analytically in each step. The number of states can be determined by Bayesian information criterion (BIC). This criterion is introduced by Schwarz and Prajogo (2011) suggests using it for determination of number of states. BIC adjusts value of likelihood by penalizing the number of parameters scaled by a function of number of observations:

$$BIC = -2\ln Q + k\ \ln T,$$

Q: likelihood calculated from Baum-Welch algorithm
k: the number of parameters
T: the number of observations.

The criterion optimizes likelihood while penalizing the number of parameters used in the model. In the Hidden Markov Model case, $k = K \cdot N + K \cdot N^2 + K^2 \cdot K$ is the number of states. N is the dimension of observations.

After learning the model in some training period, we can identify characteristics of the states and joint distributions for asset classes and tactics. Then we can use the forward-backwards algorithm to assign filtered probability.

Following Bae et al. [22], we can begin the forward algorithm by assigning:

$$\alpha(s,1) = P_{S(1)|Y(1)}\left(s \middle| y(1)\right) = \frac{P_{S(1)}(s)\, P_{Y(t)|S(t)}\left(y(1) \middle| s\right)}{\sum_{\tilde{s}\in S} P_{S(1)}(\tilde{s})\, P_{Y(t)|S(t)}\left(y(1) \middle| \tilde{s}\right)},\ \forall s \in S$$

where S is the set of all states and $\alpha(s, 1)$ is the filtered probability at time $t = 1$. From this starting point, we can smooth the distribution of states by computing:

$$P_{S(t)|Y_{1,t-1}}\left(s \middle| y_{1,t-1}\right) = \sum_{\tilde{s}\in S} P_{S(t)|S(t-1)}\left(s \middle| \tilde{s}\right) \cdot \alpha(\tilde{s}, t-1).$$

And we can also compute the joint probability of the state and current observation:

$$P_{S(t),Y_{1,t}|Y_{1,t-1}}\left(s, y(t) \middle| y_{1,t-1}\right) = P_{Y(t)|S(t)}\left(y(t) \middle| s\right) \cdot P_{S(t)|Y_{1,t-1}}\left(s \middle| y_{1,t-1}\right).$$

Next, we compute:

$$\gamma(t) = \sum_{s\in S} P_{S(t),Y_{1,t}|Y_{1,t-1}}\left(s, y(t) \middle| y_{1,t-1}\right)$$

and finally, we can update the distribution of states:

$$\alpha(s,t) = P_{S(t)|Y_{1,t}}\left(s \middle| y_{1,t}\right) = \frac{P_{S(t),Y(t)|Y_{1,t-1}}\left(s, y(t) \middle| y_{1,t-1}\right)}{\gamma(t)}$$

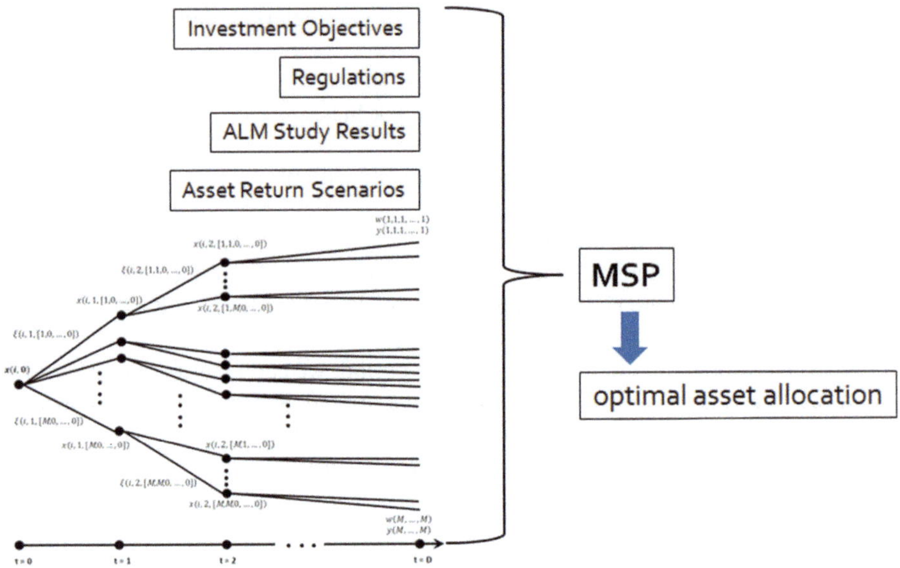

Fig. 7.14 Creating optimal portfolios via multi-stage stochastic programming

With filtered probability throughout time, we can back test our Hidden Markov Model to check whether it can forecast market conditions.

After back testing our trained Hidden Markov Model, we can formulate the stochastic programming framework:

We construct a scenario tree like Fig. 7.14 to map random elements to a set of numbers. Denote the daily growth for each asset or tactic as $\varepsilon\,(i, t, \omega) \cdot i$ is the index for each asset or tactic, ω is the random element. Random elements can be represented by a finite number of branches and each path of the tree becomes a single possible scenario b_t at time t. $b_t \in \{0, 1, \ldots, M\}$. M is some predetermined number to represent number of nodes. $t \in \{0, 1, \ldots, T\}$. T is the number of periods. Control variables are the allocations to each asset or tactic in each period, which can be represented by $x(i, t, [b_1, b_2, \ldots, b_T])$. By the same means, $\varepsilon(i, t, \omega)$ can be represented by $\varepsilon(i, t, [b_1, b_2, \ldots, b_T])$. Finally, we can represent the objective function Z by $Z(b_1, b_2, \ldots, b_T)$.

We follow the procedures below to formulate the stochastic programming framework:

1. Decide the number of children for the current node of the scenario tree. We denote $B\,(b, t) = [b_1, b_2, \ldots, b_T]$, where $b_t \in \{0, 1, \ldots, M\}$ and b_{t+1} to $b_T = 0$. Zero represents values "Not Assigned".

2. For each filtered probability $\Pr(s, t, B(b, t))$, where $s \in \{1, \ldots, N\}$. Assign b'th node to s if $\sum_{i=0}^{s-1} \Pr(i, t, B(b, t)) < \frac{b'}{M} \leq \sum_{i=0}^{s-1} \Pr(i, t, B(b, t))$. Then, $B(b, t+1) = [b_1, b_2, \ldots, b_t, b_{t+1} = b', 0, .., 0]$
3. For each child node, if state s is assigned to it, apply the mean vector μ_s and covariance matrix Σ_s from the estimated parameter set θ.
4. Generate the sample return $\mathrm{r}(s, t+1, B(b, t+1)) \sim MVN(\mu_s, \ \Sigma_s)$. So

$$\varepsilon(i, t+1, B(b, t+1)) = 1 + \mathrm{r}(s, t+1, B(b, t+1))$$

5. Calculate the filtered probability $\Pr(s', t+1, B(b, t+1))$ by using the forward algorithm given $\mathrm{r}(s, t+1, B(b, t+1))$ and θ.
6. Repeat steps for each node until t = T.

After setting up the tree, we can optimize the objective function using the formulation below:

Objective function:

$$\text{Maximize} \sum_{b_1=1}^{M} \cdots \sum_{b_T=1}^{M} \frac{1}{M^T} Z(b_1, b_2, \ldots, b_T)$$

Constraints:

Initial wealth: $\sum_i x(i, 0, [0, 0, \ldots, 0]) = W_0$

Constraints for each period $t = 1, \ldots, T$:

$$\begin{aligned}\sum_i -\varepsilon(i, t, [b_1, b_2, \ldots, b_t, 0, \ldots, 0])\, x(i, t \\ -1, [b_1, b_2, \ldots, b_{t-1}, 0, \ldots, 0]) \\ + x(i, t, [b_1, b_2, \ldots, b_t, 0, \ldots, 0]) = 0, \mathit{for}\ t \\ = 1 \ldots T-1\end{aligned}$$

Final wealth: $\sum_i -\varepsilon(i, T, [b_1, b_2, \ldots, b_T])\, x(i, T-1, [b_1, b_2, \ldots, b_{T-1}, 0]) + W(b_1, b_2, \ldots, b_T) = 0$

7.5 Conclusions

We show the current portfolio and ALM issues for institutional investors, and the limitations of asset allocation with traditional asset categories. We advocate that investors seek wider diversification in alternative categories. We illustrate the need for dynamic asset allocation and rebalancing portfolios between allocation reviews. Thus, a liquid index with good performance and low correlation with traditional assets is needed. We described a tracking index for CTA funds and showed its performance characteristics. It has good growth rate, is highly liquid and has low correlation with traditional assets, especially during contagion periods. Then, we show how to use the index in portfolio performance enhancing. The index can be used as an overlay strategy for core portfolio. Also, more sophisticated techniques can be used to construct the probabilistic laws of the index. Fundamentally understanding the constituents in the index, we can build scenario generator and Hidden Markov Model to construct the probabilistic laws. Then based on the probabilistic laws, multi-stage stochastic programming can be applied to construct optimal portfolios for various objectives.

The research points to some promising future directions. More research on "fundamental" tracking indexes may be done on other asset classes or tactics. Tracking index on REITS for real estate investments can be useful as a way to rebalance illiquid asset of real estate. For example, a tracking index for exchange fund engaging in carry trade and related tactics is DBV. In addition, a tracking portfolio for pension surplus may be very useful for pension plans in terms of duration enhancing.

Aside from tracking indexes, asset allocation model of illiquid and liquid assets also needs further discussion. In this paper, we show the approach of constructing and applying tracking indexes. Another approach is to optimize multi-period asset allocation of illiquid and liquid assets directly. Dynamics for illiquid asset retruns and illiquidity constraints must be modeled. Also, transaction or illiquidity costs need to be incorporated. In this paper, we described an index in the commodity futures market. Other tracking portfolios of illiquid assets can also be constructed and be implemented via overlays. Overlays require no dedicated capital beyond the core portfolio, providing higher risk-adjusted returns than approaches based on traditional leverage.

References

1. B.G. Malkiel, *A Random Walk Down Wall Street: The Time-Tested Strategy for Successful Investing*, 10th edn. (W. W. Norton, 2012)
2. Deutsche Insurance Asset Management, TrendLines semi-annual report on development in the insurance investment market
3. T. Watson, *Global Pension Assets Study 2013, 2012, 2011*
4. Vanguard Group, Inc. *Survey of Defined Benefit Plan Sponsors, 2012 and 2010*
5. A. Golden, *Endowment Management at Princeton University, Highlighting Some Quantitative Methods*, Lecture notes from course ORF 311
6. W. Buffet, *The Essays of Warren Buffet: Lessons for Investors and Managers* (Wiley, Singapore, 2013)
7. J.M. Mulvey, A rule based commodity index. J. Investment Manage. (2013)

8. J.M. Mulvey, K. Simsek, Z. Zhang, Improving performance for pension plans. J. Asset Manage. **7**, 93–108 (2006)
9. J.M. Mulvey, C. Ural, Z. Zhang, Improving performance for long-term investors: wide diversification. Leverage, and overlay strategies. Quant. Financ. **7**(2), 1–13 (2007)
10. J.M. Mulvey, W. Kim, Constantly rebalanced portfolio—is mean reversion necessary? in *Encyclopedia of Quantitative Finance* (Wiley, UK, 2009)
11. J.M. Mulvey, G. Gould, C. Morgan, An asset and liability management system for Towers Perrin-Tillinghast. Interfaces **30**(1), 96–114 (2000)
12. J.M. Mulvey, W. Kim, M. Bilgili, Dynamic investment strategies and rebalancing gains, in *The Euromoney Algorithmic Trading Handbook 2007/2008* (2008)
13. J.M. Mulvey, W. Kim, Y. Ma, Duration-enhancing overlay strategies for defined benefit pension plans. J. Asset Manage. **11**, 136–162 (2010)
14. J. Frankel, R. Andrew, *Determinants of Agricultural and Mineral Commodity Prices*. HKS Faculty Research Working Paper Series RWP10-038 (John F. Kennedy School of Government, Harvard University, 2010)
15. J. Gruber, R. Vigfusson, Interest rates and the volatility and correlation of commodity prices. International Finance Discussion Papers 1065 (Board of Governors of the Federal Reserve System, Washington, DC, 2012)
16. K. Tang, W. Xiong, Index investment and financialization of commodities. Financ. Anal. J. **68**(6), 54–74 (2012)
17. J.D. Hamilton, A new approach to the economic analysis of non-stationary time series and the business cycle. Econometrica **57**, 357–384 (1989)
18. C. Turner, R. Startz, C.R. Nelson, A Markov model of heteroskedasticity, risk, and learning in the stock market. J. Financ. Econ. **25**, 3–22 (1989)
19. B. Hansen, The likelihood ratio test under non-standard conditions: testing the Markov switching model of GNP. J. Appl. Econ. **7**, 61–82 (1992)
20. J.D. Hamilton, R. Susmel, Autoregressive conditional heteroskedasticity and changes in regime. J. Econ. **64**, 307–333 (1994)
21. R. Garcia, Asymptotic null distribution of the likelihood ratio test in Markov switching models. Int. Econ. Rev. **39**, 763–788 (1998)
22. G. Bae, W. Kim, J.M. Mulvey, Dynamic asset allocation for varied financial markets under regime switching framework. Eur. J. Oper. Res. (2013)
23. A.M. Fraser, *Hidden Markov Models and Dynamical Systems*. Society for Industrial and Applied Mathematics (2008)
24. M.C. Bishop, *Pattern Recognition and Machine Learning*. Information Science and Statistics (2006)

Chapter 8
Stabilizing Implementable Decisions in Dynamic Stochastic Programming

Michael A.H. Dempster, Elena A. Medova, and Yee Sook Yong

Abstract We present a novel approach to address sampling error when discretely approximating a dynamic stochastic programme with a limited finite number of scenarios to represent the underlying path probability distribution. This represents a tentative solution to the problems first identified in our companion paper (Dempster et al., A comparative study of sampling methods for stochastic programming, forthcoming). Conventional approaches to such problems have been to find the best discretization of the statistical properties of the simulated processes in terms of the *objective* of the problem based on probability metrics. Here we consider the stability of the *implementable decisions* of a stochastic programme, which is key to financial investment and asset liability management (ALM) problems, while simultaneously reducing the discretization bias resulting from small-sample scenario discretization. We tackle discretization error by reducing the degrees of freedom of the decision space in a financially meaningful way by constraining the decisions to lie within a carefully chosen subspace. This avoids overfitting the optimized decisions to the simulated in-sample scenarios which often do not generalize to unseen scenarios drawn from the same probability distribution of paths. We illustrate the application of versions of the proposed technique using a practical four-stage ALM problem previously studied in Dempster et al. (J Portf Manag 32(2):51–61, 2006. Empirical results show their effectiveness in reducing the discretization bias and improving the stability of the implementable decisions without adding much to the computational complexity of the original problem.

The views expressed here are solely those of the authors and not Credit Suisse.

M.A.H. Dempster (✉) • E.A. Medova
Department of Pure Mathematics and Statistics, Centre for Financial Research, University of Cambridge, Cambridge CB2 1AG, UK

Cambridge Systems Associates, Cambridge CB5 8AF, UK
e-mail: mahd2@cam.ac.uk; eam28@cam.ac.uk

Y.S. Yong
Credit Suisse AG, Singapore
e-mail: yeesook@cantab.net

© Springer International Publishing Switzerland 2017

G. Consigli et al. (eds.), *Optimal Financial Decision Making under Uncertainty*,
International Series in Operations Research & Management Science 245,
DOI 10.1007/978-3-319-41613-7_8

Keywords Discretization bias • Stability • Implementable decisions • Sampling error • Dynamic stochastic programming

8.1 Introduction and Background

A dynamic (multi-stage) stochastic program can be written as

$$\min_{x \in \mathcal{X}} \int_{\Omega} f(x, \omega) dP(\omega), \tag{8.1}$$

where $f(x, \omega)$ is the objective function defined in terms of both the uncertain paths ω and the decision space $\mathcal{X}$ of first stage decisions which imply feasibility in subsequent stages. Modelling of the complex practical features of such problems requires numerical computation of optimal solutions which in turn requires a discrete sampling scenario tree approximation of the continuous path probability distribution P. A vast literature has been devoted to finding the best approximation of the continuous distribution in a variety of stochastic programming problems. For example, there are efforts concentrating on having the scenario sample and underlying moments matched [3, 7, 8] or on minimizing the Monge-Kantorovich-Wasserstein and other probability metrics [4–6, 10, 11]. All these efforts aim to find an approximation that best matches certain statistical properties of the discretized sample distribution to the underlying theoretical one. However, practically more important criteria are to evaluate the impact of these methods on the stability of the optimal objective function value and to test against possible sampling bias [9]. A comparative study of the effectiveness of some of these methods can be found in our previous paper [2].

In this paper, we emphasize an additional criterion in evaluating scenario sampling methods, namely, the stability of the root-node recommended decisions. In financial portfolio allocation problems, this criterion is crucial to the practical implementation of the recommended decisions which are hedged against the sampled future scenarios and must hold until the next rebalance date. At this future date the parameters of the underlying theoretical path distributions will be recalibrated to market data which includes its update since the implemented portfolio. In practical dynamic problems, since statistical models of asset returns are well known to be prone to significant misspecification, root-node decision instability can lead to increased actual transactions costs. Even if these costs are penalized in the problem formulation, as here, instabilities of root-node decisions at recalibration of the model used can lead to excessive transaction cost accumulation from chasing after spurious profits.

This new criterion does not appear in the literature, presumably based on the argument that two different decisions that yield similar objective function values should not be penalized as being unstable. This situation will occur if the surface of the objective function is a flat plateau or has multiple optima of similar values.[1]

[1]However these cases are precluded for a dynamic (multi-stage) linear stochastic programme.

This argument in general relies heavily on the assumption that the surface of the objective function of a scenario-based stochastic programming problem is accurately represented. Sampling error usually causes this surface to vary across different simulations of scenario trees even for two stage problems. So, similarity in objective function values does not truly measure the stability of the stochastic programming problem's solution. Two distinct implementable decisions with similar objective function values might lead to very different values of the "true" objective function surface. This implies that in-sample stability should be measured with respect to *both* criteria, i.e. stability of both objectives and implementable decisions.

There are different ways of defining stability of the implementable decision vector. For example, we could take the standard deviation of the Euclidean norm, or compute the standard deviation of each initial decision component and find the maximum. However, for a financial asset liability management problem the implementable decisions are an optimal asset mix which can be characterized in terms of expected portfolio return and volatility. This provides a more convenient and familiar way of measuring stability of the implementable decisions in financial problems. For example, when we have asset classes that are highly correlated or have nearly the same risk and return characteristics, *different* asset mixes may yield similar portfolio characteristics as a whole and are therefore appropriately viewed as *similar* optimal solutions in terms of decision stability.

As noted above, stability of implementable decisions is important since these are the actions to be implemented in the real world. The empirical results from the in- and out-of-sample tests in [2] clearly highlight the problems of insufficient scenario branching. A low branching factor in the scenario tree does not truly represent the assumed underlying probability distribution, causing the exploitation of these unrepresentative scenarios by the optimization process to yield a solution which is *over-fitted* to the simulated data presented to it. In this paper, we aim to have the initial decision obtained from the optimizer *generalize* to unseen scenarios generated from the *same* underlying path probability distribution. Thus, we approach the problem from the viewpoint of having *robust optimization* with respect to sampling error interpreted as a problem of *incomplete data*.

The problem of small sample over-fitting can be tackled by reducing the degrees of freedom of the implementable decision space. First, let us write the scenario tree root node decision x_t^1 at time t as a linear combination of some *basis factors* $a_{i,t}$, $i = 1, \ldots, r$,

$$x_t^1 = c_t + \sum_{i=1}^{r} \mu_{i,t}\, a_{i,t}, \tag{8.2}$$

where $\mu_{i,t}$ is the weight at time t for the ith basis factor, denoted by $a_{i,t}$, c_t is an offsetting constant vector and r denotes the total number of factors used in the optimization process. The restriction in the degrees of freedom of the implementable decisions may be achieved via:

1. reducing the dimension of the basis factors by setting $r < d$, where d is the total dimension of the decision variables in the original optimization problem, or

2. limiting the values which the weights $\mu_{i,t}$ for the original decisions, $i = 1, \ldots, d$, can take.

For the first case, we constrain the decision to be a linear combination of a specially chosen set of factors instead of allowing the decision process to be fully flexible. For the second method, we only allow the original decisions to be within specially chosen limits. Note that when lower and upper limits on some of the weights are set to zero, the second case collapses to the first. The resulting reduced flexibility prohibits the optimizer from finding a solution which is fine-tuned to the small-sample scenarios. Two issues need to be addressed when using this proposed formulation. Namely: How are the basis factors to be chosen? Given that we know ways to find the factors, how many should we retain?

To demonstrate how stabilization of implementable decisions can be effectively adopted we use the Pioneer guaranteed return fund problem [1] as an illustrative example. The next section reviews the salient features of the dynamic stochastic optimization model employed for this problem and highlights some of the issues with small-sample approximation described in [2]. Sections 8.2 and 8.3 of the paper consider respectively the evaluation of under-estimation of portfolio risk in this problem and its possible solution. Section 8.4 treats the empirical evaluation of the proposed techniques and Sect. 8.5 concludes and suggests some directions for further research.

8.2 Review of Pioneer Guaranteed Return Funds

Pioneer Investments, the asset management division of UniCredit bank, offered to investors in the EU from the early mid-2000s a range of unit-linked guaranteed products, technically *guaranteed minimum investment benefit* (GMIB) *variable annuities*. Pre-crisis, the funds backing these products totalled about 14 billion euros. In this paper we will study the proof of concept model used to develop the models actually used by Pioneer to manage these funds, see [1] and [2] for respectively EUR and USD implementations of this initial model. The Pioneer guaranteed return fund model considers the portfolio optimization of a *closed-end* fund with a *nominal* guarantee of G per annum for all investors. The models actually implemented by Pioneer involved investor contributions to open-end funds and a variety of guarantee benchmarks including equity indices. The present model aims to maximize the performance of the fund taking into account the risk associated with falling short of the level of the guarantee. The model objective is thus a trade-off between risk control and maximizing portfolio wealth controlled by a non-negative parameter β (see the Appendix for details). A low value of β corresponds to very tight risk control. As β is increased, assets with higher return performance are chosen at the expense of greater risk.

When a small-sample scenario tree is used, for problems with wealth maximization as the objective (high β), simple (algebraic) two moment matching is able to yield results which are close to the true solution as long as the expected returns of each asset are captured correctly in the scenario tree. However, when a small sample scenario tree is used with tight risk control, two moment matching fails to alleviate a more prominent issue associated with insufficient scenario branching—underestimation of portfolio risk. The lack of a large enough number of sample scenarios causes an under-estimation of in-sample portfolio risk to result in an unjustified risk-taking investment recommendation impairing the fund's realized performance. This is the issue addressed in this paper.

8.3 Evaluating Under-estimation of Portfolio Risk

We have noted above that insufficient scenario branching causes an under-estimation of in-sample maximum shortfall, to result in an overly aggressive investment strategy. Here, we perform a theoretical calculation of the one period expected maximum shortfall of the Pioneer model to gain an insight into the possible degree of under-estimation in such a scenario tree. Let W_t and B_t denote the wealth and barrier level at time t respectively.[2] The *minimum* portfolio return required to stay above the barrier at the next time period, in other words, the immediate *target return* $\boldsymbol{\tau}$, is given by

$$\boldsymbol{\tau} := \frac{\mathbf{B}_{t+1}}{W_t} - 1. \tag{8.3}$$

Due to the stochastic nature of the barrier, the target is also stochastic. If $B_{t+1} < W_t$, then the current wealth level is already above the barrier of the next period, and any investment strategy that maintains the current wealth level (i.e. with non-negative returns) will suffice to meet the negative target. On the other hand, if $B_{t+1} > W_t$ then we have a positive target. Given an allocation strategy, the *immediate* portfolio return is normally distributed, since we model the asset returns as correlated Gaussian processes. This means that conditional on the current wealth level, the wealth distribution for the next period is normal. Note however, that this does not imply that *terminal* wealth is normally distributed, as the portfolio undergoes yearly re-balancing and the effects of compounding. Nevertheless, the immediate conditional portfolio return characteristics allow us to derive the immediate maximum shortfall as a non-negative quantity, which serves as a lower bound to the maximum shortfall over the planning horizon.

Given an allocation strategy, let the expected portfolio return and volatility be denoted by μ_p and σ_p. Therefore, conditional on the current wealth, the next period wealth is normally distributed as

[2]Here we use boldface to denote (conditionally) stochastic entities, and inequalities and equations between such entities are assumed to hold almost surely.

$$W_{t+1|t} \sim N\left(W_t(1+\mu_p), (W_t\sigma_p)^2\right). \tag{8.4}$$

Note that with a Gaussian returns model we might end up with negative wealth, but this is not possible in the real world since the maximum loss that we might incur is the total amount of the initial investment. However, since this relates to the tail of the normal distribution, which in this context occurs with low probability, we can ignore this inconsistency in the following calculation. The next period *expected maximum shortfall* (EMS_{t+1}) is given by

$$EMS_{t+1} = \frac{1}{\sqrt{2\pi}W_t\sigma_p}\int_{-\infty}^{B_{t+1}} (W_{t+1} - B_{t+1}) e^{-\frac{\left(W_{t+1} - W_t(1+r_p)\right)^2}{2(W_t\sigma_p)^2}} dW_{t+1}. \tag{8.5}$$

Changing variables by letting $r := \frac{W_{t+1}}{W_t} - 1$, we have

$$\begin{aligned}
\overline{EMS}_{t+1} := \frac{EMS_{t+1}}{W_t} &= \frac{1}{\sqrt{2\pi}\sigma_p}\int_{-\infty}^{\tau} (r-\tau) e^{-\frac{\left(r-\mu_p\right)^2}{2\sigma_p^2}} dr \qquad (8.6)\\
&= \frac{1}{\sqrt{2\pi\sigma_p^2}}\int_{-\infty}^{\tau} (r-\mu_p) e^{-\frac{(r-\mu_p)^2}{2\sigma_p^2}} dr \\
&\quad + \frac{\mu_p - \tau}{\sqrt{2\pi\sigma_p^2}}\int_{-\infty}^{\tau} e^{-\frac{(r-\mu_p)^2}{2\sigma_p^2}} dr
\end{aligned}$$

$$\begin{aligned}
&= -\frac{\sigma_p}{\sqrt{2\pi}} e^{-\frac{(\tau-\mu_p)^2}{2\sigma_p^2}} + (\mu_p - \tau) Prob\left(\mathbf{z} < \frac{\tau - \mu_p}{\sigma_p}\right) \qquad (8.7)\\
&= -\frac{\sigma_p}{\sqrt{2\pi}} e^{-\frac{(\tau-\mu_p)^2}{2\sigma_p^2}} + \frac{\mu_p - \tau}{2}\left(1 + erf\left(\frac{(\tau-\mu_p)}{\sqrt{2}\sigma_p}\right)\right).
\end{aligned}$$

This theoretical EMS calculation is a non-linear function of the decision variables and thus is difficult to implement as a constraint in the Pioneer model. Nonetheless, (8.7) can give us an indication of the range of values of portfolio returns and volatilities for a given EMS and a target value. Figure 8.1 plots the contour of the *normalized* (by wealth) EMS, denoted by $\overline{EMS}$, for different portfolio returns and volatilities, from which we see that the mean-variance efficient portfolio yields the least expected maximum shortfall for a specified return.

Given a limit c on $\overline{EMS}$ and a target τ, define A to be the set of assets/portfolios with risk/return characteristics yielding immediate (one-period) normalized EMS less than the limit, i.e.

$$A(c,\tau) = \{(\mu,\sigma) : \overline{EMS}(\mu,\sigma,\tau) \leq c\}, \tag{8.8}$$

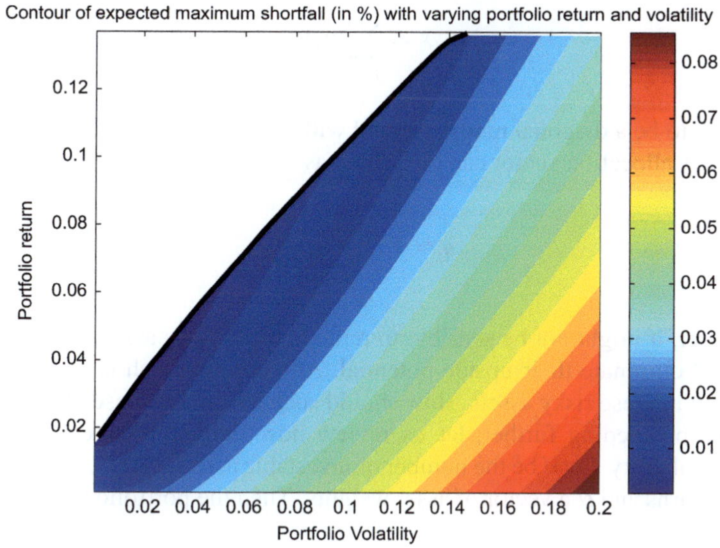

Fig. 8.1 Contour of normalized expected maximum shortfall

and let $\sigma_{\max}$ be the maximum volatility of the assets from this set, i.e.

$$\sigma_{\max}(c, \tau) := \arg\max_{\sigma}\{(\mu, \sigma) : \overline{EMS}(\mu, \sigma, \tau) \leq c\}. \tag{8.9}$$

Due to the concave nature of the normalized EMS contour, $\sigma_{\max}$ may be found by the intersection of the corresponding contour with the mean-variance efficient frontier. In the next section, we will see how we can incorporate the information obtained from Eqs. (8.8) and (8.9) into the appropriate choice of the basis factors defined in (8.2).

8.3.1 Position Limits Based on a Volatility Constraint

Let the reference to the initial model period t be implicit in the sequel and let 1_j be a $d \times 1$ column vector with all elements set to zero except the jth element, which is set to one. From Fig. 8.1, we can see that if the volatility of an asset is greater than $\sigma_{\max}$ then irrespective of its expected return, a 100 % investment in this asset will result in an immediate EMS that violates the given limit. An easy way to address this problem is to remove altogether such risky assets from the set of investable assets.

This removal may be formulated in terms of (8.2) for suitable r by setting μ_j to zero and a_j to 1_j for all j such that $\{\sigma_j > \sigma_{\max}\}$. The allocation vector (in dollar amounts) can thus be written as

$$x = \sum_{j=1}^{r} \mu_j 1_j. \tag{8.10}$$

If we let P to be a diagonal matrix formed with current asset prices on its diagonal, the *quantity* allocation vector can be written as

$$q = P^{-1} \sum_{j=1}^{r} \mu_j 1_j. \tag{8.11}$$

However, it might not be sensible to remove these risky assets altogether, since this would eliminate their upside potential. Another approach is to limit the total investment in these risky assets. How should such a limit be chosen?

Before proceeding further, let us review how individual assets contribute to portfolio volatility. Let d be the number of investable assets, C the covariance matrix of their returns and α a vector of asset weights. Then the portfolio volatility σ_p may be written as

$$\sigma_p = \sqrt{\sum_{j=1}^{d} \alpha_j^2 C_{jj} + \sum_{j=1}^{d} \sum_{i=1}^{d} \alpha_i \alpha_j C_{ij}} \tag{8.12}$$

with $\alpha_i \geq 0, i = 1, \ldots, d$, when no short-selling is allowed. The diversification process is clearly illustrated by (8.12); a reduction in the portfolio volatility is achieved when the second term is negative and the portfolio volatility is bounded from above by the maximum volatility σ_{max} of the assets under consideration. However, due to the interaction/diversification among the assets, it is difficult to ascertain limits on each asset which yield portfolio volatility in the appropriate range. Therefore we let the maximum proportion invested in asset i be heuristically determined by

$$\alpha_i^{\text{hi}} := \frac{\sigma_{\max}}{\sqrt{C_{ii}}}, \tag{8.13}$$

as holdings above this limit (irrespective of holdings in the other assets) may result in a portfolio volatility that is too aggressive. Therefore, setting $\alpha_i^{\text{lo}} = 0$and $\alpha_i^{\text{hi}} = \frac{\sigma_{\max}}{\sqrt{C_{ii}}}$, we add the following constraints to the initial model optimization problem at each rebalance period

$$x = \sum_{j=1}^{d} \mu_j \; 1_j, \tag{8.14}$$

where

$$W\alpha_i^{\text{lo}} \leq \mu_i \leq W\alpha_i^{\text{hi}} \tag{8.15}$$

and W is the current wealth level. The position limits (8.15) based on (8.13) can be used by risk managers and regulators to avoid fund managers putting too much money into speculative investments, thereby exposing their fund to excessive risk. With these position limits, we can reduce the initial decision space to a region of suitable risk, avoiding the optimization process overfitting the small scenario tree sample used in the problem.

However, the heuristic we have used to bound the position limits in (8.13) ignores the diversification benefits that the assets bring to the portfolio as a whole. An asset, though risky, might be negatively correlated with other assets which, when combined, yield lower portfolio risk. Therefore we next investigate how the return covariance matrix *diagonalization* procedure transforms the asset returns into *uncorrelated* space, which allows a more straightforward determination of position limits. Assuming that the assets are uncorrelated, the portfolio volatility collapses to

$$\sigma_p^2 = \sum_{i=1}^{d} \alpha_i^2 C_{ii} = \alpha_k^2 C_{kk} + \sum_{i \neq k} \alpha_i^2 C_{ii}. \tag{8.16}$$

Suppose we have a constraint $\sigma_p \leq \sigma_{\max}$. Then, combining this with (8.16) gives

$$\alpha_k^2 \leq \frac{\sigma_{\max}^2}{C_{kk}} - \frac{\sum_{i \neq k} \alpha_i^2 C_{ii}^2}{C_{kk}} \leq \frac{\sigma_{\max}^2}{C_{kk}}. \tag{8.17}$$

Because the second term is difficult to determine, we shall use

$$\alpha_k \leq \frac{\sigma_{\max}}{\sqrt{C_{kk}}} \tag{8.18}$$

as the appropriate bound. Note the similarity of (8.18) to the heuristic we used in (8.13). However, the position limit in (8.18) is in the uncorrelated return space.

For completeness we describe the standard algebraic procedure for transforming the asset returns from the original space to the uncorrelated space. Let $\mathbf{r}$ be the vector of normally distributed returns of the d assets with $\mathbf{r} \sim N(\mu, C)$. We will assume C to be a full covariance matrix since assets are usually not perfectly correlated with one another. Performing the eigenvalue decomposition of C, we obtain

$$C = BDB^{-1}, \tag{8.19}$$

where B is the matrix formed by the eigenvectors b_i as its columns and D is a diagonal matrix of the eigenvalues. Instead of the standard L_2 normalization of the eigenvectors, we will use the normalization which requires the elements of each the eigenvectors to sum to unity. Note that the orthogonal property of the eigenvectors gives

$$b_i^T b_j = \begin{cases} f_{ii} & i = j \\ 0 & i \neq j. \end{cases} \tag{8.20}$$

These equations may also be represented in matrix form as $B^T B = F$, where F is a diagonal matrix. This diagonalization process transforms the returns in the original space to a space in which they are uncorrelated. It may be performed by multiplying the original return vector $\mathbf{r}$ with the matrix B^T. Then the return vector in the transformed space becomes $r^* = B^T r$ and the resulting diagonal covariance matrix of the transformed asset returns is given by

$$\begin{aligned} Cov\left(r^* r^{*T}\right) &= B^T \mathbb{E}\left[(r-\mu)(r-\mu)^T\right] B \\ &= B^T \left(BDB^{-1}\right) B = FD. \end{aligned} \tag{8.21}$$

From the point of view of asset management, the transformation B^{-1} bundles the original assets into *uncorrelated portfolios* with the weights of each portfolio given by certain coefficients. We henceforth refer to these uncorrelated portfolios as the original portfolio basis factors, with weights for each factor in which negative weights correspond to short-selling the corresponding portfolio. The resulting variance of each portfolio basis factor is given by the corresponding eigenvalue D_{ii} multiplied by the normalizing constant f_{ii} as shown in (8.21). Each of these portfolio basis factors is uncorrelated with the others. Thus we may constrain each of them *independently*.

Next, let us see how an allocation to the portfolio basis factors relates to the assets in the original space. Let i be the index of the assets in the original space and j be the index of portfolio basis factors. Assume that a single dollar portfolio investment is decomposed into an α_j proportion assigned to the jth portfolio basis factor, $j = 1, \ldots, d$, with $\sum_{j=1}^{d} \alpha_j = 1$. The quantity of asset i held in the original portfolio, denoted by q_i, is given by

$$q_i = \sum_{j=1}^{d} \alpha_j q_i^{(j)}, \tag{8.22}$$

where $q_i^{(j)}$ is the quantity invested in asset i in the portfolio basis j. Recalling that for $B = (b_{ij})$ of (8.19), the proportion of the dollar unit $q_i^{(j)}$ invested in portfolio basis j can be found to be $\frac{b_{ij}}{P_i}$ and given that we invest W dollars of financial wealth in the original portfolio, the quantity q_i of asset i held in this is given by

$$q_i = \frac{W}{P_i} \sum_{j=1}^{d} \alpha_j b_{ij}. \tag{8.23}$$

We have thus derived the relationship between portfolio decisions in the uncorrelated space and the original space. This transformation does not alter the optimization problem, so the allocation may be performed equivalently in either the transformed or the original space.

However, as noted above, once we have this transformation into the uncorrelated space, constraining the decisions by choosing only the ones with appropriate risk/return characteristics becomes much more straightforward, since we can consider each of the uncorrelated basis portfolios independently. Removal of risky assets in the uncorrelated return space is performed by setting

$$\alpha_j = 0 \quad \text{for} j \text{ suchthat } \sqrt{D_{jj} f_{jj}} > \sigma_{\max} \tag{8.24}$$

which leads to

$$q_i = \frac{W}{P_i} \sum_{j=1}^{r} \alpha_j \, b_{ij}, \quad i = 1, \ldots, d, \tag{8.25}$$

for suitable r in the form of (8.2). Given the weights α_j set equal to zero in the uncorrelated return space (and assuming for simplicity that the indices are relabelled so that these are $j = r + 1, \ldots, d$), the constraints (8.24) can be formulated equivalently, using the orthogonal properties of the basis factors, as

$$b_k^T q^{\text{original}} = \operatorname{diag}\left(\frac{W}{P}\right) b_k^T \left(\sum_{j=1}^{d} \alpha_j b_j\right) = 0 \quad \forall k > r, \tag{8.26}$$

where b_k denotes the kth eigenvector and q^{original} is the initial quantity vector in the original space, since

$$b_k^T \left(\sum_{j=1}^{r} \alpha_j b_j\right) = \begin{cases} 0 & \text{if } j \neq k \\ \alpha_k f_{kk} & \text{if } j = k. \end{cases} \tag{8.27}$$

From the implementation point of view, the constraints (8.26) offer an advantage in that they do not involve introducing a new set of initial decision variables α into the optimization problem. The added constraints can thus be viewed as either: (1) limiting the initial decision space to the subspace formed by linear combinations of the appropriately chosen basis factors, or (2) restricting the optimization process to only finding decisions that satisfy the orthogonal properties of the remaining basis factors.

We could also set the position limits on the uncorrelated portfolios to preserve the upside of the risky assets similarly to (8.15) by replacing $\sqrt{C_{jj}}$ with the volatilities $\sqrt{D_{jj} f_{jj}}$ of the portfolio basis factors. Short selling is possible in the uncorrelated space, but not in the original space. Therefore, we obtain

$$-\frac{\sigma_{\max}}{\sqrt{f_{jj}D_{jj}}} \leq \alpha_j \leq \frac{\sigma_{\max}}{\sqrt{f_{jj}D_{jj}}}. \tag{8.28}$$

8.3.2 Position Limits Based on Asset Returns and Volatility Proportional Constraints

The previous section derived position limits by implicitly assuming that the individual investable assets are near the mean-variance efficient frontier. This allowed the use of volatilities only in our considerations. When an asset is not close to the efficient frontier, the maximum proportion in the asset allowable before breaching the EMS limit is an *overestimate*. Such mean-variance inefficient assets may nevertheless contribute to optimal portfolios in the presence of additional portfolio constraints, for example, limiting overall portfolio drawdown in each period of the model. This section therefore looks at an approach to relaxing the (approximate) mean-variance efficient assumption on asset returns.

With this approach the maximum proportion α_i^{hi} for a particular asset is determined as a proportional investment in that asset beyond which the required EMS target will not be satisfied even if the remaining capital is entirely invested in the *zero-coupon bond* (ZCB) forming the guarantee barrier (see the Appendix for details). Here we treat the ZCB as completely riskless, i.e. using its characteristic at the problem planning horizon. Therefore, the upper limit of the proportion for asset j is given by

$$\alpha_j^{\text{hi}} = \arg\max\{0 \leq \alpha_j \leq 1 : \overline{EMS}\left((1-\alpha_j)r_{zcb} + \alpha_j\mu_j, \alpha_j\sigma_j, \tau\right), \tag{8.29}$$

where r_{zcb} is the *annualized* return of the zero coupon bond. The risk/return profile of an investment in asset i and the ZCB forms a straight line passing through $(r_{zcb}, 0)$ and (μ_j, σ_j) and α_i is the *proportion* at which this straight line intersects the corresponding EMS contour. Similar to the other approaches described above, for each asset we here neglect the contribution of the other assets towards portfolio risk and return.

8.3.3 Summary

In this section we have introduced three alternative methods for mitigating the under-estimation of portfolio risk from small sample scenario trees in the context of the Pioneer guaranteed return fund DSP problem with diffusion return processes calibrated to monthly USD data. We measure initial portfolio risk in terms of one-period conditional expected shortfall relative to the current cost of ensuring the guarantee at maturity and its conditional small sample stability in terms of the usual portfolio volatility risk and expected return. The fundamental idea is to restrict the dimension of the initial portfolio to a subspace spanned by a limited number of factors, here by placing restrictions on portfolio weights to exclude excessively risky assets using approximate risk bounds, but more familiar financial risk factors could in principle also be employed.

Position limits based on asset returns and volatility proportional constraints are specific to the guaranteed fund problem case study of Sect. 8.4 and we shall see (Fig. 8.2) that it is important in the asset restriction process that bounds take account of asset return covariance. We therefore summarize here the steps of that approach which is based upon rotating the Gaussian asset return space to consist of uncorrelated portfolios composed from the original assets in which short selling is allowed:

Step 1. Perform the eigenvalue decomposition of the return vector conditional covariance matrix (3.17).

Step 2. Apply the asset risk bounds (3.22) to remove risky assets in the uncorrelated portfolio return space and transform to quantities in the original asset space using (3.23).

Step 3. Alternatively, express the original asset restriction constraints in terms of linear constraints which involve the conditional covariance matrix eigenvectors given by (3.24) with dimension the number of risky assets excluded in the uncorrelated portfolio return space.

Although the case study of the next section involves Gaussian conditional asset returns, this standard linear algebra based procedure is applicable to general conditional asset return distributions with two moments finite, provided that small sample initial portfolio stability is measured by initial conditional portfolio risk and return. The effectiveness of this procedure in more general settings, with other asset return distributions and portfolio risk measures different from the guaranteed expected maximum shortfall used for the Pioneer problem in the sequel, is an empirical matter for future research.

8.4 Empirical Results

We now investigate the effectiveness of the $\sigma_{\max}$ and proportional constraints described in the previous section in stabilizing and reducing discretization error in the implementable decisions of the 4-stage Pioneer guaranteed return fund model. Only the root node implementable decisions are so constrained. We perform in and out-of-sample tests using the same tree structures employed in [2], i.e. 7-7-7-7, 7-7-7, 7-7 and 7 at each roll forward period, so that a comparison with our earlier results can be made to ascertain the effectiveness of our methods.

Table 8.1 shows the in-sample terminal wealth, expected maximum shortfall and hitting probability for the different stabilization methods employed. The uncorrelated space $\sigma_{\max}$ position limits and the proportional limits yield lower in-

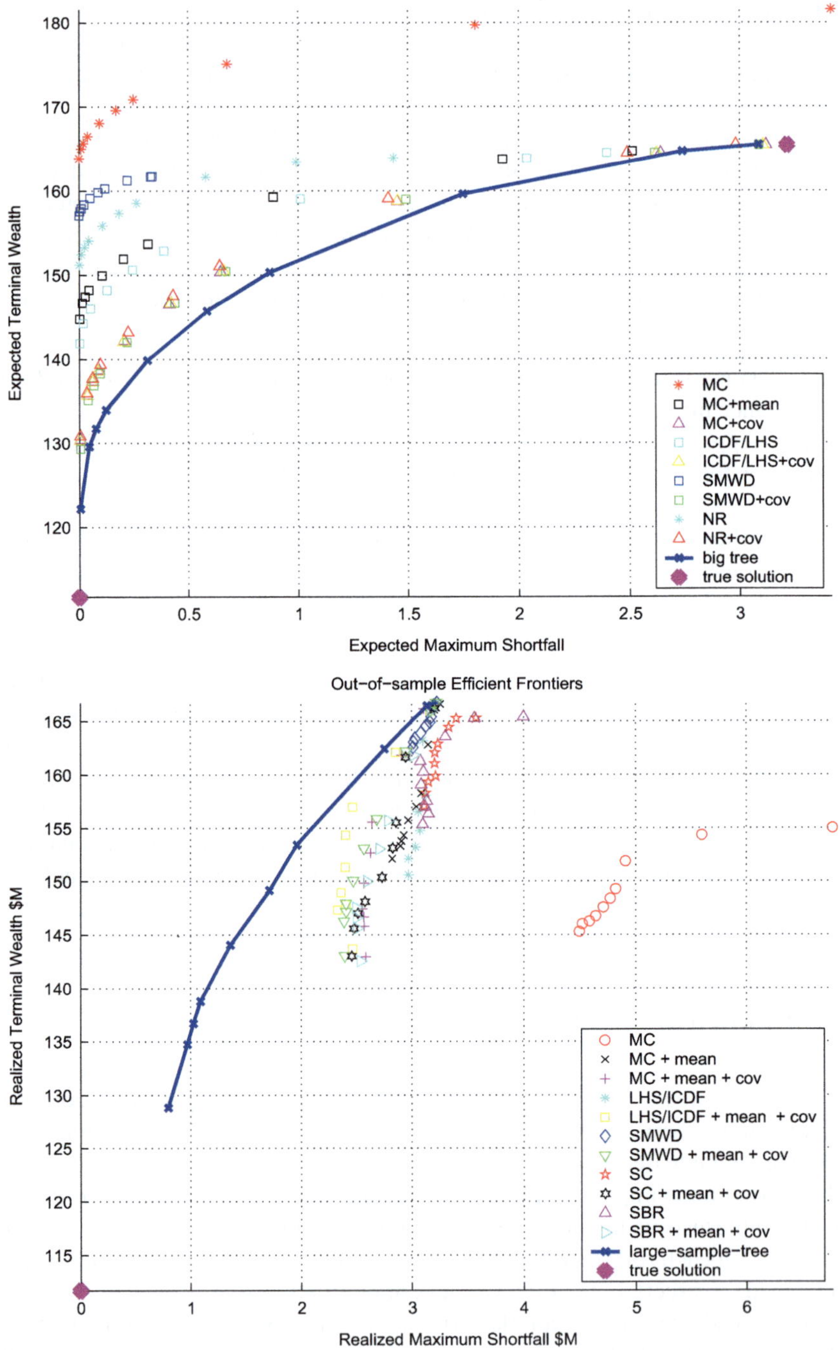

Fig. 8.2 In- and out-of-sample results for the original guaranteed fund 4-stage Pioneer problem [2]

Table 8.1 In-sample stability results with stabilization of root node decisions

β	Method	Terminal wealth		Expected maximum shortfall		Hitting probability	
0.0075	$\sigma_{\max}$ (original)	129.33	(1.60)	0.03	(0.01)	0.03	(0.01)
	$\sigma_{\max}$ (uncorrelated)	127.45	(1.31)	0.02	(0.01)	0.02	(0.01)
	Proportional	127.58	(1.35)	0.02	(0.01)	0.02	(0.01)
	Mean + cov	137.33	(5.80)	0.06	(0.03)	0.05	(0.02)
0.6	$\sigma_{\max}$ (original)	165.44	(0.04)	3.12	(0.53)	0.33	(0.07)
	$\sigma_{\max}$ (uncorrelated)	165.44	(0.04)	3.12	(0.53)	0.33	(0.07)
	Proportional	165.44	(0.04)	3.12	(0.53)	0.33	(0.07)
	Mean + cov	165.44	(0.04)	3.12	(0.53)	0.33	(0.07)

Table 8.2 In-sample portfolio allocation stability results for 4-stage barrier problem with stabilization of root node decision

		Total equities		Total bonds		Portfolio return		Portfolio volatility	
β	Method	Avg	Stdev	Avg	Stdev	Avg	Stdev	Avg	Stdev
0.0075	$\sigma_{\max}$ (original)	10.2	(2.1)	89.8	(2.1)	3.80	(0.15)	3.59	(0.39)
	$\sigma_{\max}$ (uncorrelated)	8.1	(0.5)	91.9	(0.5)	3.73	(0.07)	3.38	(0.20)
	Proportion	8.6	(1.3)	91.4	(1.3)	3.67	(0.09)	3.39	(0.16)
	Mean + cov	34.8	(22.5)	65.2	(22.5)	6.37	(2.26)	6.27	(2.74)
0.6	$\sigma_{\max}$ (original space)	100.0	(0.0)	0.0	(0.0)	13.70	(0.00)	14.80	(0.00)
	$\sigma_{\max}$ (uncorrelated)	100.0	(0.0)	0.0	(0.0)	13.70	(0.00)	14.80	(0.00)
	Proportion	100.0	(0.0)	0.0	(0.0)	13.70	(0.00)	14.80	(0.00)
	Mean + cov	100.0	(0.0)	0.0	(0.0)	13.70	(0.00)	14.80	(0.00)

sample terminal wealth than the other two position limits in the tight risk control region.

Table 8.2 shows the return and volatility of the recommended portfolios at the root node. A more detailed breakdown of the allocations is tabulated in Tables 8.3 and 8.4. Table 8.3 shows the effects of limits imposed on the assets in the original space for different values of β. With very tight risk control, only B1 of Table 8.3 and the ZCB are allowable assets in which to invest. The recommended portfolio at $\beta = 0.0075$ invests heavily (about 90 %) in bonds. As β increases, a higher proportion is invested in the more risky assets.

Comparing the portfolio recommendations with and without stabilization to mean and covariance matching only (mean+cov), shows the success of all stabilization schemes in reducing the in-sample over-fitting with a limited number of scenarios by employing a much more conservative portfolio and increasing the stability of the implementable decisions. The stabilization constraint does not affect optimization at high β due to the dominating wealth maximization term.

We now investigate the effect of the stabilization constraints on crucial *out-of-sample* performance. Figure 8.3 plots the realized efficient frontier, in which all stabilization methods under consideration successfully trace out the curve derived

Table 8.3 Asset position limits

	$\beta = 0.001$	$\beta = 0.075$	$\beta = 0.60$
S&P	0.58	3.75	79.24
EAFE	0.65	4.15	87.63
R2K	0.47	3.01	63.65
REIT	0.74	4.74	100.00
B1	100.00	100.00	100.00
B2	10.26	65.90	100.00
B3	5.06	32.46	100.00
B4	3.54	22.75	100.00
B5	2.83	18.14	100.00
B10	1.61	10.35	100.00
B30	1.03	6.59	100.00

from a 10,000 scenario problem in the low risk region. Comparing this to the results obtained using standard moment matching shown in Fig. 8.2[3] indicates that these methods lead to higher realized terminal wealth at lower risk. Comparing the different stabilization methods employed, we see that position limits derived in the uncorrelated return space and with the proportional approach yield better out-of-sample performance. The improved performance is due to the implicit portfolio volatility constraints that induce more conservative portfolios in the tight risk control region. These portfolios reduce the occurrence of out-of-sample portfolio losses resulting from exploitation of the spurious profit opportunities present in small-sample scenario trees.

The more conservative portfolios induced also result in a lower ZCB barrier hitting probability as shown in Table 8.5. Comparing the barrier hitting probability at $\beta = 0.0075$ for methods with and without stabilization constraints shows that this hitting probability is reduced from 33 % to about 13.4 %. Note however that the in-sample risk is still under-estimated (by about an order of magnitude) due to the low scenario branching factor remaining still somewhat prone to over-fitting. However, comparing this degree of under-estimation with that resulting from only second order matching, we have nevertheless successfully decreased this under-estimation by about a factor of 4 through reduction in the degrees of freedom of the initial decisions. The in-sample estimate of EMS for $\beta = 0.0075$ when using only covariance matching with random scenario sampling is 2.47, whereas the EMS realized with stabilized decisions in the correlated return space value is 0.06. By contrast, the in-sample estimate of EMS for position limits in the uncorrelated space is 0.02 and its out-of-sample value is 0.11 (see Tables 8.1 and 8.5). The out-of-sample efficient frontier in Fig. 8.3 is much closer to that of the large sample scenario tree than is that of the original problem presented in [2].

[3]See [2] for an explanation of the alternative methods reported in the figure including simple Monte Carlo (MC) and MC with first and second moment matching (MC+mean, MC+mean+cov).

Table 8.4 In-sample portfolio allocation stability results with stabilization of decisions

β	Asset	Position limits		Position limits (uncorrelated)		Proportion limits	
0.0075	S&P	3.4	(0.7)	0.2	(0.4)	2.4	(0.3)
	EAFE	0.7	(1.2)	0.1	(0.2)	0.6	(0.8)
	R2K	1.4	(1.2)	0.0	(0.0)	1.1	(0.7)
	REIT	4.7	(0.1)	7.8	(0.9)	4.5	(0.1)
	B1	0.0	(0.0)	0.0	(0.0)	0.0	(0.1)
	B2	0.8	(3.2)	0.0	(0.3)	1.2	(3.4)
	B3	17.2	(13.9)	3.0	(7.4)	7.2	(4.6)
	B4	21.9	(2.8)	5.9	(7.2)	8.3	(0.0)
	B5	14.5	(5.9)	3.6	(4.9)	6.9	(0.1)
	B10	5.9	(4.5)	0.4	(1.2)	3.6	(1.2)
	B30	4.2	(2.8)	2.3	(2.4)	2.0	(1.0)
	ZCB	25.2	(14.2)	76.7	(7.5)	62.0	(6.1)
	Total equities	10.2	(2.1)	8.1	(0.5)	8.6	(1.3)
	Total bonds	89.8	(2.1)	91.9	(0.5)	91.4	(1.3)
	Portfolio returns	3.8	(0.2)	3.7	(0.1)	3.7	(0.1)
	Portfolio volatility	3.6	(0.4)	3.4	(0.2)	3.4	(0.2)
0.6	S&P	0.0	(0.0)	0.0	(0.0)	0.0	(0.0)
	EAFE	0.0	(0.0)	0.0	(0.0)	0.0	(0.0)
	R2K	0.0	(0.0)	0.0	(0.0)	0.0	(0.0)
	REIT	100.0	(0.0)	100.0	(0.0)	100.0	(0.0)
	B1	0.0	(0.0)	0.0	(0.0)	0.0	(0.0)
	B2	0.0	(0.0)	0.0	(0.0)	0.0	(0.0)
	B3	0.0	(0.0)	0.0	(0.0)	0.0	(0.0)
	B4	0.0	(0.0)	0.0	(0.0)	0.0	(0.0)
	B5	0.0	(0.0)	0.0	(0.0)	0.0	(0.0)
	B10	0.0	(0.0)	0.0	(0.0)	0.0	(0.0)
	B30	0.0	(0.0)	0.0	(0.0)	0.0	(0.0)
	ZCB	0.0	(0.0)	0.0	(0.0)	0.0	(0.0)
	Total equities	100.0	(0.0)	100.0	(0.0)	100.0	(0.0)
	Total bonds	0.0	(0.0)	0.0	(0.0)	0.0	(0.0)
	Portfolio returns	13.7	(0.0)	13.7	(0.0)	13.7	(0.0)
	Portfolio volatility	14.8	(0.0)	14.8	(0.0)	14.8	(0.0)

Among the stabilization approaches implemented, the general method given by position limits in the uncorrelated return space yields the best results although the volatility proportional constraints give very similar results. Adding these constraints does not add much complexity to the model and they are therefore an effective method to reduce bias in root node implementable decisions. The much lower computational costs of employing these stabilization schemes with small-sample scenario trees makes them effective practical methods to suppress sampling error.

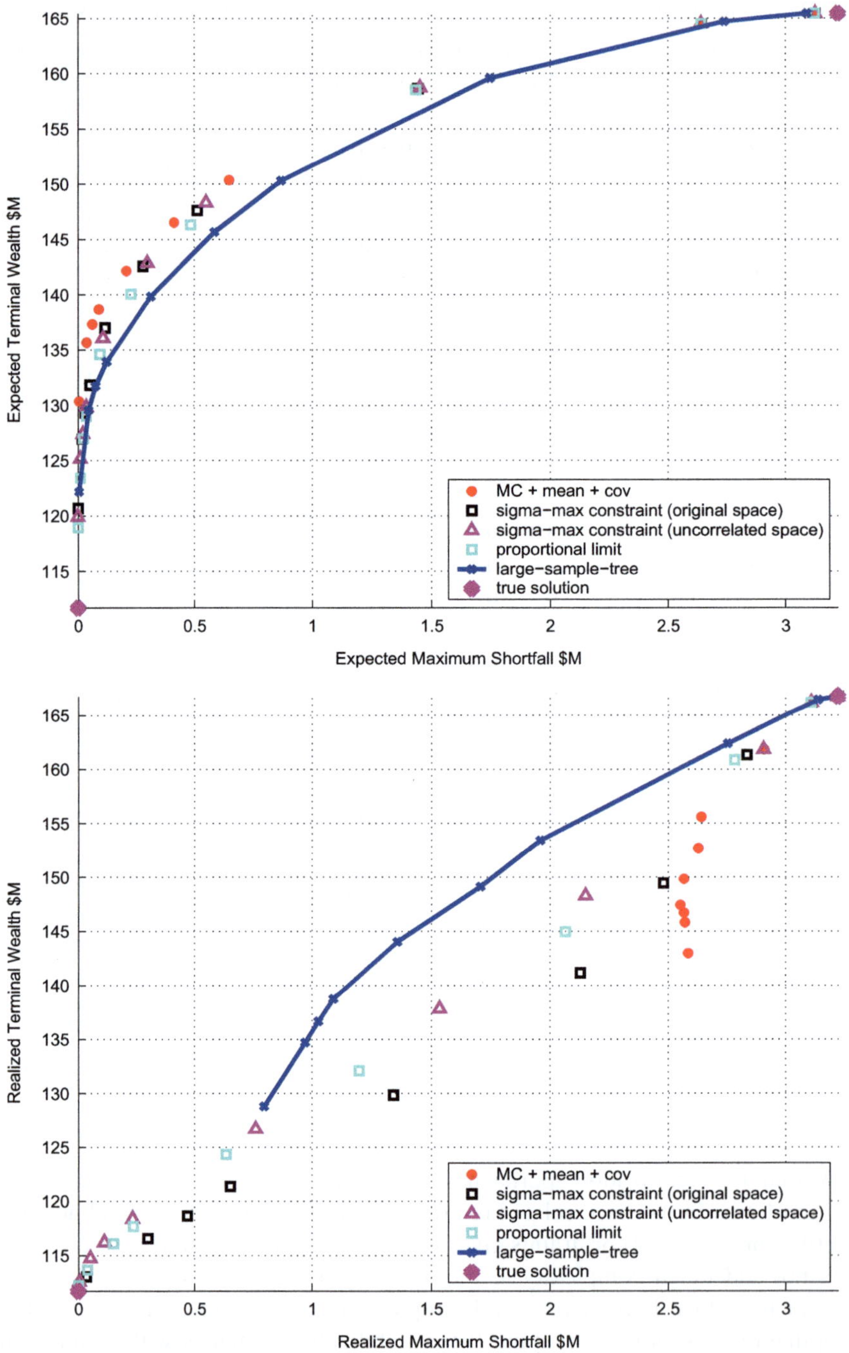

Fig. 8.3 In- and out-sample results for 4-stage Pioneer guaranteed fund problem with stabilization of decisions

Table 8.5 Out-of-sample results for the stabilization method

β	Method	Terminal wealth		Expected maximum shortfall		Hitting probability
0.0075	$\sigma_{\max}$ (original space)	118.63	(6.06)	0.47	(1.15)	0.24
	$\sigma_{\max}$ (uncorrelated space)	116.23	(3.03)	0.11	(0.43)	0.1340
	Proportion limit	116.62	(3.70)	0.20	(0.65)	0.1700
	Mean+cov	146.69	(36.75)	2.57	(4.93)	0.39
0.6	$\sigma_{\max}$ (original space)	166.75	(45.95)	3.22	(6.85)	0.3140
	$\sigma_{\max}$ (uncorrelated space)	166.75	(45.95)	3.22	(6.85)	0.3140
	Proportion limit	166.75	(45.95)	3.22	(6.85)	0.3140
	Mean+cov	166.75	(45.95)	3.22	(6.85)	0.31

8.5 Conclusions and Future Directions

We have examined alternative methods to address the under-estimation of portfolio risk in small scenario sample dynamic stochastic programming financial models. We investigated methods to suppress discretization error and implementable decision sampling variability by reducing the degrees of freedom of the initial decision space. This may be carried out by limiting the implementable decisions to be a linear combination of a properly chosen basis, thus forcing the decisions to lie within an appropriate subspace. We explore how these bases may be optimally chosen in the context of under-estimation of maximum shortfall in the Pioneer guaranteed return fund model. An approximation to a theoretical calculation of one-period maximum shortfall is used to add an indirect constraint on the portfolio volatility through the use of position limits. Different schemes to determine the appropriate limits have been investigated, and it was found using limits in the uncorrelated return space yields the best out-of-sample performance. Discretization error is reduced and the stability of the root node decision is improved, especially for tight risk control.

The area of stabilization of implementable decisions is a new area in dynamic stochastic programming, where to our knowledge limited research has so far been carried out. Though the stability of objective function values is important, for financial applications of dynamic stochastic programming the stability of the implementable decisions is essential to the successful application of these models in the real world, since these are the recommendations actually carried out. Discretization bias in the implementable decisions may thus erode the full potential of incorporating stochasticity into dynamic decision modelling in the face of risk.

The results from our first investigation of limiting the degrees of freedom of the implementable decisions space in an effort to reduce discretization error and stabilize these decisions are promising. They indicate that this is a fruitful area for further research. An obvious extension is to limit the degrees of freedom using the methods described in this paper for *all* prospective forward decisions in financial applications of dynamic stochastic programming models. More extensive investigation should also be carried out to evaluate the effectiveness of these methods in improving in-sample risk estimation with more general asset return distributions, especially for risk measures which are highly dependent on the lower tail of portfolio return distributions such as Value at Risk and Expected Shortfall.

Acknowledgements We would like to thank an anonymous referee and the Editor whose comments materially improved the paper.

Appendix: Pioneer Guaranteed Return Fund Model Formulation [2]

The Pioneer guaranteed return fund model [1] is a portfolio optimization of a *closed-end* fund with a *nominal* return guarantee of G per annum. At each period the model aims to maximize the performance of the fund taking into account the risk associated with falling short of the guaranteed value level. The formulation presented here follows closely [1] except that we only impose transactions costs on the change in portfolio holdings. Selling the *off-the-run* bonds and replacing them with *on-the-run* bonds are assumed not to incur any transactions cost. The model parameter and variable definitions are given in Table 8.6.

Objective

$$\max \quad \sum_{w \in \Omega} p(\omega) \left(\beta \sum_{t \in T^d} \hat{W}_t(\omega) - (1-\beta) H(\omega) \right) \tag{8.30}$$

Cash Balance Constraints

- Running cash balance constraints

$$\sum_{a \in A} P_{t,a}^{(buy)}(w) q_{t,a}(w) + \sum_{a \in A} \left(\tau_a P_{t,a}^{(buy)}(w) s_{t,a}^{+}(w) + \tau_a P_{t,a}^{(sell)}(w) s_{t,a}^{-}(w) \right)$$

$$= \sum_{a \in A} \left(P_{t,a}^{(sell)}(w) q_{t-1,a}(w) + D_{t,a}(w) q_{t-1,a}(w) \right) \tag{8.31}$$

$$\forall t \in T^d \backslash \{1\} \quad \forall w \in \Omega$$

Table 8.6 Model parameters and variables

Sets definition	
$T^d = \{1, \dots, T+1\}$	Set of decision/simulation times
A	Set of all assets
Ω	Set of scenarios
Parameter definitions	
β	Risk aversion attitude ($\beta = 1$ corresponding to risk loving)
Q_a	Initial asset holdings (in units) of asset $a \in A$
c_1	Initial cash amount
Stochastic parameter definitions	
$P_{t,a}^{(buy)}(w)/P_{t,a}^{(sell)}$	Buy/sell price of asset $a \in A$ at time t in scenario w
$D_{t,a}(w)$	Annual coupon of bond $a \in A$ paid (in arrears) at time t in scenario w
$Z_t(w)$	Zero coupon bond price at time t in scenario w
$L_t(w) = W_1(1+G)^T Z_t(w)$	Barrier level at time t of guarantee G per annum in scenario w
$p(w)$	Probability of scenario w
Decision variable definitions	
$q_{t,a}(w)$	Quantity held in asset $a \in A$ over period $[t, t+1)$ in scenario w
$q_{t,a}^{+}(w)/q_{t,a}^{-}(w)$	Quantity bought/sold in asset $a \in A$ at time t in scenario w
$s_{t,a}^{+}(w)/s_{t,a}^{-}(w)$	Increment/decrement (in units) of asset $a \in A$ at time t in scenario w
$W_t(w)$	Financial wealth *before* portfolio rebalancing at time t in scenario w
$\hat{W}_t(w)$	Financial wealth *after* portfolio rebalancing at time t in scenario w
$h_t(w) = \max(0, \hat{W}_t(w) - L_t(w))$	Shortfall at time t in scenario w
$H(w)$	Maximum shortfall in scenario w

- Initial cash balance constraints

$$\sum_{a \in A} P_{1,a}^{(buy)}(w) q_{1,a}(w) + \sum_{A} \left(\tau_a P_{1,a}^{(buy)}(w) s_{1,a}^{+}(w) + \tau_a P_{1,a}^{(sell)}(w) s_{1,a}^{-}(w) \right)$$

$$= c_1 + \sum_{A} \left(P_{1,a}^{(sell)}(w) Q_a + D_{1,a}(w) Q_a \right) \tag{8.32}$$

Quantity Balance Constraints

- Running quantity balance constraints

$$q_{t,a}(w) = q_{t-1,a}(w) + q^{+}_{t,a}(w) - q^{-}_{t,a}(w) \tag{8.33}$$

$$\forall t \in T^{d}\backslash\{1\} \quad \forall a \in A \quad \forall w \in \Omega$$

- Initial quantity balance constraint

$$q_{1,a}(w) = Q_a + q^{+}_{1,a}(w) - q^{-}_{1,a}(w) \quad \forall a \in A \quad \forall w \in \Omega \tag{8.34}$$

Annual Bond Roll-Over Constraints The *off-the-run* bonds are sold and the new *on-the-run* bonds are bought. Note that we do not incur transaction costs on buying and selling resulting from annual rolling. Transaction costs are only incurred on changes in asset holdings.

$$\begin{aligned} q^{-}_{t,a}(w) &= q_{t-1,a}(w) \quad \forall t \in T^{d}\backslash\{1\} \quad \forall a \in A \quad \forall w \in \Omega \\ q^{-}_{1,a}(w) &= Q_a \qquad\qquad \forall a \in A \quad \forall w \in \Omega \end{aligned} \tag{8.35}$$

This constraint implies that

$$q_{t,a}(w) = q^{+}_{t,a}(w) \quad \forall t \in T^{d} \quad \forall a \in A \quad \forall w \in \Omega \tag{8.36}$$

Liquidation Constraints The financial portfolio is liquidated in cash at the final horizon for at least the guarantees to be paid to the clients.

$$q_{T,a}(w) = 0 \qquad \forall a \in A \quad \forall w \in \Omega \tag{8.37}$$

This equation implies that

$$\begin{aligned} s^{+}_{T,a}(w) &= 0 \qquad\qquad \forall a \in A \quad \forall w \in \Omega \\ s^{-}_{T,a}(w) &= q_{T-1,a}(w) \quad \forall a \in A \quad \forall w \in \Omega \end{aligned} \tag{8.38}$$

Wealth Accounting Constraints

- Wealth before rebalancing

$$W_t(w) = \sum_{A} \left(P^{(sell)}_{t,a}(w) q_{t-1,a}(w) + D_{t,a}(w) q_{t-1,a}(w) \right) \tag{8.39}$$

$$\forall t \in T^{d}\backslash\{1\} \quad \forall w \in \Omega$$

$$W_1(w) = \sum_{A} \left(P^{(sell)}_{1,a}(w) Q_a + D_{1,a}(w) Q_a \right) + c_1 \tag{8.40}$$

- Wealth after rebalancing

$$\hat{W}_t(w) = \sum_A P^{buy}_{t,a}(w) q_{t,a}(w) \quad \forall t \in T^d \backslash \{T+1\} \quad \forall w \in \Omega \tag{8.41}$$

$$\hat{W}_T(w) = \sum_A \left(P^{(sell)}_{T,a}(w)(q_{T-1,a}(w) - \tau_a s^-_{T,a}(w)) + D_{T,a}(w) q_{T-1,a}(w) \right)$$

$$\forall w \in \Omega \tag{8.42}$$

Portfolio Change Constraints We calculate the portfolio change (in units) through the following constraints:

- Decrement in asset position

$$q^+_{t,a}(w) - q^-_{t,a}(w) + s^-_{t,a}(w) \geq 0 \quad \forall t \in T^d \quad \forall a \in A \quad \forall w \in \Omega \tag{8.43}$$

- Increment in asset position

$$q^-_{t,a}(w) - q^+_{t,a}(w) + s^+_{t,a}(w) \geq 0 \quad \forall t \in T^d \quad \forall a \in A \quad \forall w \in \Omega \tag{8.44}$$

Barrier Constraints We use the wealth *after* rebalance to evaluate whether it is above the barrier. The wealth after rebalancing is used because in the real world where the product is sold to the client, the fund manager will need to liquidate the financial portfolio in cash to pay the clients at least the amount they are guaranteed. Taking transaction costs into consideration, this will drive the portfolio strategies to be more conservative.

- Shortfall constraint

$$h_t(w) + \hat{W}_t(w) \geq L_t(w) \quad \forall t \in T^d \quad \forall w \in \Omega \tag{8.45}$$

- Maximum shortfall constraint

$$H(w) \geq h_t(w) \quad \forall t \in T^d \quad \forall w \in \Omega \tag{8.46}$$

Non-anticipativity Constraints The non-anticipativity of the decision variables is implicit once we represent the stochastic processes using the scenario tree format. Therefore, no additional constraints are required.

References

1. M.A.H. Dempster, M. Germano, E.A. Medova, M.I. Rietbergen, F. Sandrini, M. Scrowston, Managing guarantees. J. Portf. Manag. **32**(2), 51–61 (2006)

2. M.A.H. Dempster, E.A. Medova, Y.S. Yong, A comparative study of sampling methods for stochastic programming, in *Stochastic Optimization Methods in Finance and Energy*. International series in operations research & management science, **163** (Springer, New York) 389–425 (2011)
3. N. Gulpinar, B. Rustem, R. Settergren, Simulation and optimization approaches to scenario tree generation. J. Econ. Dyn. Control **28**(7), 1291–1315 (2004)
4. H. Heitsch, W. Romisch, Scenario reduction algorithms in stochastic programming. Comput. Optim. Appl. **24**, 187–206 (2003)
5. H. Heitsch, W. Romisch, Generation of multivariate scenario trees to model stochasticity in power management, in *Proceedings of IEEE St. Petersburg Power Tech* (2005)
6. R. Hochreiter, G. Pflug, Financial scenario generation for stochastic multi-stage decision processes as facility location problems. AURORA Technical Report, European Centre for Parallel Computing at Vienna, 2002
7. K. Hoyland, S.W. Wallace, Generating scenario trees for multi-stage decision problems. Manag. Sci. **47**, 295–307 (2001)
8. K. Hoyland, M. Kaut, S.W. Wallace, A heuristic for moment-matching scenario generation. Comput. Optim. Appl. **47**, 295–307 (2003)
9. M. Kaut, S. Wallace, Evaluation of scenario-generation methods for stochastic programming. Stochastic Programming E-Print Series 14, 2003. http://www.speps.org
10. G.C. Pflug, Scenario tree generation for multi-period financial optimization by optimal discretization. Math. Program. **89**, 251–271 (2001)
11. W. Romisch, Stability of stochastic programming problems, in *Stochastic Programming*. Handbooks in Operations Research and Management Science, vol. 10 (Elsevier, Amsterdam, 2003)

Chapter 9
The Growth Optimal Investment Strategy Is Secure, Too

László Györfi, György Ottucsák, and Harro Walk

Abstract This paper is a revisit of discrete time, multi period and sequential investment strategies for financial markets showing that the log-optimal strategies are secure, too. Using exponential inequality of large deviation type, the rate of convergence of the average growth rate is bounded both for memoryless and for Markov market processes. A kind of security indicator of an investment strategy can be the market time achieving a target wealth. It is shown that the log-optimal principle is optimal in this respect.

Keywords Portfolio selection • Growth rate • Log-optimality • Sequential investment strategies

9.1 Introduction

This paper gives some additional features of the investment strategies in financial stock markets inspired by the results of information theory, non-parametric statistics and machine learning. Investment strategies are allowed to use information collected from the past of the market and determine, at the beginning of a trading period, a portfolio, that is, a way to distribute their current capital among the available assets. The goal of the investor is to maximize his wealth in the long run without knowing the underlying distribution generating the stock prices. Under this assumption the asymptotic rate of growth has a well-defined maximum which can be achieved in full knowledge of the underlying distribution generated by the stock prices.

L. Györfi (✉) • G. Ottucsák
Department of Computer Science and Information Theory, Budapest University of Technology and Economics, Stoczek u.2, 1521 Budapest, Hungary
e-mail: gyorfi@cs.bme.hu; oti@cs.bme.hu

H. Walk
Department of Mathematics, University of Stuttgart, Pfaffenwaldring 57, 70569 Stuttgart, Germany
e-mail: walk@mathematik.uni-stuttgart.de

© Springer International Publishing Switzerland 2017

G. Consigli et al. (eds.), *Optimal Financial Decision Making under Uncertainty*, International Series in Operations Research & Management Science 245,
DOI 10.1007/978-3-319-41613-7_9

In Sect. 9.2, under memoryless assumption on the underlying process generating the asset prices, the log-optimal portfolio achieves the maximal asymptotic average growth rate, that is, the expected value of the logarithm of the return for the best fix portfolio vector. Using exponential inequality of large deviation type, the rate of convergence of the average growth rate to the optimum growth rate is bounded. Consider a security indicator of an investment strategy, which is the market time achieving a target wealth. The log-optimal principle is optimal in this respect, too.

In Sect. 9.3, for generalized dynamic portfolio selection, when asset prices are generated by a stationary and ergodic process, there are universally consistent (empirical) methods that achieve the maximal possible growth rate. If the market process is a first order Markov process, then the rate of convergence of the average growth rate is obtained more generally.

Consider a market consisting of d assets. The evolution of the market in time is represented by a sequence of price vectors $\mathbf{S}_1, \mathbf{S}_2, \ldots \in \mathbb{R}_+^d$, where

$$\mathbf{S}_n = (S_n^{(1)}, \ldots, S_n^{(d)})$$

such that the jth component $S_n^{(j)}$ of $\mathbf{S}_n$ denotes the price of the jth asset on the nth trading period.

Let us transform the sequence of price vectors $\{\mathbf{S}_n\}$ into the sequence of return (relative price) vectors $\{\mathbf{X}_n\}$ as follows:

$$\mathbf{X}_n = (X_n^{(1)}, \ldots, X_n^{(d)})$$

such that

$$X_n^{(j)} = \frac{S_n^{(j)}}{S_{n-1}^{(j)}}.$$

Thus, the jth component $X_n^{(j)}$ of the return vector $\mathbf{X}_n$ denotes the amount obtained after investing a unit capital in the jth asset on the nth trading period.

9.2 Constantly Rebalanced Portfolio Selection

The dynamic portfolio selection is a multi-period investment strategy, where at the beginning of each trading period the investor rearranges the wealth among the assets. A representative example of the dynamic portfolio selection is the constantly rebalanced portfolio (CRP). The investor is allowed to diversify his capital at the beginning of each trading period according to a portfolio vector $\mathbf{b} = (b^{(1)}, \ldots b^{(d)})$. The jth component $b^{(j)}$ of $\mathbf{b}$ denotes the proportion of the investor's capital invested in asset j. Throughout the paper it is assumed that the portfolio vector $\mathbf{b}$ has nonnegative components with $\sum_{j=1}^d b^{(j)} = 1$. The fact that $\sum_{j=1}^d b^{(j)} = 1$ means that the investment strategy is self financing and consumption of capital is excluded. The non-negativity of the components of $\mathbf{b}$ means that short selling and buying stocks on margin are not permitted. The simplex of possible portfolio vectors is denoted by Δ_d.

Let S_0 denote the investor's initial capital. Then at the beginning of the first trading period $S_0 b^{(j)}$ is invested into asset j, and it results in return $S_0 b^{(j)} x_1^{(j)}$, therefore at the end of the first trading period the investor's wealth becomes

$$S_1 = S_0 \sum_{j=1}^{d} b^{(j)} X_1^{(j)} = S_0 \langle \mathbf{b}\, , \, \mathbf{X}_1 \rangle \, ,$$

where $\langle \cdot\, , \, \cdot \rangle$ denotes inner product. For the second trading period, S_1 is the new initial capital

$$S_2 = S_1 \cdot \langle \mathbf{b}\, , \, \mathbf{X}_2 \rangle = S_0 \cdot \langle \mathbf{b}\, , \, \mathbf{X}_1 \rangle \cdot \langle \mathbf{b}\, , \, \mathbf{X}_2 \rangle \, .$$

By induction, for the trading period n the initial capital is S_{n-1}, therefore

$$S_n = S_{n-1} \langle \mathbf{b}\, , \, \mathbf{X}_n \rangle = S_0 \prod_{i=1}^{n} \langle \mathbf{b}\, , \, \mathbf{X}_i \rangle \, .$$

The asymptotic average growth rate of this portfolio selection is

$$\begin{aligned} \lim_{n\to\infty} \frac{1}{n} \ln S_n &= \lim_{n\to\infty} \left(\frac{1}{n} \ln S_0 + \frac{1}{n} \sum_{i=1}^{n} \ln \langle \mathbf{b}\, , \, \mathbf{X}_i \rangle \right) \\ &= \lim_{n\to\infty} \frac{1}{n} \sum_{i=1}^{n} \ln \langle \mathbf{b}\, , \, \mathbf{X}_i \rangle \, , \end{aligned}$$

therefore without loss of generality one can assume in the sequel that the initial capital $S_0 = 1$.

If the market process $\{\mathbf{X}_i\}$ is memoryless, i.e., it is a sequence of independent and identically distributed (i.i.d.) random return vectors then we show that the best constantly rebalanced portfolio (BCRP) is the log-optimal portfolio:

$$\mathbf{b}^* := \arg\max_{\mathbf{b}\in\Delta_d} \mathbb{E}\{\ln \langle \mathbf{b}\, , \, \mathbf{X}_1 \rangle\}.$$

This optimality was formulated as follows:

Proposition 1 (Kelly [30], Latané [32], Breiman [11], Finkelstein and Whitley [19], Barron and Cover [8]). *If $S_n^* = S_n(\mathbf{b}^*)$ denotes the capital after day n achieved by a log-optimal portfolio strategy $\mathbf{b}^*$, then for any portfolio strategy $\mathbf{b}$ with finite $\mathbb{E}\{\ln \langle \mathbf{b}\, , \, \mathbf{X}_1 \rangle\}$ and with capital $S_n = S_n(\mathbf{b})$ and for any memoryless market process $\{\mathbf{X}_n\}_{-\infty}^{\infty}$,*

$$\lim_{n\to\infty} \frac{1}{n} \ln S_n \le \lim_{n\to\infty} \frac{1}{n} \ln S_n^* \quad \textit{almost surely (a.s.)} \tag{9.1}$$

and maximal asymptotic average growth rate is

$$\lim_{n\to\infty} \frac{1}{n} \ln S_n^* = W^* := \mathbb{E}\{\ln \langle \mathbf{b}^* , \mathbf{X}_1 \rangle\} \quad a.s.$$

Proof. This optimality is a simple consequence of the strong law of large numbers. Introduce the notation

$$W(\mathbf{b}) = \mathbb{E}\{\ln \langle \mathbf{b} , \mathbf{X}_1 \rangle\}.$$

Then

$$\begin{aligned}
\frac{1}{n} \ln S_n &= \frac{1}{n} \sum_{i=1}^{n} \ln \langle \mathbf{b} , \mathbf{X}_i \rangle \\
&= \frac{1}{n} \sum_{i=1}^{n} \mathbb{E}\{\ln \langle \mathbf{b} , \mathbf{X}_i \rangle\} + \frac{1}{n} \sum_{i=1}^{n} (\ln \langle \mathbf{b} , \mathbf{X}_i \rangle - \mathbb{E}\{\ln \langle \mathbf{b} , \mathbf{X}_i \rangle\}) \\
&= W(\mathbf{b}) + \frac{1}{n} \sum_{i=1}^{n} (\ln \langle \mathbf{b} , \mathbf{X}_i \rangle - \mathbb{E}\{\ln \langle \mathbf{b} , \mathbf{X}_i \rangle\}) .
\end{aligned}$$

Kolmogorov's strong law of large numbers implies that

$$\frac{1}{n} \sum_{i=1}^{n} (\ln \langle \mathbf{b} , \mathbf{X}_i \rangle - \mathbb{E}\{\ln \langle \mathbf{b} , \mathbf{X}_i \rangle\}) \to 0 \quad \text{a.s.,}$$

therefore

$$\lim_{n\to\infty} \frac{1}{n} \ln S_n = W(\mathbf{b}) = \mathbb{E}\{\ln \langle \mathbf{b} , \mathbf{X}_1 \rangle\} \quad \text{a.s.}$$

Similarly,

$$\lim_{n\to\infty} \frac{1}{n} \ln S_n^* = W^* := W(\mathbf{b}^*) = \max_{\mathbf{b}} W(\mathbf{b}) \quad \text{a.s.}$$

■

In [31] the log-optimal portfolio selection was studied for a continuous time model, where the main question of interest is the choice of sampling frequency such that the rebalancing is done at sampling time instances. They assumed that the assets' prices are cross-correlated geometric motions and therefore the return vectors of sampled price processes are memoryless. For high sampling frequency, the log-optimal strategy is a special case of mean-variance rule, called semi-log-optimal strategy (cf. [23, 36, 37]).

There is an obvious question here: how secure a growth optimal portfolio strategy is? The strong law of large numbers has another interpretation. Put

$$R_n := \inf_{n \le m} \frac{1}{m} \ln S^*_m,$$

then e^{nR_n} is a lower exponential envelope for S^*_n, i.e.,

$$e^{nR_n} \le S^*_n.$$

Moreover,

$$R_n \uparrow W^* \quad \text{a.s.},$$

which means that for an arbitrary $R < W^*$, we have that

$$e^{nR} \le S^*_n$$

for all n after a random time N large enough.

In the sequel we bound N, i.e., derive a rate of convergence of the strong law of large numbers. Assume that there exist $0 < a_1 < 1 < a_2 < \infty$ such that

$$a_1 \le X^{(j)} \le a_2 \tag{9.2}$$

for all $j = 1, \dots, d$. For the New York Stock Exchange (NYSE) daily data, this condition is satisfied with $a_1 = 0.7$ and with $a_2 = 1.2$. $a_1 = 0.7$ means that the worst that happened in a single day was 30 % drop, while $a_2 = 1.2$ corresponds to 20 % increase within a day. (cf. [18, 28].) Figure 9.1 shows the histogram of Coca Cola's daily logarithmic relative prices such that most of the days the relative prices are in the interval [0.95, 1.05] from 1962 to 2006. Here are some statistical data:

$$\begin{aligned}
\text{minimum} &= -0.2836\\
\text{1st qu.} &= -0.0074\\
\text{median} &= 0.0000\\
\text{mean} &= 0.00053\\
\text{3rd qu.} &= 0.0083\\
\text{maximum} &= 0.1796.
\end{aligned}$$

Theorem 1. *If the market process* $\{\mathbf{X}_i\}$ *is memoryless and the condition (9.2) is satisfied, then for an arbitrary* $R < W^*$*, we have that*

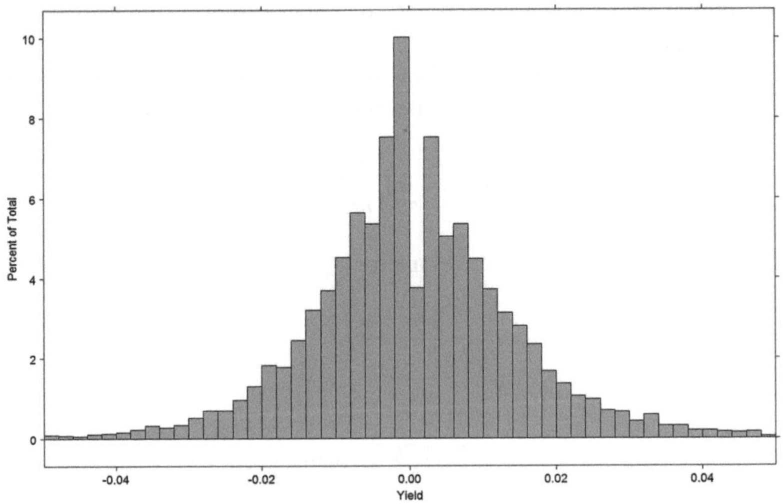

Fig. 9.1 The histogram of log-returns for Coca Cola

$$\mathbb{P}\left\{e^{nR} > S_n^*\right\} \le e^{-2n\frac{(W^*-R)^2}{(\ln a_2 - \ln a_1)^2}}.$$

Proof. We have that

$$\begin{aligned}\mathbb{P}\left\{e^{nR} > S_n^*\right\} &= \mathbb{P}\left\{R > \frac{1}{n}\ln S_n^*\right\} \\ &= \mathbb{P}\left\{R - W^* > \frac{1}{n}\sum_{i=1}^{n}\left(\ln\langle \mathbf{b}^*, \mathbf{X}_i\rangle - \mathbb{E}\{\ln\langle \mathbf{b}^*, \mathbf{X}_i\rangle\}\right)\right\}.\end{aligned}$$

Apply the Hoeffding [27] inequality: Let $X_1, \ldots, X_n$ be independent random variables with $X_i \in [c, c+K]$ with probability one. Then, for all $\epsilon > 0$,

$$\mathbb{P}\left\{\frac{1}{n}\sum_{i=1}^{n}(X_i - \mathbb{E}\{X_i\}) < -\epsilon\right\} \le e^{-2n\frac{\epsilon^2}{K^2}}.$$

Because of the condition,

$$\ln a_1 \le \ln\langle \mathbf{b}^*, \mathbf{X}_i\rangle \le \ln a_2,$$

therefore the theorem follows from the Hoeffding inequality for the correspondences

$$\epsilon = W^* - R$$

and

$$X_i = \ln \langle \mathbf{b}^* , \mathbf{X}_i \rangle$$

and

$$K = \ln a_2 - \ln a_1 .$$

■

Using Theorem 1, we can bound the probability that after n there is a time instant m such that $e^{mR} > S_m^*$:

Corollary 1. *If the market process $\{\mathbf{X}_i\}$ is memoryless and the condition (9.2) is satisfied, then for an arbitrary $R < W^*$, we have that*

$$\mathbb{P}\left\{\cup_{m=n}^{\infty}\{e^{mR} > S_m^*\}\right\} \le e^{-2n\frac{(W^*-R)^2}{K^2}} \frac{e^{2\frac{(W^*-R)^2}{K^2}}}{e^{2\frac{(W^*-R)^2}{K^2}} - 1}. \tag{9.3}$$

Proof. From Theorem 1 we get that

$$\begin{aligned}\mathbb{P}\left\{\cup_{m=n}^{\infty}\{e^{mR} > S_m^*\}\right\} &\le \sum_{m=n}^{\infty} \mathbb{P}\left\{e^{mR} > S_m^*\right\} \\ &\le \sum_{m=n}^{\infty} e^{-2m\frac{(W^*-R)^2}{(\ln a_2 - \ln a_1)^2}} \\ &= e^{-2n\frac{(W^*-R)^2}{(\ln a_2 - \ln a_1)^2}} \frac{1}{1 - e^{-2\frac{(W^*-R)^2}{(\ln a_2 - \ln a_1)^2}}}.\end{aligned}$$

■

Theorem 1 and Corollary 1 are about the probability of underperformance depending on a_1 and a_2. Using central limit theorem (CLT), one can derive modifications of Theorem 1 and Corollary 1. The advantage of the CLT is that the resulted formula does not depend on a_1 and a_2, it depends only of the variance of the log-returns. However, in contrast to large deviation bounds, the CLT is only an approximation.

An additional hard open problem is how to construct empirical strategies taking into account proportional transaction cost (see, for example, [20, 21]).

When it comes to security, the small-sample behavior should be more interesting. Consider the relative amount of times j between 1 and n, for which S_j^* is below e^{jR} for $R < W^*$ near to W^*, say $R = R_n = W^* - \frac{m}{\sqrt{n}}\sigma$ for fixed $m > 0$ with $\sigma^2 = \mathbf{Var}(\ln \langle \mathbf{b}^* , \mathbf{X}_1 \rangle)$ assumed to be positive and finite. For $0 \le x \le 1$ we have

$$\mathbb{P}\left\{\frac{1}{n}\sum_{j=1}^{n}\mathbb{I}_{\{S_j^*<e^{jR}\}}\leq x\right\}$$

$$=\mathbb{P}\left\{\frac{1}{n}\sum_{j=1}^{n}\mathbb{I}_{\{\frac{1}{j}\sum_{i=1}^{j}(\ln\langle\mathbf{b}^*,\mathbf{X}_i\rangle-\mathbb{E}\{\ln\langle\mathbf{b}^*,\mathbf{X}_i\rangle\})<R-W^*\}}\leq x\right\}$$

$$=\mathbb{P}\left\{\frac{1}{n}\sum_{j=1}^{n}\mathbb{I}_{\{\frac{1}{\sqrt{n}\sigma}\sum_{i=1}^{j}(\ln\langle\mathbf{b}^*,\mathbf{X}_i\rangle-\mathbb{E}\{\ln\langle\mathbf{b}^*,\mathbf{X}_i\rangle\})+m\frac{j}{n}<0\}}\leq x\right\}$$

$$\to\mathbb{P}\left\{\int_0^1\mathbb{I}_{\{W(u)+mu\leq 0\}}du\leq x\right\}$$

with standard Brownian motion W, by Donsker's functional central limit theorem (see [9]) for the functional $f\to\int_0^1\mathbb{I}_{\{f(u)+mu\leq 0\}}du$.

By the generalized arc-sine law of Takács [41] the right hand side equals

$$F_m(x)$$

$$:=2\int_0^x\left[\frac{\varphi(m\sqrt{1-u})}{\sqrt{1-u}}+m\Phi(m\sqrt{1-u})\right]\left[\frac{\varphi\left(-m\sqrt{u}\right)}{\sqrt{u}}-m\Phi\left(-m\sqrt{u}\right)\right]du$$

for $0\leq x\leq 1$, where $F_m(1)=1$, and φ and Φ are the standard normal density and distribution functions, respectively. We have a non-degenerate limit distribution. Here for $m\to\infty$ and also for the case $R=R_n'$ with $(W^*-R)\sqrt{n}\to\infty$, especially a constant $R_n'<W^*$, we have degeneration to the Dirac distribution concentrated at 0. The proof of these assertions can be as follows: For each $0<\epsilon<1/2$, on $[\epsilon,1-\epsilon]$ the uniformly bounded integrand uniformly converges to 0 for $m\to\infty$, thus $F_m(1-\epsilon)-F(\epsilon)\to 0$. Further $F_m(0)=0$ and $F_m(1)=1$ for each m, and $F_m(x)$ is non-decreasing for each $0\leq x\leq 1$. Thus, $F_m(x)\to 1$ for each $0<x\leq 1$. Finally one notices that $m<\sqrt{n}(W^*-R_n')\to\infty$ $(n\to\infty)$ implies

$$\liminf_n\mathbb{P}\left\{\frac{1}{n}\sum_{j=1}^{n}\mathbb{I}_{\{\frac{1}{\sqrt{n}\sigma}\sum_{i=1}^{j}(\ln\langle\mathbf{b}^*,\mathbf{X}_i\rangle-\mathbb{E}\{\ln\langle\mathbf{b}^*,\mathbf{X}_i\rangle\})+\sqrt{n}(W^*-R_n')\frac{j}{n}<0\}}\leq x\right\}$$

$$\geq\lim_n\mathbb{P}\left\{\frac{1}{n}\sum_{j=1}^{n}\mathbb{I}_{\{\frac{1}{\sqrt{n}\sigma}\sum_{i=1}^{j}(\ln\langle\mathbf{b}^*,\mathbf{X}_i\rangle-\mathbb{E}\{\ln\langle\mathbf{b}^*,\mathbf{X}_i\rangle\})+m\frac{j}{n}<0\}}\leq x\right\},$$

for each m. It should be mentioned that under the assumption (9.2) the latter of the assertions is also a consequence of Theorem 1 for $R=R_n'$.

In the literature there is a discussion on good and bad properties of log-optimal investment (see [34], Sects. 30 and 39, with references). Beside

$$\limsup \frac{1}{n} \log(S_n/S_n^*) \leq 0$$

almost surely (see (9.1) and (9.4) below, good long-run performance) one has

$$\mathbb{E}\{S_n/S_n^*\} \leq 1$$

for all n (good short-term performance). Both properties were established by Algoet and Cover [3] in the much more general context of a stationary and ergodic process of daily returns $\mathbf{X}_n$ and conditionally log-optimal investment (here regarding past returns, but nothing more: myopic policy). Leaving the concept of a logarithmic utility function induced by the multiplicative structure of investment, Samuelson [38] in his critics pointed out that maximizing the expected return $\mathbb{E}\{\langle \mathbf{b}, \mathbf{X}_i \rangle\}$ instead of expected logarithmic return, with in this sense optimal portfolio choice $\mathbf{b}^{**}$ and corresponding wealth S_n^{**}, leads to $\mathbb{E}\{S_n^{**}\}/\mathbb{E}\{S_n^*\} \to \infty$, see also the comments of Markowitz [35]. But under the risk aspect of the deviation of a random variable from its expectation, use of logarithm is more advantageous. The log transform is a special case of the Box–Cox [10] transforms introduced in view of stabilization and widely used in science, e.g., in medical science. Nevertheless there is the question whether the risk aversion of log utility is big enough to save an investor with very high probability from large terminal losses for medium time horizon. Simulation studies discussed by MacLean, Thorp, Zhao and Ziemba in MacLean et al. [34], Sect. 38, show that in a minority of scenarios such events occur. These effects depend on time horizon and distribution of the daily return, which allows a "proper use in the short and medium run" provided one has a good knowledge of the distribution. Corollary 1 allows for small $\epsilon > 0$ to obtain a lower bound N for the time horizon having a probability $\geq 1 - \epsilon$ that after this time the investor's wealth is for ever at least the unit starting capital: on the right-hand side of (9.3) set $R = 0$ and then choose N as the lowest integer n such that the right-hand side is at most ϵ. Here as in the following, $W^* > 0$ is assumed.

The good long-run and short-run performance of the various strategies are discussed in the literature, but usually the corresponding results concern only the expectation. Both in financial theory and practice, people care about the distribution as well. For the log-optimal strategy, there are almost sure statements, too (cf. Proposition 1).

Besides the growth rate of an investment strategy, one may consider the market time achieving a target wealth. We consider only strategies $\mathbf{b}$ with $\mathbb{E}\{\ln \langle \mathbf{b}, \mathbf{X}_1 \rangle\} > 0$. Again, $S_n^* = S_n(\mathbf{b}^*)$ denotes the capital after day n applying log-optimum portfolio strategy $\mathbf{b}^*$, and $S_n = S_n(\mathbf{b})$ the capital using the portfolio strategy $\mathbf{b}$. For a target wealth $\bar{s}$, introduce the market times

$$\tau(\bar{s}) := \min\{m; S_m \geq \bar{s}\}$$

and similarly

$$\tau^*(\bar{s}) := \min\{m; S_m^* \geq \bar{s}\}.$$

There are some studies how to minimize the expected market time $\mathbb{E}\{\tau(\bar{s})\}$ for large $\bar{s}$ [5, 6, 11, 26, 29], where Ethier [16] established an asymptotic median log-optimality of the (mean) log-optimal investment strategy. Breiman [11] conjectured that, for large $\bar{s}$, the asymptotically best strategy is the growth optimal one such that we apply the growth optimal strategy until we reach a neighborhood of $\bar{s}$.

Using the representation

$$\{S_m \geq \bar{s}\} = \left\{\sum_{i=1}^{m} \ln \langle \mathbf{b}\,, \mathbf{X}_i \rangle \geq \ln \bar{s}\right\}$$

the renewal theory for extended renewal processes, i.e., random walks with drift (see, for instance, [12, 17]), yields

Proposition 2 (Breiman [11]). *One has that*

$$\frac{\tau(\bar{s})}{\ln \bar{s}} \to \frac{1}{\mathbb{E}\{\ln \langle \mathbf{b}\,, \mathbf{X}_1 \rangle\}}$$

a.s.,

$$\frac{\mathbb{E}\{\tau(\bar{s})\}}{\ln \bar{s}} \to \frac{1}{\mathbb{E}\{\ln \langle \mathbf{b}\,, \mathbf{X}_1 \rangle\}},$$

especially

$$\frac{\tau^*(\bar{s})}{\ln \bar{s}} \to \frac{1}{W^*}$$

a.s.,

$$\frac{\mathbb{E}\{\tau^*(\bar{s})\}}{\ln \bar{s}} \to \frac{1}{W^*}$$

($\bar{s} \to \infty$).

In this sense the growth optimal strategy has another optimality property. This result has been refined by Breiman [11] and can be extended to

$$\begin{aligned}
&\frac{\ln \bar{s}}{\mathbb{E}\{\ln \langle \mathbf{b}^*\,, \mathbf{X}_1 \rangle\}} - \frac{\ln \bar{s}}{\mathbb{E}\{\ln \langle \mathbf{b}\,, \mathbf{X}_1 \rangle\}} + \frac{\mathbb{E}\{((\ln \langle \mathbf{b}^*\,, \mathbf{X}_1 \rangle)_+)^2\}}{(\mathbb{E}\{\ln \langle \mathbf{b}^*\,, \mathbf{X}_1 \rangle\})^2} \\
&\geq \mathbb{E}\{\tau^*(\bar{s})\} - \mathbb{E}\{\tau(\bar{s})\} \\
&\geq \frac{\ln \bar{s}}{\mathbb{E}\{\ln \langle \mathbf{b}^*\,, \mathbf{X}_1 \rangle\}} - \frac{\ln \bar{s}}{\mathbb{E}\{\ln \langle \mathbf{b}\,, \mathbf{X}_1 \rangle\}} - \frac{\mathbb{E}\{((\ln \langle \mathbf{b}\,, \mathbf{X}_1 \rangle)_+)^2\}}{(\mathbb{E}\{\ln \langle \mathbf{b}\,, \mathbf{X}_1 \rangle\})^2}
\end{aligned}$$

by Lorden's [33] upper bound for excess result.

Next we bound the tail distribution of $\tau^*(\bar{s})$ in case of large $\bar{s} = e^{nR}$, where $R < W^*$. We get that

$$\mathbb{P}\{\tau^*(e^{nR}) > n\} = \mathbb{P}\left\{\cap_{m=1}^{n}\{S_m^* < e^{nR}\}\right\} \le \mathbb{P}\left\{S_n^* < e^{nR}\right\},$$

therefore Theorem 1 implies that

$$\mathbb{P}\{\tau^*(e^{nR}) > n\} \le e^{-2n\frac{(W^*-R)^2}{(\ln a_2 - \ln a_1)^2}}.$$

9.3 Time Varying Portfolio Selection

For a general dynamic portfolio selection, the portfolio vector may depend on the past data. As before, $\mathbf{X}_i = (X_i^{(1)}, \ldots X_i^{(d)})$ denotes the return vector on trading period i. Moreover, denote the segment $\mathbf{X}_1, \ldots, \mathbf{X}_i$ by $\mathbf{X}_1^i$. Let $\mathbf{b} = \mathbf{b}_1$ be the portfolio vector for the first trading period. For initial capital S_0, we get that

$$S_1 = S_0 \cdot \langle \mathbf{b}_1 \,, \mathbf{X}_1 \rangle \,.$$

For the second trading period, S_1 is new initial capital, the portfolio vector is $\mathbf{b}_2 = \mathbf{b}(\mathbf{X}_1)$, and

$$S_2 = S_0 \cdot \langle \mathbf{b}_1 \,, \mathbf{X}_1 \rangle \cdot \langle \mathbf{b}(\mathbf{X}_1) \,, \mathbf{X}_2 \rangle \,.$$

For the nth trading period, a portfolio vector is $\mathbf{b}_n = \mathbf{b}(\mathbf{X}_1, \ldots, \mathbf{X}_{n-1}) = \mathbf{b}(\mathbf{X}_1^{n-1})$ and

$$S_n = S_0 \prod_{i=1}^{n} \left\langle \mathbf{b}(\mathbf{X}_1^{i-1}) \,, \mathbf{X}_i \right\rangle = S_0 e^{nW_n(\mathbf{B})}$$

with the average growth rate

$$W_n(\mathbf{B}) = \frac{1}{n} \sum_{i=1}^{n} \ln \left\langle \mathbf{b}(\mathbf{X}_1^{i-1}) \,, \mathbf{X}_i \right\rangle.$$

The fundamental limits, determined in [3], and in [1, 2], reveal that the so-called *log-optimum portfolio* $\mathbf{B}^* = \{\mathbf{b}^*(\cdot)\}$ is the best possible choice.

Proposition 3 (Algoet and Cover [3]). *On trading period n let $\mathbf{b}^*(\cdot)$ be such that*

$$\mathbb{E}\left\{\ln \left\langle \mathbf{b}^*(\mathbf{X}_1^{n-1}) \,, \mathbf{X}_n \right\rangle \middle| \mathbf{X}_1^{n-1}\right\} = \max_{\mathbf{b}(\cdot)} \mathbb{E}\left\{\ln \left\langle \mathbf{b}(\mathbf{X}_1^{n-1}) \,, \mathbf{X}_n \right\rangle \middle| \mathbf{X}_1^{n-1}\right\}.$$

If $S_n^* = S_n(\mathbf{B}^*)$ denotes the capital achieved by a log-optimum portfolio strategy $\mathbf{B}^*$, after n trading periods, then for any other investment strategy $\mathbf{B}$ with capital $S_n = S_n(\mathbf{B})$ and with

$$\sup_n \mathbb{E}\left\{(\ln\left\langle \mathbf{b}_n(\mathbf{X}_1^{n-1}), \mathbf{X}_n\right\rangle)^2\right\} < \infty,$$

and for any stationary and ergodic process $\{\mathbf{X}_n\}_{-\infty}^{\infty}$,

$$\limsup_{n\to\infty} \frac{1}{n} \ln \frac{S_n}{S_n^*} \leq 0 \quad a.s. \tag{9.4}$$

and

$$\lim_{n\to\infty} \frac{1}{n} \ln S_n^* = W^* \quad a.s., \tag{9.5}$$

where

$$W^* := \mathbb{E}\left\{\max_{\mathbf{b}(\cdot)} \mathbb{E}\left\{\ln\left\langle \mathbf{b}(\mathbf{X}_{-\infty}^{-1}), \mathbf{X}_0\right\rangle \middle| \mathbf{X}_{-\infty}^{-1}\right\}\right\}$$

is the maximal possible growth rate of any investment strategy.

Note that for memoryless markets $W^* = \max_{\mathbf{b}} \mathbb{E}\{\ln\langle \mathbf{b}, \mathbf{X}_0\rangle\}$ which shows that in this case the log-optimal portfolio is a constantly rebalanced portfolio.

Proof. For martingale difference sequences, there is a strong law of large numbers: If $\{Z_n\}$ is a martingale difference sequence with respect to $\{X_n\}$ and

$$\sum_{n=1}^{\infty} \frac{\mathbb{E}\{Z_n^2\}}{n^2} < \infty$$

then

$$\lim_{n\to\infty} \frac{1}{n} \sum_{i=1}^{n} Z_i = 0 \text{ a.s.}$$

(cf. [13], see also [40, Theorem 3.3.1]). Introduce the decomposition

$$\begin{aligned}
\frac{1}{n} \ln S_n &= \frac{1}{n} \sum_{i=1}^{n} \ln\left\langle \mathbf{b}(\mathbf{X}_1^{i-1}), \mathbf{X}_i\right\rangle \\
&= \frac{1}{n} \sum_{i=1}^{n} \mathbb{E}\{\ln\left\langle \mathbf{b}(\mathbf{X}_1^{i-1}), \mathbf{X}_i\right\rangle \mid \mathbf{X}_1^{i-1}\} \\
&+ \frac{1}{n} \sum_{i=1}^{n} \left(\ln\left\langle \mathbf{b}(\mathbf{X}_1^{i-1}), \mathbf{X}_i\right\rangle - \mathbb{E}\{\ln\left\langle \mathbf{b}(\mathbf{X}_1^{i-1}), \mathbf{X}_i\right\rangle \mid \mathbf{X}_1^{i-1}\}\right).
\end{aligned}$$

The last average is an average of martingale differences, so it tends to zero a.s. Similarly,

$$\frac{1}{n}\ln S_n^* = \frac{1}{n}\sum_{i=1}^{n}\mathbb{E}\{\ln\left\langle \mathbf{b}^*(\mathbf{X}_1^{i-1}), \mathbf{X}_i\right\rangle \mid \mathbf{X}_1^{i-1}\}$$
$$+\frac{1}{n}\sum_{i=1}^{n}\left(\ln\left\langle \mathbf{b}^*(\mathbf{X}_1^{i-1}), \mathbf{X}_i\right\rangle - \mathbb{E}\{\ln\left\langle \mathbf{b}^*(\mathbf{X}_1^{i-1}), \mathbf{X}_i\right\rangle \mid \mathbf{X}_1^{i-1}\}\right).$$

Because of the definition of the log-optimal portfolio we have that

$$\mathbb{E}\{\ln\left\langle \mathbf{b}(\mathbf{X}_1^{i-1}), \mathbf{X}_i\right\rangle \mid \mathbf{X}_1^{i-1}\} \le \mathbb{E}\{\ln\left\langle \mathbf{b}^*(\mathbf{X}_1^{i-1}), \mathbf{X}_i\right\rangle \mid \mathbf{X}_1^{i-1}\},$$

and the proof of (9.4) is finished. In order to prove (9.5) we have to show that

$$\frac{1}{n}\sum_{i=1}^{n}\mathbb{E}\{\ln\left\langle \mathbf{b}^*(\mathbf{X}_1^{i-1}), \mathbf{X}_i\right\rangle \mid \mathbf{X}_1^{i-1}\} \to W^*$$

a.s. Introduce the notations

$$\mathbf{b}^*_{-k}(\mathbf{X}_{n-k}^{n-1}) = \arg\max_{\mathbf{b}(\cdot)}\mathbb{E}\left\{\ln\left\langle \mathbf{b}(\mathbf{X}_{n-k}^{n-1}), \mathbf{X}_n\right\rangle \mid \mathbf{X}_{n-k}^{n-1}\right\}$$

$(1 \le k < n)$ and

$$\mathbf{b}^*_{-\infty}(\mathbf{X}_{-\infty}^{n-1}) = \arg\max_{\mathbf{b}(\cdot)}\mathbb{E}\left\{\ln\left\langle \mathbf{b}(\mathbf{X}_{-\infty}^{n-1}), \mathbf{X}_n\right\rangle \mid \mathbf{X}_{-\infty}^{n-1}\right\}.$$

Obviously,

$$\mathbb{E}\{\ln\left\langle \mathbf{b}^*_{-k}(\mathbf{X}_{i-k}^{i-1}), \mathbf{X}_i\right\rangle \mid \mathbf{X}_{i-k}^{i-1}\} \le \mathbb{E}\{\ln\left\langle \mathbf{b}^*(\mathbf{X}_1^{i-1}), \mathbf{X}_i\right\rangle \mid \mathbf{X}_1^{i-1}\}$$

$(i > k)$ and

$$\mathbb{E}\{\ln\left\langle \mathbf{b}^*(\mathbf{X}_1^{i-1}), \mathbf{X}_i\right\rangle \mid \mathbf{X}_1^{i-1}\} \le \mathbb{E}\{\ln\left\langle \mathbf{b}^*_{-\infty}(\mathbf{X}_{-\infty}^{i-1}), \mathbf{X}_i\right\rangle \mid \mathbf{X}_{-\infty}^{i-1}\}.$$

Thus, the ergodic theorem implies that

$$W^*_{-k} := \mathbb{E}\left\{\max_{\mathbf{b}(\cdot)}\mathbb{E}\left\{\ln\left\langle \mathbf{b}(\mathbf{X}_{-k}^{-1}), \mathbf{X}_0\right\rangle \middle| \mathbf{X}_{-k}^{-1}\right\}\right\}$$
$$= \lim_n \frac{1}{n}\sum_{i=1}^{n}\mathbb{E}\{\ln\left\langle \mathbf{b}^*_{-k}(\mathbf{X}_{i-k}^{i-1}), \mathbf{X}_i\right\rangle \mid \mathbf{X}_{i-k}^{i-1}\}$$
$$\le \liminf_n \frac{1}{n}\sum_{i=1}^{n}\mathbb{E}\{\ln\left\langle \mathbf{b}^*(\mathbf{X}_1^{i-1}), \mathbf{X}_i\right\rangle \mid \mathbf{X}_1^{i-1}\}$$

a.s. and

$$\limsup_n \frac{1}{n}\sum_{i=1}^n \mathbb{E}\{\ln\langle \mathbf{b}^*(\mathbf{X}_1^{i-1}),\mathbf{X}_i\rangle \mid \mathbf{X}_1^{i-1}\}$$

$$\le \lim_n \frac{1}{n}\sum_{i=1}^n \mathbb{E}\{\ln\langle \mathbf{b}^*_{-\infty}(\mathbf{X}_{-\infty}^{i-1}),\mathbf{X}_i\rangle \mid \mathbf{X}_{-\infty}^{i-1}\} = W^*.$$

a.s. Using martingale argument one can check that

$$W^*_{-k} \uparrow W^*,$$

and so (9.5) is proved. ■

Put

$$\epsilon = \frac{W^* - R}{2}. \tag{9.6}$$

Concerning the rate of convergence we have that

Theorem 2. *If the market process* $\{\mathbf{X}_i\}$ *is stationary, ergodic and the condition (9.2) is satisfied, then for an arbitrary* $R < W^*$, *we have that*

$$\mathbb{P}\left\{e^{nR} > S_n^*\right\} \le e^{-n\frac{(W^*-R)^2}{2(\ln a_2 - \ln a_1)^2}} + \mathbb{P}\Big\{R + \epsilon > \frac{1}{n}\sum_{i=1}^n \mathbb{E}\{\ln\langle \mathbf{b}^*(\mathbf{X}_1^{i-1}),\mathbf{X}_i\rangle \mid \mathbf{X}_1^{i-1}\}\Big\}.$$

Proof. Apply the previous decomposition:

$$\begin{aligned}
&\mathbb{P}\left\{e^{nR} > S_n^*\right\}\\
&= \mathbb{P}\left\{R > \frac{1}{n}\ln S_n^*\right\}\\
&= \mathbb{P}\Big\{R + \epsilon - \epsilon > \frac{1}{n}\sum_{i=1}^n \mathbb{E}\{\ln\langle \mathbf{b}^*(\mathbf{X}_1^{i-1}),\mathbf{X}_i\rangle \mid \mathbf{X}_1^{i-1}\}\\
&\quad + \frac{1}{n}\sum_{i=1}^n \left(\ln\langle \mathbf{b}^*(\mathbf{X}_1^{i-1}),\mathbf{X}_i\rangle - \mathbb{E}\{\ln\langle \mathbf{b}^*(\mathbf{X}_1^{i-1}),\mathbf{X}_i\rangle \mid \mathbf{X}_1^{i-1}\}\right)\Big\}\\
&\le \mathbb{P}\Big\{R + \epsilon > \frac{1}{n}\sum_{i=1}^n \mathbb{E}\{\ln\langle \mathbf{b}^*(\mathbf{X}_1^{i-1}),\mathbf{X}_i\rangle \mid \mathbf{X}_1^{i-1}\}\Big\}\\
&\quad + \mathbb{P}\Big\{-\epsilon > \frac{1}{n}\sum_{i=1}^n \left(\ln\langle \mathbf{b}^*(\mathbf{X}_1^{i-1}),\mathbf{X}_i\rangle - \mathbb{E}\{\ln\langle \mathbf{b}^*(\mathbf{X}_1^{i-1}),\mathbf{X}_i\rangle \mid \mathbf{X}_1^{i-1}\}\right)\Big\}
\end{aligned}$$

For the second term of the right hand side, we apply the Hoeffding [27], Azuma [7] inequality: Let $X_1, X_2, \dots$ be a sequence of random variables, and assume that $V_1, V_2, \dots$ is a martingale difference sequence with respect to $X_1, X_2, \dots$. Assume, furthermore, that there exist random variables $Z_1, Z_2, \dots$ and nonnegative constants $c_1, c_2, \dots$ such that for every $i > 0$, Z_i is a function of $X_1, \dots, X_{i-1}$, and

$$Z_i \le V_i \le Z_i + c_i \quad \text{a.s.}$$

Then, for any $\epsilon > 0$ and n,

$$\mathbb{P}\left\{\sum_{i=1}^n V_i \ge \epsilon\right\} \le e^{-2\epsilon^2/\sum_{i=1}^n c_i^2}$$

and

$$\mathbb{P}\left\{\sum_{i=1}^n V_i \le -\epsilon\right\} \le e^{-2\epsilon^2/\sum_{i=1}^n c_i^2}.$$

Thus

$$\begin{aligned}
&\mathbb{P}\Big\{-\epsilon > \frac{1}{n}\sum_{i=1}^n \left(\ln\left\langle \mathbf{b}^*(\mathbf{X}_1^{i-1}), \mathbf{X}_i\right\rangle - \mathbb{E}\{\ln\left\langle \mathbf{b}^*(\mathbf{X}_1^{i-1}), \mathbf{X}_i\right\rangle \mid \mathbf{X}_1^{i-1}\}\right)\Big\} \\
&\le e^{-2n\frac{\epsilon^2}{(\ln a_2 - \ln a_1)^2}} \\
&= e^{-n\frac{(W^*-R)^2}{2(\ln a_2 - \ln a_1)^2}}.
\end{aligned}$$

■

If the market process is just stationary and ergodic, then it is impossible to get rate of convergence of the term

$$\mathbb{P}\Big\{R + \epsilon > \frac{1}{n}\sum_{i=1}^n \mathbb{E}\{\ln\left\langle \mathbf{b}^*(\mathbf{X}_1^{i-1}), \mathbf{X}_i\right\rangle \mid \mathbf{X}_1^{i-1}\}\Big\}.$$

In order to find conditions, for which a rate can be derived, one possibility is that for $i > k$

$$\begin{aligned}
\mathbb{E}\{\ln\left\langle \mathbf{b}^*(\mathbf{X}_1^{i-1}), \mathbf{X}_i\right\rangle \mid \mathbf{X}_1^{i-1}\} &= \max_{\mathbf{b}(\cdot)} \mathbb{E}\{\ln\left\langle \mathbf{b}(\mathbf{X}_1^{i-1}), \mathbf{X}_i\right\rangle \mid \mathbf{X}_1^{i-1}\} \\
&= \max_{\mathbf{b}} \mathbb{E}\{\ln\left\langle \mathbf{b}, \mathbf{X}_i\right\rangle \mid \mathbf{X}_1^{i-1}\} \\
&\ge \max_{\mathbf{b}} \mathbb{E}\{\ln\left\langle \mathbf{b}, \mathbf{X}_i\right\rangle \mid \mathbf{X}_{i-k}^{i-1}\},
\end{aligned}$$

and so we may increase the above probability. We expected that the density of

$$\max_{\mathbf{b}} \mathbb{E}\{\ln \langle \mathbf{b} , \mathbf{X}_{k+1} \rangle \mid \mathbf{X}_1^k\}$$

has a small support, which moves to the right, when k increases.

We made an experiment verifying this conjecture empirically. At the web page http://www.szit.bme.hu/~oti/portfolio there are two benchmark data sets from NYSE:

- The first data set consists of daily data of 36 stocks with length 22 years (5651 trading days ending in 1985). More precisely, the data set contains the daily price relatives, that was calculated from the nominal values of the *closing prices* corrected by the dividends and the splits for all trading day. This data set has been used for testing portfolio selection in [15, 22–25, 39].
- The second data set is an extended version of the first one. It was augmented by 22 years and covers 44 years period from 1962 to 2006 containing 11,178 trading days. As opposed to the first data set it contains only 19 stocks out of the 36 stocks due to the fact that 4 illiquid and 13 bankrupted stocks were left out. In the analysis of financial time series there often happens a censoring, which means that the time series is terminated (bankrupt, merging, withdraw from the stock market, etc.). If one takes into account only the non-censored time series, then the survivals cause a bias in the statistical inference, called survival bias. Thus, the leaving out the bankrupted stocks adds survival bias to the simulation. However in case of actively managed portfolio strategies as re-balancing or online portfolio selection the effect of the survival bias is less important than the liquidity of the traded stocks. For example, if IROQU and KINAR (a bankrupted and a small capitalization stock) were not left out then the achieved wealth would be unrealistically high (cf. [22]). Based on the above argument the following 4 illiquid stocks were excluded from the data set: SHERW, KODAK, COMME and KINAR. Further benchmark data sources are available at http://www.cais.ntu.edu.sg/~chhoi/olps/datasets.html. Clearly, the distributions of the market process were not the same over the past 44 years. The empirical strategies applied are not sensitive with respect to the changes of the distributions.

As in [25], we considered the kernel based portfolio strategies $\mathbf{B}^{(k)} = \{\mathbf{b}^{(k)}(\cdot)\}$, where the window size $k = 1, \ldots, 5$ and the corresponding radius is

$$r_k^2 = 0.00035 \cdot d \cdot k.$$

According to the kernel based rule, the portfolio vector for day n is selected such that one searches for similar patterns to the near past segment $\mathbf{X}_{n-k}^{n-1}$ and design a portfolio to the subsequence of return vectors followed the similarities. For $n > k + 1$, define the random variable $Z_{n,k}$ by

$$Z_{n,k} = \frac{\max_{\mathbf{b} \in \Delta_d} \sum_{\{k<i<n: \|\mathbf{X}_{i-k}^{i-1} - \mathbf{X}_{n-k}^{n-1}\| \le r_k\}} \ln \langle \mathbf{b} , \mathbf{X}_i \rangle}{\left|\{k < i < n : \|\mathbf{X}_{i-k}^{i-1} - \mathbf{X}_{n-k}^{n-1}\| \le r_k\}\right|},$$

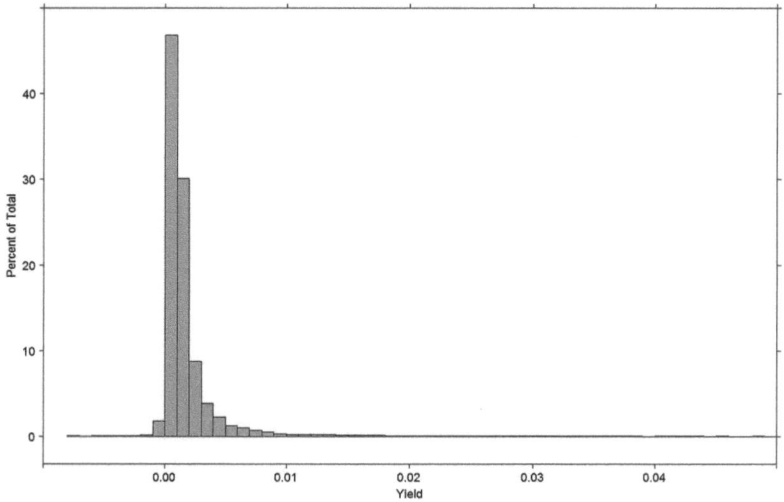

Fig. 9.2 The histogram of the maximum of the conditional expectations for $k = 1$

if the sum is non-void. Then the histogram of $\{Z_{n,k}, n = k+1, \ldots N\}$ can be an approximation of the density of $\max_{\mathbf{b}} \mathbb{E}\{\ln \langle \mathbf{b}\, , \, \mathbf{X}_{k+1} \rangle \mid \mathbf{X}_1^k\}$.

For $k = 1, \ldots, 5$, we generated the five histograms of the maximum of these empirical conditional expectations. The main observation was that these histograms do not depend on k, therefore one can assume that the market process is a first order Markov process. Figure 9.2 shows a histogram out of the five, which corresponds to $k = 1$. Surprisingly, this histogram has a small support. Here are some data:

$$\begin{aligned} \text{minimum} &= -0.008 \\ \text{1st qu.} &= 0.00061 \\ \text{median} &= 0.0010 \\ \text{mean} &= 0.0019 \\ \text{3rd qu.} &= 0.0018 \\ \text{maximum} &= 0.1092. \end{aligned}$$

An important feature of this histogram is that it has a positive skewness, which means that the right hand side tail is larger than the left hand side one. The reason of this property is that $\max_{\mathbf{b}} \mathbb{E}\{\ln \langle \mathbf{b}\, , \, \mathbf{X}_{k+1} \rangle \mid \mathbf{X}_1^k\}$ is the maximum of (dependent) random variables.

For the kernel based portfolio we generated the histogram of the log-return, too. The elementary portfolio is defined by

$$\mathbf{b}^{(k)}(\mathbf{x}_1^{n-1}) = \arg\max_{\mathbf{b} \in \Delta_d} \sum_{\{k<i<n: \|\mathbf{x}_{i-k}^{i-1} - \mathbf{x}_{n-k}^{n-1}\| \le r_k\}} \ln \langle \mathbf{b}\, , \, \mathbf{x}_i \rangle \, ,$$

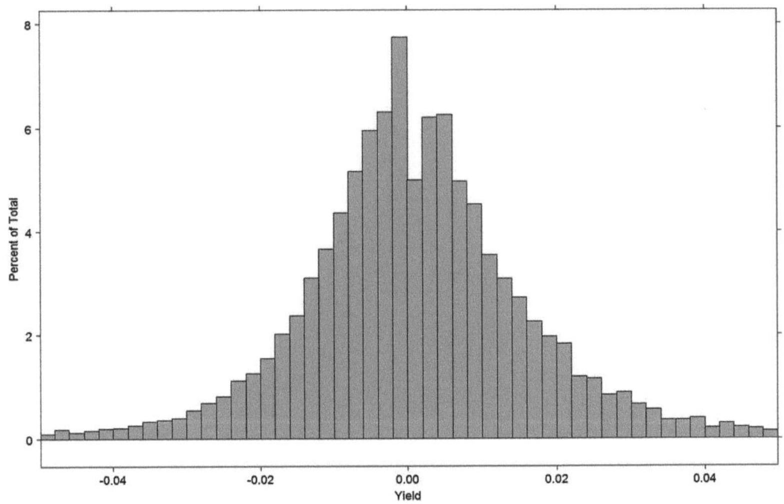

Fig. 9.3 The histogram of the log-returns for an empirical portfolio strategy

if the sum is non-void, and $\mathbf{b}_0 = (1/d, \dots, 1/d)$ otherwise. Define the random variable $Z'_{n,k}$ by

$$Z'_{n,k} = \ln \left\langle \mathbf{b}^{(k)}(\mathbf{X}_1^{n-1}) \, , \, \mathbf{X}_n \right\rangle ,$$

which is the daily log-return for day n. For $k = 1$, we generated the histogram of $\{Z'_{n,k}, n = k+1, \dots N\}$. Figure 9.3 shows the histogram of the log-return for the empirical portfolio strategy $\mathbf{B}^{(1)}$. Here are the corresponding data:

$$\begin{aligned} \text{minimum} &= -0.1535 \\ \text{1st qu.} &= -0.0077 \\ \text{median} &= 0.0003 \\ \text{mean} &= 0.00118 \\ \text{3rd qu.} &= 0.0093 \\ \text{maximum} &= 0.1522. \end{aligned}$$

Comparing the Figs. 9.1 and 9.3, one can observe that the shape and the quantiles of the histograms are almost the same. The main difference is in the mean. Since these data sets contains the relative prices for trading days only, and 1 year consists of 250 trading days, therefore in terms of average annual yields (AAY) the mean= 0.00118 in Fig. 9.3 corresponds to AAY 34 %, while the mean= 0.00118 for the Coca Cola corresponds to AAY 14 %.

Based on these empirical observations, in the following we assume that the market process $\{\mathbf{X}_i\}$ is a first-order stationary Markov process. In this case the log-optimum portfolio choice has the form $\mathbf{b}^*(\mathbf{X}_{n-1})$ (instead of $\mathbf{b}^*(\mathbf{X}_1^{n-1})$) maximizing $\mathbb{E}\{\ln\langle\mathbf{b}, \mathbf{X}_n\rangle \mid \mathbf{X}_{n-1}\}$ such that

$$\mathbb{E}\{\ln\langle\mathbf{b}^*(\mathbf{X}_{n-1}), \mathbf{X}_n\rangle\} = W^*.$$

We assume that $\mathbf{X}_i$ has a denumerable state space $S \subset [a_1, a_2]^d$, which is realistic because the values of the components of $\mathbf{X}_i$ are quotients of integer valued prices. Further we assume that the Markov process is irreducible and aperiodic. Finally, suppose that the Markov kernel $\nu(H \mid \mathbf{x})$ defined by

$$\nu(H \mid \mathbf{x}) := \mathbb{P}\{\mathbf{X}_2 \in H \mid \mathbf{X}_1 = \mathbf{x}\}$$

($\mathbf{x} \in S, H \subset S$) is continuous in total variation, i.e.,

$$V(\mathbf{x}, \mathbf{x}') := \sup_{H \subset S} |\nu(H \mid \mathbf{x}) - \nu(H \mid \mathbf{x}')| \to 0 \tag{9.7}$$

if $\mathbf{x}' \to \mathbf{x}$. Notice that by Scheffé's theorem

$$V(\mathbf{x}, \mathbf{x}') := \frac{1}{2} \sum_{\mathbf{x}^* \in S} |\nu(\{\mathbf{x}^*\} \mid \mathbf{x}) - \nu(\{\mathbf{x}^*\} \mid \mathbf{x}')|.$$

The following theorem with $R < W^*$ gives exponential bounds for the probability that $e^{nR} > S_n^*$ and for the probability that after n there is a time instant m such that $e^{mR} > S_m^*$.

Theorem 3. *Let the market process $\{\mathbf{X}_i\}$ be a first-order stationary denumerable Markov chain, which is irreducible and aperiodic, satisfies (9.2) and (9.7). Then for arbitrary $R < W^*$, there exist $c, C, c^*, C^* \in (0, \infty)$ depending on $W^* - R$, $\ln a_2 - \ln a_1$ and the ergodic behavior of $\{\mathbf{X}_i\}$ such that for all n*

$$\mathbb{P}\left\{e^{nR} > S_n^*\right\} \le e^{-n\frac{(W^*-R)^2}{2(\ln a_2 - \ln a_1)^2}} + Ce^{-cn}, \tag{9.8}$$

and

$$\mathbb{P}\left\{\cup_{m=n}^{\infty}\{e^{mR} > S_m^*\}\right\} \le C^* e^{-c^* n}. \tag{9.9}$$

Proof. With the notation (9.6), Theorem 2 implies that

$$\mathbb{P}\left\{e^{nR} > S_n^*\right\} \le e^{-n\frac{(W^*-R)^2}{2(\ln a_2 - \ln a_1)^2}} + \mathbb{P}\left\{R + \epsilon > \frac{1}{n}\sum_{i=1}^{n} \mathbb{E}\{\ln\langle\mathbf{b}^*(\mathbf{X}_{i-1}), \mathbf{X}_i\rangle \mid \mathbf{X}_{i-1}\}\right\}.$$

By stationarity, the distribution μ of $\mathbf{X}_i$ does not depend on i and satisfies

$$\int \nu(\cdot \mid \mathbf{x})\mu(d\mathbf{x}) = \mu,$$

i.e.,

$$\sum_{\mathbf{x}\in S} \nu(\{\mathbf{x}^*\} \mid \mathbf{x})\mu(\{\mathbf{x}\}) = \mu(\{\mathbf{x}^*\}). \tag{9.10}$$

It is well known from the theory of denumerable Markov chains (see, e.g., [17]), that (9.10) together with irreducibility and aperiodicity of $\{\mathbf{X}_i\}$ implies that $\{\mathbf{X}_i\}$ is positive recurrent with mean recurrence time $1/\mu(\{\mathbf{x}\}) < \infty$ and weak convergence of $P_{\mathbf{X}_n|\mathbf{X}_1=\mathbf{x}}$ to μ. Thus, by Scheffé and Riesz-Vitali theorems, even

$$\begin{aligned}
&\sup_{H\subset S} |\mathbb{P}\{\mathbf{X}_n \in H \mid \mathbf{X}_1 = \mathbf{x}\} - \mu(H)| \\
&= \frac{1}{2}\sum_{\mathbf{x}^*\in S} |\mathbb{P}\{\mathbf{X}_n = \mathbf{x}^* \mid \mathbf{X}_1 = \mathbf{x}\} - \mu(\{\mathbf{x}^*\})| \\
&\to 0
\end{aligned}$$

$(n \to \infty)$ for each $\mathbf{x} \in S$. Further for each integer n

$$\begin{aligned}
&\sup_{H\subset S} |\mathbb{P}\{\mathbf{X}_n \in H \mid \mathbf{X}_1 = \mathbf{x}\} - \mathbb{P}\{\mathbf{X}_n \in H \mid \mathbf{X}_1 = \mathbf{x}'\}| \\
&= \frac{1}{2}\sum_{\mathbf{x}^*\in S} |\mathbb{P}\{\mathbf{X}_n = \mathbf{x}^* \mid \mathbf{X}_1 = \mathbf{x}\} - \mathbb{P}\{\mathbf{X}_n = \mathbf{x}^* \mid \mathbf{X}_1 = \mathbf{x}'\}| \\
&= \frac{1}{2}\sum_{\mathbf{x}^*\in S} |\sum_{\mathbf{y}\in S} \mathbb{P}\{\mathbf{X}_n = \mathbf{x}^* \mid \mathbf{X}_2 = \mathbf{y}\}(\mathbb{P}\{\mathbf{X}_2 = \mathbf{y} \mid \mathbf{X}_1 = \mathbf{x}\} - \mathbb{P}\{\mathbf{X}_2 = \mathbf{y} \mid \mathbf{X}_1 = \mathbf{x}'\})| \\
&\leq \frac{1}{2}\sum_{\mathbf{x}^*\in S}\sum_{\mathbf{y}\in S} \mathbb{P}\{\mathbf{X}_n = \mathbf{x}^* \mid \mathbf{X}_2 = \mathbf{y}\}|\mathbb{P}\{\mathbf{X}_2 = \mathbf{y} \mid \mathbf{X}_1 = \mathbf{x}\} - \mathbb{P}\{\mathbf{X}_2 = \mathbf{y} \mid \mathbf{X}_1 = \mathbf{x}'\}| \\
&= \frac{1}{2}\sum_{\mathbf{y}\in S} |\mathbb{P}\{\mathbf{X}_2 = \mathbf{y} \mid \mathbf{X}_1 = \mathbf{x}\} - \mathbb{P}\{\mathbf{X}_2 = \mathbf{y} \mid \mathbf{X}_1 = \mathbf{x}'\}| \\
&= \sup_{H\subset S} |\mathbb{P}\{\mathbf{X}_2 \in H \mid \mathbf{X}_1 = \mathbf{x}\} - \mathbb{P}\{\mathbf{X}_2 \in H \mid \mathbf{X}_1 = \mathbf{x}'\}| \\
&\to 0
\end{aligned}$$

$(\mathbf{x}' \to \mathbf{x})$ by (9.7). Therefore even

$$\sup_{H\subset S,\mathbf{x}\in S} |\mathbb{P}\{\mathbf{X}_n \in H \mid \mathbf{X}_1 = \mathbf{x}\} - \mu(H)| \to 0.$$

Thus, the process $\{\mathbf{X}_i\}$ is φ-mixing. Also the sequence

$$\{\mathbb{E}\{\ln \langle \mathbf{b}^*(\mathbf{X}_{i-1}), \mathbf{X}_i\rangle \mid \mathbf{X}_{i-1}\}\}$$

is φ-mixing with mixing coefficients $\varphi_m \to 0$. Now we can apply Collomb's exponential inequality (p. 449 in [14]) with $d = \delta = \sqrt{D} = \frac{1}{n}(\ln a_2 - \ln a_1)$. For $m \in \{1, \ldots, n\}$ we obtain

$$\mathbb{P}\left\{R + \epsilon > \frac{1}{n}\sum_{i=1}^{n} \mathbb{E}\{\ln \langle \mathbf{b}^*(\mathbf{X}_{i-1}), \mathbf{X}_i\rangle \mid \mathbf{X}_{i-1}\}\right\}$$
$$\leq \exp\left\{\frac{n}{m}\left(3\sqrt{e}\varphi_m + \frac{3}{8}\frac{1 + 4\sum_{i=1}^{m}\varphi_i}{m} - \frac{\epsilon}{4(\ln a_2 - \ln a_1)}\right)\right\}.$$

Suitable choice of $m = M(\epsilon)$ with $n \geq N(\epsilon)$ leads to the second term on the right hand side of (9.8) as a bound for all n. Finally, from (9.8) we obtain (9.9) as in the proof of Corollary 1. ■

Remark. Theorem 3 can be extended to the case of a Harris-recurrent, strongly aperiodic Markov chain, not necessarily being stationary or having denumerable state space; compare in a somewhat other context Theorem 2 in [20], where Theorem 4.1 (i) of [4] and Collomb's inequality are used.

Acknowledgements This work was partially supported by the European Union and the European Social Fund through project FuturICT.hu (grant no.: TAMOP-4.2.2.C-11/1/KONV-2012-0013).

References

1. P. Algoet, Universal schemes for prediction, gambling, and portfolio selection. Ann. Probab. **20**, 901–941 (1992)
2. P. Algoet, The strong law of large numbers for sequential decisions under uncertainty. IEEE Trans. Inf. Theory **40**, 609–634 (1994)
3. P. Algoet, T.M. Cover, Asymptotic optimality asymptotic equipartition properties of log-optimum investments. Ann. Probab. **16**, 876–898 (1988)
4. K.B. Athreya, P. Ney, A new approach to the limit theory of recurrent Markov chains. Trans. Am. Math. Soc. **245**, 493–501 (1978)
5. D.C. Aucamp, An investment strategy with overshoot rebates which minimizes the time to attain a specified goal. Manag. Sci. **23**, 1234–1241 (1977)
6. D.C. Aucamp, On the extensive number of plays to achieve superior performance with the geometric mean strategy. Manag. Sci. **39**, 1163–1172 (1993)
7. K. Azuma, Weighted sums of certain dependent random variables. Tohoku Math. J. **68**, 357–367 (1967)
8. A.R. Barron, T.M. Cover, A bound on the financial value of information. IEEE Trans. Inf. Theory **34**, 1097–1100 (1988)
9. R. Billingsley, *Convergence of Probability Measures*, 2nd edn. (Wiley, New York, 1999)

10. G.E.P. Box, D.R. Cox, An analysis of transformations. J. R. Stat. Soc. Ser. B **26**, 211–252 (1964)
11. L. Breiman, Optimal gambling systems for favorable games, in *Proceedings of the Fourth Berkeley Symposium on Mathematical Statistics and Probability*, vol. 1 (University of California Press, Berkeley, 1961), pp. 65–78
12. L. Breiman, *Probability* (Addison-Wesley, Reading, MA, 1968)
13. Y.S. Chow, Local convergence of martingales and the law of large numbers. Ann. Math. Stat. **36**, 552–558 (1965)
14. G. Collomb, Propriétés de convergence presque complète du prédicteur à noyau. Z. Wahrscheinlichkeitstheorie verw. Geb. **66**, 441–460 (1984)
15. T. Cover, Universal portfolios. Math. Financ. **1**, 1–29 (1991)
16. S.N. Ethier, The Kelly system maximizes median fortune. J. Appl. Probab. **41**, 1230–1236 (2004)
17. W. Feller, *An Introduction to Probability Theory and Its Applications*, vol. 1, 3rd edn., vol. 2, 2nd edn. (Wiley, New York, 1968/1971)
18. E.R. Fernholz, *Stochastic Portfolio Theory* (Springer, New York, 2000)
19. M. Finkelstein, R. Whitley, Optimal strategies for repeated games. Adv. Appl. Probab. **13**, 415–428 (1981)
20. L. Györfi, H. Walk, Empirical portfolio selection strategies with proportional transaction cost. IEEE Trans. Inf. Theory **58**, 6320–6331 (2012)
21. L. Györfi, H. Walk, Log-optimal portfolio selection strategies with proportional transaction costs, in *Machine Learning for Financial Engineering*, ed. by L. Györfi, G. Ottucsák, H. Walk (Imperial College Press, London, 2012), pp. 119–152
22. L. Györfi, G. Lugosi, F. Udina, Nonparametric kernel based sequential investment strategies. Math. Financ. **16**, 337–357 (2006)
23. L. Györfi, A. Urbán, I. Vajda, Kernel-based semi-log-optimal empirical portfolio selection strategies. Int. J. Theor. Appl. Financ. **10**, 505–516 (2007)
24. L. Györfi, F. Udina, H. Walk, Nonparametric nearest-neighbor-based sequential investment strategies. Stat. Decis. **26**, 145–157 (2008)
25. L. Györfi, G. Ottucsák, A. Urbán, Empirical log-optimal portfolio selections: a survey, in *Machine Learning for Financial Engineering*, ed. by L. Györfi, G. Ottucsák, H. Walk (Imperial College Press, London, 2012), pp. 81–118
26. T.P. Hayes, How Not to Win a Million Dollars: A Counterexample to a Conjecture of L. Breiman (2011). http://arxiv.org/pdf/1112.0829v1.pdf
27. W. Hoeffding, Probability inequalities for sums of bounded random variables. J. Am. Stat. Assoc. **58**, 13–30 (1963)
28. M. Horváth, A. Urbán, Growth optimal portfolio selection with short selling and leverage, in *Machine Learning for Financial Engineering*, ed. by L. Györfi, G. Ottucsák, H. Walk (Imperial College Press, London, 2012), pp. 153–178
29. C. Kardaras, E. Platen, Minimizing the expected market time to reach a certain wealth level. SIAM J. Financ. Math. **1**, 16–29 (2010)
30. J.L. Kelly, A new interpretation of information rate. Bell Syst.Tech. J. **35**, 917–926 (1956)
31. D. Kuhn, D.G. Luenberger, Analysis of the rebalancing frequency in the log-optimal portfolio selection. Quant. Financ. **10**, 221–234 (2010)
32. H.A. Latané, Criteria for choice among risky ventures. J. Polit. Econ. **67**, 145–155 (1959)
33. G. Lorden, On excess over the boundary. Ann. Math. Stat. **41**, 520–527 (1970)
34. L.C. MacLean, E.O. Thorp, W.T. Ziemba (eds.), *The Kelly Capital Growth Investment Criterion: Theory and Practice* (World Scientific, New Jersey, 2011)
35. H.M. Markowitz, Investment for the long run: new evidence for an old rule. J. Financ. **31**, 1273–1286 (1976)
36. L.B. Pulley, Mean-variance approximation to expected logarithmic utility. Oper. Res. **40**, 685–696 (1983)
37. R. Roll, Evidence on the growth-optimum model. J. Financ. **28**, 551–566 (1973)

38. P.A. Samuelson, The "fallacy" of maximizing the geometric mean in long sequences of investment or gambling. Proc. Natl. Acad. Sci. U. S. A. **68**, 2493–2496 (1971)
39. Y. Singer, Switching portfolios. Int. J. Neural Syst. **8**, 445–455 (1997)
40. W.F. Stout, *Almost Sure Convergence* (Academic Press, New York, 1974)
41. L. Takács, On a generalization of the arc-sine law. Ann. Appl. Probab. **6**, 1035–1040 (1996)

Chapter 10
Heuristics for Portfolio Selection

Manfred Gilli and Enrico Schumann

Thus computing is, or at least should be, intimately bound up with both the source of the problem and the use that is going to be made of the answers—it is not a step to be taken in isolation from reality.
Richard W. Hamming, An Essay on Numerical Methods

Abstract Portfolio selection is about combining assets such that investors' financial goals and needs are best satisfied. When operators and academics translate this actual problem into optimisation models, they face two restrictions: the models need to be empirically meaningful, and the models need to be soluble. This chapter will focus on the second restriction. Many optimisation models are difficult to solve because they have multiple local optima or are 'badly-behaved' in other ways. But on modern computers such models can still be handled, through so-called heuristics. To motivate the use of heuristic techniques in finance, we present examples from portfolio selection in which standard optimisation methods fail. We then outline the principles by which heuristics work. To make that discussion more concrete, we describe a simple but effective optimisation technique called Threshold Accepting and how it can be used for constructing portfolios. We also summarise the results of an empirical study on hedge-fund replication.

Keywords Portfolio optimisation • Heuristics • Financial modelling • Model risk • Model errors • Hedge fund replication

M. Gilli (✉)
University of Geneva and Swiss Finance Institute, Geneva, Switzerland
e-mail: manfred.gilli@unige.ch

E. Schumann
AQ Investment AG, Zurich, Switzerland
e-mail: es@enricoschumann.net

© Springer International Publishing Switzerland 2017
G. Consigli et al. (eds.), *Optimal Financial Decision Making under Uncertainty*, International Series in Operations Research & Management Science 245,
DOI 10.1007/978-3-319-41613-7_10

10.1 Introduction

Before Markowitz, there was nothing.

Such a statement may exaggerate, but not much. Before Markowitz [41], investments were good or bad solely according to how much profit they promised. Though investors and academic writers had an intuition that investing in several assets was better than putting 'all eggs in the one basket', the overall goal was to select assets with high returns. Assets were not added to portfolios for risk reasons alone.

Portfolio theory changed that. Markowitz showed that adding ever more assets to a portfolio can typically not completely remove return-variation, but that a careful choice of assets can reduce risk. Thus, managing risk became an explicit part of investment management.

Markowitz's model is quite simple: assets are held for a fixed period of time; which assets are chosen depends on two properties of the resulting portfolio, reward and risk. Markowitz equated reward with expected return and risk with return variance. In this chapter, we will look at such Markowitz-type models. There exist more complex models for sure (e.g., spanning several time periods), but the one-period setting still provides enough empirical and computational difficulties that it warrants discussion. Thus, we shall stay in the one-period setting; we will, however, deviate from Markowitz's model in our definition of risk and reward.

In fact, already back in the 1950s Markowitz thought that there might be better specifications for risk. Downside semi-variance seemed more appealing than variance [42] because it does not penalise upside return-variation. Eventually, Markowitz rejected the idea because at that time it was too difficult to compute optimal portfolios. But with heuristics—the methods that we explain in this chapter—we can solve models without restrictions on the functional form of the selection criterion or the constraints; thus, downside-risk specifications can easily be handled.

However, we will not only discuss computation. After all, this chapter's topic is the selection of financial portfolios, and so we need to explain how this application is special, and how a prescription to optimise portfolios differs from a general discussion of numerical optimisation. We shall argue that the challenges to the practitioner of portfolio construction—we call him 'the analyst'—can be grouped into three topics :

1. The modelling. The analyst needs to decide what financial goals and necessities exist and should be put into the model. And, of course, he cannot speak of abstract quantities such as 'risk'—everything has to be made precise.
2. The forecasting. Portfolio selection is about choosing assets today in the hope they do well tomorrow. Thus, the analyst can only use quantities in his models that he can forecast sufficiently well.
3. The computation. Given the model and forecasts, the analyst needs to solve the model. A 'good' model cannot be good if we cannot compute its solution. (The

proof is simple: if we never could solve it, we could never test it. So how could we tell it was good in the first place?)

In fact, computational restrictions are much less of an obstacle to working with portfolio models than is sometimes thought. The computational power that the analyst has at his disposal today—literally in his hands, given the developments in tablets and other handheld computers—is such that many models that appear difficult in theory turn out to be very tractable practically.

But with power comes responsibility. In the past, a financial economist could well concentrate on theory, only to leave the lowly computational work to the specialist who, in turn, understood nothing of finance and economics. Today these roles have merged.[1] To be sure, such a separation was always bad, but at least it could serve as an excuse for ignoring certain aspects of the portfolio selection process. Today, the excuse has disappeared. It is the analyst's job to understand and manage the process of portfolio selection at all stages—modelling, forecasting, computation.

Two principles will guide us throughout this chapter (borrowed from Gilli et al. [27]):

(i) the application matters, and
(ii) go experiment.

Principle (i) means that what matters is to select well-performing portfolios. Any computational technique that helps us to reach that goal is fine; likewise, any apparent computational difficulty needs to be judged by how much it impedes us reaching our goal.

Principle (ii) is particularly relevant to computation. During the process of portfolio selection, many situations arise in which we are faced with choices: many small decisions that have to be taken while writing down a model, preparing data or setting up an algorithm. We can rarely give general advice on such decisions, but that is not a problem: just go experiment and find out for yourself.

These two principles may also be read as kind of a warning: we will focus on the practical application of portfolio optimisation. We care little whether what we say fits into theoretical paradigms. Neither do we care about the kind of 'practical use' to which many results in academic finance and operations research are put—to end up in financial institutions' marketing departments, or those of software vendors. We describe optimisation methods that allow the analyst to specify models as he likes, but only to test those models and to discard those that do not work—which is, unfortunately, most of them. We will offer our judgement on what we deem important, and what is not. In our opinion; you are welcome to disagree.

The chapter is structured as follows. In the next section, we shall discuss problems and models (throughout this chapter, you will find us strictly distinguishing between those terms). Then, in Sect. 10.3, we will describe the principles of heuristics. As we said, these methods allow the analyst to handle models without

[1] Such a merging of roles did not only happen in computational finance; it also took place in publishing and data analysis in general.

restrictions on the functional form of the objective function and the constraints. Of course, a price must be paid for that, and so heuristics come with their own difficulties. In particular, almost all heuristics are stochastic algorithms, and hence solutions need to be evaluated with some care. We shall address these difficulties and suggest ways to handle them practically. Two example sections follow: in Sect. 10.4, we detail one heuristic, Threshold Accepting, and how it can be applied to portfolio selection. Then, in Sect. 10.5, we present a summary of an empirical study that used Threshold Accepting. Section 10.6 concludes.

10.2 Of Problems, Models and Methods

We start with a brief discussion of a one-period optimisation model. These descriptions should be familiar to most readers. We will step back then and reflect on this model; we shall zoom in again and discuss how to solve the model only in the next section.

10.2.1 A One-Period Investment Model

We want to invest a budget v_0 into assets from a universe of n_A assets. These assets are then held for a period of length T. Let p_0 be a vector of current (known) prices, then the budget constraint can be written as

$$p_0' x \leq v_0\,, \tag{10.1}$$

in which x is a vector of numbers of contracts we hold. At time T, prices will have evolved to p_T and the portfolio value will be

$$v_T = p_T' x\,. \tag{10.2}$$

To rank different choices for x, we need an objective function ϕ that maps v_T into a real number. In fact, the function may also look at the path $\{v\}_0^T$ that the portfolio value takes between portfolio inception and T. We remain in the one-period framework, since we do not trade between time 0 and T. Thus, the goal becomes

$$\underset{x}{\text{minimise}}\ \phi\left(\{v\}_0^T\right) \tag{10.3}$$

subject, at least, to the budget constraint. (In order to maximise, we minimise $-\phi$.) Possible specifications for ϕ could be moments (also partial or conditional moments), quantiles, or in essence any function that can be evaluated for $\{v\}_0^T$. See Gilli et al. [27, Chap. 13] for examples and a detailed discussion.

In many cases, we add further constraints. We may have legal or regulatory restrictions on how to invest. For instance, many institutional investors cannot legally hold more than a specific amount of assets from one counter party. Conceptually-simple restrictions can be difficult for standard optimisation methods; see for instance Scozzari et al. [50].

Empirically, restrictions can provide safeguards against over-fitting. There is much evidence that limiting position sizes improves performance. The classic reference is Frost and Savarino [15]; see also Jagannathan and Ma [32].

Constraints can also help to make estimated parameters interpretable. Variances cannot be negative and probabilities must lie between zero and one. But with unconstrained numerical procedures, there is no guarantee that such restrictions hold, and we may need to impose them to get meaningful results.

10.2.2 Reality to Model, and Back

What we described in the last section is a model, not the actual problem. The problem is finding assets that give, loosely speaking, much reward with little risk. The model we described assumed a simple investment process (buy-and-hold) and fixed the notions of reward and risk through our choice of ϕ.

10.2.2.1 Sources of Error

Modelling is the process of putting the actual problem into a form that can be understood by a computer. We have to make vague notions precise, and we often need to simplify and approximate. This, in turn, will introduce errors. The word error must not be understood in the sense that something did not work as expected. Approximation errors originate from the very practice of modelling.

Following a classic discussion in von Neumann and Goldstine [55]—expanded in Morgenstern [43]—we group these errors into two categories: empirical errors (a.k.a. model errors) and numerical errors. The analyst's job is not only to acknowledge such errors, but to actually evaluate them. In this respect, we are fortunate in finance since we can often measure the magnitude of errors in meaningful units, namely euros (or dollars, francs or whatever your favourite currency). Some errors are simply bigger than others and, thus, matter more. True, such interpretation is often difficult and imprecise, but a carefully exploring, quantifying and discussing the effects of model choices etc. is always preferred to dismissing such a discussion as 'out-of-scope'.

Let us start with model errors. A portfolio in Markowitz' model is, in essence, a return distribution, which looks good or bad according to the objective function. How we define this objective function has considerable impact on what portfolio we choose. The word risk for instance is often used almost synonymously with return variance. But there are other ways to define risk. A typical objection against variance is that it penalises upside as well as downside. And indeed, already more than half a century ago, Markowitz thought about using downside semi-variance instead, which corresponds much better to the financial practitioner's notion of risk [42]. To quantify how relevant these differences in model specification are, we need to empirically compare[2] different models with respect to how they help to solve our actual problem.

There are other factors than the objective function that affect the quality of a model. Transactions costs for instance can be relevant for specific asset classes, and hence a model that includes them may be better than a model that does not. That does not mean that the analyst should try to put every possible detail into the model. Every Unix user knows that `less` is more powerful than `more`, and the same often holds in modelling. In some cases, simple back-of-the-envelope calculations make clear that certain aspects cannot matter. But more often, whether a certain aspect should be modelled or not, is not clear from the start. Principle (ii) tells us what to do in such cases: we run experiments. We should stress that if we decide to ignore certain aspects in the model, it will be because we consider them unimportant, perhaps even harmful, in an economic or empirical sense, or because we cannot reliably implement them (think of mean return forecasts). But it will not be because we could not handle them computationally.

Once a model is established, it needs to be connected to reality. We need to input forecasts and expectations. We can, for instance, only minimise variance if we have a variance–covariance matrix. Again, such inputs may be good or bad, and we have another source of error. The difficulties in forecasting the required variables are well-established, see Brandt [5] for an overview. And it is not only the forecasting problem: results are often extremely sensitive to seemingly minor setup variations, for instance, the chosen time horizon [1, 21, 36]. That makes it difficult to reject bad models.

The focus of this chapter is not the empirics of portfolio selection models, but their numerical solution. Nevertheless, we chose to review these problems to put this part of the portfolio selection process into perspective. From now on, we will assume that the model and its input have been fixed, so our task remains to solve the model, for which we use a computer. There are two sources of error. Round-off error, because we cannot represent every real number within finite memory; and truncation error, because all computations that 'go to the limit' (be it zero or infinity) must stop before the limit is reached.

Round-off error should rarely be a concern in financial optimisation (see also Trefethen [52]). It can cause trouble, for sure, but its impact, when compared with model error, is many orders of magnitude smaller.

Truncation error is more relevant for our discussion. In principle, we could solve any optimisation model through random-sampling. If we sampled more and more candidate solutions, we should—in principle—come arbitrarily close to the

[2]In an empirically sound way, which essentially means careful data analysis and replication. See, for example, Cohen [7].

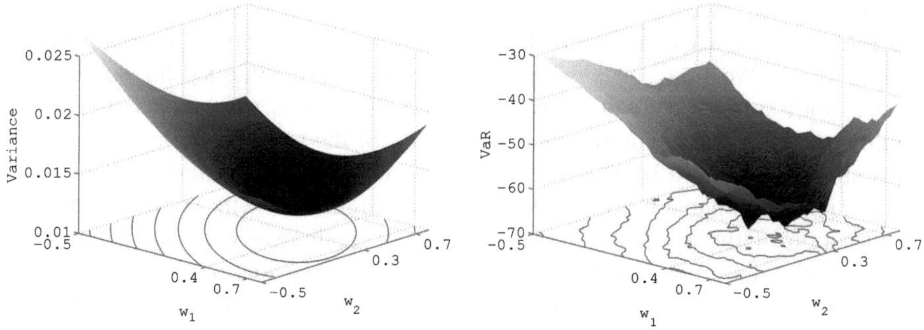

Fig. 10.1 Objective function values for a portfolio selection model with three assets. x- and y-axis show weights for two assets; the third weight is fixed through the budget constraint. *Left panel*: objective function is variance. *Right panel*: objective function is Value-at-Risk

model's solution. But clearly, in most cases that would be an extremely inefficient way to handle a model.

With heuristics, we face a variant of this truncation error. We have not actually detailed so far what heuristics actually are, but it should suffice to say here that they are iterative methods. The truncation error comes in here because heuristics only provide a stochastic approximation to the optimal solution. In other words, if we run a heuristic once, the result can be considered the realisation of random variable. The distribution of this random variable is a function of the computational effort we make. More precisely, more effort (e.g., more iterations), better solutions.

Of course, only obtaining a stochastic approximation of a solution is not really satisfactory from the standpoint of optimisation theory. After all, a model's solution is the optimum; theoretically, there are no better or worse solutions, only *the* solution and everything else.

But such approximate solutions are still useful, simply because better *models* can be used, models that would be too difficult to solve with a classical method. And it turns out that in portfolio selection most models are difficult to solve. As an example, Fig. 10.1 shows, in its left panel, the variance of a portfolio consisting of three assets. (Actually the squareroot of the variance. The third asset's weight is fixed through the budget constraint.) This is Markowitz's objective function. In the right panel we use the same dataset, but this time the objective function is Value-at-Risk, a quantile of the return distribution. The function for Value-at-Risk is not smooth, and a classic method that uses the gradient may become stuck in a local minimum (if the gradient provides useful guidance at all). Heuristics were specially designed to overcome such local minima, as we will discuss in the next section.

But let us summarise the discussion first. Selecting financial portfolios is much more than running an optimisation algorithm. The modelling process goes from actual problem to a model to the model's solution. (And, finally, we may want to implement the model-solution.)

To see whether we have solved the actual problem—in our case, selecting 'good' portfolios—or at least improved the current status, we follow the chain, but in reverse order. We first check whether our numerical techniques for solving the model work sufficiently well. Then we check if there is a relation between the quality of an in-sample solution and the quality of the out-of-sample solution; see Gilli and Schumann [22]. This is a very important test of the quality of a model, since it is unreasonable to assume that 'only the minimum' will work well; a small perturbation of the optimal solution should not qualitatively change the results. For example, if we generated a large number of random solutions (see Burns [6], for a discussion) and sorted these random solutions by in-sample quality, then we would like to see a positive relation between in-sample and out-of-sample quality. In fact, such tests can also help to evaluate the required precision for an optimisation algorithm. If we cannot empirically establish this relation, there is little point in optimising. Finally, we can compare different models with one another [23].

10.3 Heuristics

10.3.1 What Are Heuristics?

Different people mean different things when they speak of 'heuristics'. Very loosely, a heuristic is a decision rule (or *modus operandi*) that (1) typically helps to solve a problem or to a make good decision, and that (2) is simple. This is roughly the definition of Pearl [45], and it coincides with a definition that many computer scientists and programmers employ: heuristics as simple rules that provide good answers to problems in typical cases.

In mathematics, a heuristic is a line of reasoning that cannot be formally proved but leads to correct conclusions nonetheless [46]. This idea deserves repetition, because it is relevant for practical optimisation, too: we may not be able to prove that something works, but we can have empirical evidence that it does. Or in other words: not being able to prove that something works does not imply that it does not work.

In psychology a heuristic is a rule-of-thumb, a simple prescription, for decision making. While D. Kahneman and A. Tversky's heuristics-and-biases programme gave the term a rather bad reputation [53], a more favourable re-interpretation of their results has gained influence more recently; see for example Gigerenzer [17, 18].

In fact, there is something fascinating about simplicity when it comes to predicting and operating under uncertainty. Studies in fields such as econometrics, psychology, marketing research, machine learning or forecasting in general document that while simple methods lose out against more sophisticated ones in stylised settings, they yield excellent results in realistic situations [3, 8, 9, 29, 37–39]. More specifically, simple methods work well in the presence of noise and uncertainty—which is exactly what we have in finance.

But within their respective disciplines, those studies represent no more than niches. The basic idea—providing evidence that simple is sufficiently good—can also be found in the literature on portfolio optimisation in the 1970s; see for instance Elton and Gruber [13] or Elton et al. [14]. The common thread throughout these papers was the justification of simplified financial models. The problem back then was to reduce computational cost, and the authors tried to empirically justify simpler techniques. Today, complicated models are feasible, but they are still not necessarily better. (Again, Principle (i): the application matters.)

We will define heuristics in a narrower, more-precise sense: as a class of numerical methods for solving optimisation models. The typical model is

$$\underset{x}{\text{minimise}}\, \phi(x, \text{data})$$

in which ϕ is a scalar-valued function and x is a vector of decision variables. (As we already said, by putting a minus in front of ϕ we can make it a maximisation model.) In many cases, we will add constraints to the model.

We find it helpful to not think in terms of a mathematical description, but rather to replace ϕ by something like

$$\texttt{solutionQuality} = \texttt{function(x, data)}\,.$$

That is, we need to be able to program a mapping from a solution to its quality, given the data. There is no need for a closed-form mathematical description of the function.[3] Indeed, in many applied disciplines there are no closed-form objective functions. The function ϕ could include an experimental setup, with x the chosen treatment and $\phi(x)$ the desirability of its outcome. Or evaluating ϕ might require a complicated stochastic simulation, such as an agent-based model.

A number of requirements describe an optimisation heuristic further ([4, 57, 58] list similar criteria):

- The method should give a 'good' stochastic approximation of the true optimum, with 'goodness' measured in computing time or solution quality.
- The method should be robust when we change the model—for instance, when we modify the objective function or add a constraint—and also when we increase the problem size. Results should not vary too much for different parameter settings for the heuristic.
- The technique should be easy to implement.

[3]Mathematically a function is nothing but a mapping, so there is no contradiction here. But when people see $\phi(x)$ they intuitively often think of something like $\phi(x) = \sqrt{x} + x^2$. We would prefer they thought of a programme, not a formula.

- Implementation and application of the technique should not require subjective elements.

Such a definition is not unambiguous, but it is a start. Actually, we think that users can only gain intuition about heuristics through studying examples—which we will do in the next section. But for now, we shall go on dwelling on principles.

In a broad sense, we can differentiate between two classes of heuristics, constructive methods and iterative-search methods. In this chapter, we shall concentrate on the latter type, so let us give a quick example for constructive methods and then not mention them any further. For a constructive method, an algorithm starts with an empty solution and adds components step-by-step; the procedure terminates when it has completed one solution. An example: a reasonable low-variance equity portfolio of cardinality N can be constructed by (1) obtaining forecasts for the marginal variances of all eligible assets, (2) sort the assets by forecast variance and (3) keep the N assets with the lowest forecast variance in the portfolio (equally-weighted); see Schumann [49].

For iterative search methods the algorithm moves from solution to solution, that is, a complete existing solution is modified to obtain a new solution. Such a new solution may be quite different from previous ones, as some methods, such as Genetic Algorithms, create new solutions in a rather discontinuous ways. But still, a new solution will share characteristics with its predecessor (if that was not the case, we would be doing random-sampling).

10.3.2 Principles

The following pseudocode should make the idea of an iterative method more precise.

```
generate initial solution x^c
while stopping condition not met do
    create new solution x^n = N(x^c)
    if A(φ, x^n, x^c, ...) then x^c = x^n
end while
return x^c
```

In words: we start with a solution x^c, typically randomly chosen. Then, in each iteration, the function N ('neighbour') makes a copy of x^c and modifies this copy; thus, we get a new candidate solution x^n. The function A ('accept') decides whether x^n replaces x^c, typically by comparing the objective function values of the solutions. The process is repeated until a stopping condition is satisfied; finally, x^c is returned.

To implement such a method, we need to specify

- how we represent a solution x,
- how we evaluate a solution (the function ϕ),
- how we change a solution (the function N),
- how to decide whether to accept a solution (the function A),
- when to stop.

These building blocks would still apply to a classical method. For example, for a gradient-based method x would be a numeric vector; N would evaluate the gradient at x^c and then move minus the gradient with a specified stepsize; A would evaluate x^c and x^n, and replace x^c only if x^n is better; if not, the search is stopped.

Heuristics use other, often simpler, mechanisms. In fact, two characteristics will show up in almost all methods. (1) Heuristics will not insist on the best possible moves. A heuristic may accept a new solution x^n even if it is worse than the current solution. (2) Heuristics typically have random elements. For instance, a heuristic may change x^c randomly (instead of locally-optimally as in a gradient search). These characteristics make heuristics inefficient for well-behaved models. But for difficult models (for instance, such with many local optima as in Fig. 10.1), they enable heuristics to move away from local optima.[4]

Let us give a concrete example, namely the problem we already used earlier: we want to select N assets, equally-weighted, out of a large number of assets, such that the resulting portfolio has a small variance. We assume that we have a forecast for the variance–covariance matrix available. Then a simple method for getting a very good solution to this model is a local search. For a local search,

- the solution x is a list of the included assets;
- the objective function ϕ is a function that computes the variance forecast for a portfolio x;
- the function N picks one neighbour by randomly removing one asset from the portfolio and adding another one;
- the function A compares $\phi(x^c)$ and $\phi(x^n)$, and if x^n is not worse, accepts it;
- the stopping rule is to quit after a fixed number of iterations.

Note that local search is still greedy in a sense, since it will not accept a new solution that is worse than the previous one. Thus, if the search arrives at a solution that is better than all its neighbours, it can never move away from it—even if this solution is only a local optimum. Heuristic methods that build on local search thus employ additional strategies for escaping such local optima.

And indeed, with a small—but important—variation we arrive at Simulated Annealing [35]. We use a different acceptance rule A: If the new solution is better, accept it. If it is worse, do still accept it, but only with a specific probability. This probability in turn depends on the new solution's quality: the worse it is, the less likely it is the solution is accepted. Also, the probability of acceptance is typically lower in later iterations (that is, the algorithm becomes pickier). In many implementations, the probability at later stages is essentially zero; thus, Simulated Annealing turns into a local search.

[4]In principle, because of such mechanisms a heuristic could drift farther and farther off a good solution. But practically, that is very unlikely because every heuristic has a bias towards good solutions. In Threshold Accepting, the method that we describe in Sect. 10.4, that bias comes into effect because a better solution is always accepted, a worse one only if it is not too bad. Since we repeat this creating of new candidate solutions thousands of times, we can be very certain that the scenario of drifting-off a good solution does practically not occur.

10.3.3 *Constraints*

Nothing in the pseudocode that we showed above ensures that a constraint on a solution x is observed. But it is often constraints that make models realistic and difficult. Several strategies exist for including restrictions into heuristics.

10.3.3.1 Throw Away

If our model has only few constraints that are not often hit, the simplest approach is to 'throw away' infeasible new solutions. That is, if a neighbour solution violates a constraint, we just select another neighbour. Note that this means that we include the constraints in the acceptance function A.

10.3.3.2 Include Constraint in N

We can directly use the constraint to create new, feasible solutions. In portfolio selection models we usually have a budget constraint; that is, we require that all asset weights sum to one. This constraint can be enforced when we compute new solutions by increasing some weights and decreasing others such that the sum of all weight changes is zero.

10.3.3.3 Transform x

An older but still used idea is to transform variables. This approach sometimes works for constraints that require that the elements of x lie in certain ranges; see the discussion in Powell [47]. For instance, $\sin(x)$ will map any real x to the range $[-1, 1]$; $\alpha\,(\sin(x))^2$ will give a mapping to $[0, \alpha]$. But such transformations come with their own problems; see Gill et al. [24, Sect. 7.4]; in particular, it may become difficult to change a problem later on or to handle multiple constraints.

10.3.3.4 Repair x

We can introduce mechanisms to correct solutions that violate constraints. For example, if a solution x holds the portfolio weights, then dividing every element in x by the sum of the elements of x ensures that all weights sum to unity.

10.3.3.5 Penalise x

Finally, we can penalise infeasible solutions. Whenever a constraint is violated, we add a penalty term to the objective function and so downgrade the quality of the solution. In essence, we change the problem to an unconstrained one for which we can use the heuristic. The penalty is often made an increasing function of the magnitude of violation. Thus, the algorithm may move through infeasible areas of the search space, but will have guidance to return to feasible areas. The penalty approach is the most generic strategy to include constraints; it is convenient since the computational architecture needs hardly be changed. Penalties create soft constraints since the algorithm could in principle always override a penalty; practically, we can set the penalty so high that we have hard constraints.

10.3.4 Random Solutions

The most common objection against using heuristics is the fact that, since heuristics explicitly rely on random mechanisms, their solutions are also random. This randomness, it is argued, makes it difficult to evaluate the quality of solutions computed by such algorithms. (The discussion in this section builds on Gilli et al. [27].)

10.3.4.1 Randomness

A naïve approach to solving an optimisation model could be this: randomly generate a large number of candidate solutions, evaluate all solutions and pick the best one. This best solution is our overall solution.

If we repeated the whole procedure a second time, our overall solution would probably be a different one. Thus, the solution x we obtain through our sampling strategy is stochastic. The difference between our solution and the actual optimum would be a kind of truncation error, since if we sampled more and more, we should in theory come arbitrarily close to the optimum. Importantly, the variability of the solution stems from our numerical technique; it has nothing to do with the error terms that we may have in models to account for uncertainty. Stochastic solutions may even occur with non-stochastic methods: think of search spaces like those we showed in Fig. 10.1. Even if we used a deterministic method like a gradient search, the many local minima would make sure that repeated runs from different starting points result in different solutions.

We can treat the result of a stochastic algorithm as a random variable with some distribution D. What exactly the 'result' of a restart is depends on our setting. We will want to look at the objective function value (i.e., the solution quality), but we may also look at the decision variables given by a solution, that is, the portfolio weights. In any case, we collect all the quantities of interest in a vector ϱ. The result ϱ_j of a restart j is a random draw from D.

The trouble is that we do not know what D looks like. But fortunately, there is a simple way to find out for a given model. We run a reasonably large number of restarts, each time store ϱ_j, and finally compute the empirical distribution function of the $\varrho_j, j = 1, \ldots, \text{number} - \text{of} - \text{restarts}$ as an estimate for D. For a given model or model class, the shape of the distribution D will depend on the chosen method. Some techniques will be more appropriate than others and give less variable and on average better results. And D will often depend on the particular settings of the method, in particular the number of iterations—the search time—that we allow for.

Unlike classical optimization techniques, heuristics can walk away from local minima; they will not necessarily get trapped. So if we let the algorithm search for longer, we can hope to find better solutions. For minimization problems, when we increase the number of iterations, the mass of D will move to the left and the distribution will become less variable. Ideally, when we let the computing time grow ever longer, D should degenerate into a single point, the global minimum. There exist proofs of this convergence to the global minimum for many heuristic methods (see Gelfand and Mitter [16], for Simulated Annealing; Rudolph [48], for Genetic Algorithms; Gutjahr [31], Stützle and Dorigo [51], for Ant Colony Optimisation; Bergh and Engelbrecht [54], for Particle Swarm Optimisation).

Unfortunately, these proofs are not much help for practical applications. They often rely on asymptotic arguments; and many such proofs are nonconstructive (e.g., Althöfer and Koschnick [2], for Threshold Accepting): they demonstrate that parameter settings exist that lead (asymptotically) to the global optimum. Yet, practically, there is no way of telling whether the chosen parameter setting is correct in this sense; we are never guaranteed that D really degenerates to the global optimum as the number of iterations grows.

Fortunately, we do not need these proofs to make meaningful statements about the performance of specific methods. For a given model class, we can run experiments. Such experiments also help investigate the sensitivity of the solutions with respect to different parameter settings for the heuristic. Experimental results are of course no proof of the general appropriateness of a method, but they are evidence of how a method performs for a given class of models; often this is all that is needed for practical applications.

10.4 An Example: Threshold Accepting

In this section, we will discuss one heuristic, Threshold Accepting, in more detail. The algorithm is a simplified variant of Simulated Annealing and was first proposed by Dueck and Scheuer [11] and Moscato and Fontanari [44]. As far as we know, it was also the first optimisation heuristic used for portfolio selection [12]. For an overview of the application of other heuristics, such as Ant Systems or Simulated Annealing, to portfolio selection models, we recommend [40].

10.4.1 The Algorithm

Threshold Accepting (TA) builds on local search. A local search starts with a random feasible solution x^c (in our case, a random portfolio), which we call the 'current solution' since it represents the best we have so far. Then, a new solution x^n close to x^c is randomly chosen. We will discuss 'close to x^c' shortly. We call x^n a neighbour to the current solution, or simply a neighbour. If x^n is better than x^c, then x^n replaces x^c (i.e., the neighbour becomes the current solution); if not, another neighbour is selected. Algorithm 1 gives this procedure in pseudocode. The stopping criterion here is a number of iterations n_{Steps}, fixed in advance by the analyst.

Algorithm 1 Local Search

---

set n_{Steps}
randomly generate current solution x^c
for $i = 1 : n_{\text{Steps}}$ **do**
    generate $x^n = N(x^c)$ and compute $\Delta = \phi(x^n) - \phi(x^c)$
    if $\Delta < 0$ **then** $x^c = x^n$
end for
return x^c

All that the method requires is that the objective function can be evaluated for a given portfolio x; there is no need for the objective function to be continuous or differentiable. Of course, for problems with many local minima, a local search will stop at the first local optimum it finds.

TA adds a simple escape strategy for local minima: it will not only accept new solutions that are better, but also allow uphill moves, as long as the deterioration of ϕ does not exceed a fixed threshold (thus its name). Over time, that threshold decreases to zero; so eventually TA turns into a local search. The whole procedure is summarised in Algorithm 2.

Algorithm 2 Threshold Accepting

1: set n_{Steps} and n_{Rounds}
2: compute threshold sequence τ
3: randomly generate current solution x^c
4: **for** $k = 1 : n_{\text{Rounds}}$ **do**
5:     **for** $i = 1 : n_{\text{Steps}}$ **do**
6:         generate $x^n = N(x^c)$ and compute $\Delta = \phi(x^n) - \phi(x^c)$
7:         **if** $\Delta < \tau_k$ **then** $x^c = x^n$
8:     **end for**
9: **end for**
10: return x^c

Compared with local search, the changes are actually small. We add an outer loop that controls the thresholds τ. In Statement 7, the acceptance criterion is changed from '$\Delta < 0$' to '$\Delta < \tau_k$', i.e., from 'if improvement' to 'if not worse than τ_k'. For an actual implementation, we need a way to represent a solution, an objective function ϕ, the neighbourhood function N, the thresholds τ and the stopping criterion (see Sect. 10.3.2 above). The first and the last choice are quickly made: a solution is a numeric vector of positions (or perhaps weights); the stopping criterion is simply a fixed number of iterations.[5]

10.4.2 Implementation

10.4.2.1 The Objective Function

In Markowitz's mean–variance model, we need forecasts of the means, variances and correlations of the assets. With these inputs, we can easily compute forecasts of mean and variance for any specific portfolio.

Unfortunately, this approach rarely generalises to other specifications of risk and reward, or only in ways that are computationally very costly [34, Chap. 9]. For instance, even if we had forecasts of the lower partial moments of all the assets in the portfolio, we could not aggregate these to the lower partial moment of the whole portfolio [30].

But we do actually not require such an aggregation. Instead, we will do scenario optimisation [10]. The simplest case is to regard every historical return observation as one scenario. However, using actual forecasts (e.g., creating 'artificial' scenarios through resampling) can help to improve the out-of-sample performance of portfolios [20, 28].

Suppose we have n_A assets and n_S return scenarios, all collected in a matrix R of size $n_S \times n_A$. We can equivalently work with price scenarios, computed as

$$P = (1 + R) \times \text{diag}(p_0)$$

in which 1 is a matrix of ones of size $n_S \times n_A$, and 'diag' is an operator that transforms a vector into a diagonal matrix. Note that the columns of P (or R) are not time series. Every row of P holds the prices for one possible future scenario that might occur, given initial prices p_0. In fact, for many objective functions (such as partial moments), it is not relevant whether the scenarios are sorted in time, since such criteria only capture the cross-section of returns. The portfolio values v in these scenarios are given by the product Px.

But there are selection criteria that need time series, for instance drawdown. Resampling is still possible to create path scenarios: we may, for instance, use models that capture serial dependencies and then resample from their residuals, or use a block bootstrap method. Assume we have only a single scenario of paths of the assets, and arrange the prices in a matrix P^{ts} of size $(T + 1) \times n_A$, where each column holds the prices of one asset. The portfolio values over time are then given

[5]The number of iterations depends on the problem. Here, again, Principle (ii) tells us how to proceed: small-scale experiments will quickly provide us with a reasonable idea of how many iterations are needed. See Gilli et al. [27]; in particular Chaps. 11 and 12.

by $v = P^{\text{ts}}x$. Note that Px gives a sample of portfolio values over the cross-section of scenarios, while $P^{\text{ts}}x$ gives one path of the portfolio value from time 0 to time T. For both scenarios and paths, given a vector v it is easy to evaluate an objective function.

Working with scenarios in this way is not restrictive: if we preferred a parametric model, we could always calibrate it to the scenarios (e.g., compute a variance–covariance matrix from the scenarios). The approach offers another advantage, namely that we can make the time required to evaluate the objective function independent of the number of assets. Assume we work with the matrix P of price scenarios (the same holds for P^{ts}). This matrix is often fairly large, with thousands of scenarios for hundreds or thousands of assets. TA started with an initial portfolio x^{c} and now has to evaluate x^{n}. Hence the product Px^{c} has already been computed. As will be discussed below, a new portfolio will be created by a small perturbation of the original portfolio, hence

$$x^{\text{n}} = x^{\text{c}} + x^{\Delta}$$

where x^{Δ} is a vector with few nonzero elements (usually only two). Then

$$Px^{\text{n}} = P(x^{\text{c}} + x^{\Delta}) = \underbrace{Px^{\text{c}}}_{\text{known}} + Px^{\Delta}\,.$$

Many elements of x^{Δ} are zero, and hence only a few columns of P are necessary for the matrix multiplication. So we create a matrix P_* that only holds the columns where x^{Δ} is nonzero and a vector x^{Δ}_* that contains only the nonzero elements of x^{Δ}; then we replace Px^{Δ} by $P_* x^{\Delta}_*$.

10.4.2.2 The Neighbourhood Function

To move from one solution to the next, we need to define a neighbourhood function N that creates new candidate solutions. For portfolio selection problems, we have a natural way to create neighbour solutions: pick one asset in the portfolio randomly, 'sell' a small quantity of this asset, and 'invest' the amount obtained in another asset. If short positions are allowed, the chosen asset to be sold does not have to be in the portfolio. The 'small quantity' may either be a random number or a small fixed fraction such as 0.1 %. Experiments suggest that, for practical purposes, both methods give similar results; a fixed fraction may even be preferred.

10.4.2.3 The Threshold Sequence

The threshold sequence  is an ordered vector of positive numbers that decrease to zero or at least become very small. The neighbourhood definition and the thresholds are tightly connected. Larger neighbourhoods, with larger changes from one candidate portfolio to the next, need generally be accompanied by larger initial threshold values, et vice versa.

Winker and Fang [56][6] suggest a data-driven method to obtain the thresholds: generate a large number of random solutions; select one neighbour of each solution according to the given neighbourhood definition; compute the difference between the objective function values for each pair. The thresholds are then a number of decreasing quantiles of these differences; see Algorithm 3.

Algorithm 3 Computing the Threshold Sequence—Variant 1

set n_{Rounds} (# of thresholds), n_{Deltas} (# of random solutions)
for $i = 1$ to n_{Deltas} **do**
  randomly generate current solution x^c
  generate $x^n = N(x^c)$
  compute $\Delta_i = |\phi(x^n) - \phi(x^c)|$
end for
sort $\Delta_1 \leqslant \Delta_2 \leqslant \cdots \leqslant \Delta_{n_{\text{Deltas}}}$
set $\tau = \Delta_{n_{\text{Rounds}}}, \ldots, \Delta_1$

The number of thresholds n_{Rounds} with this approach is typically large, hence few steps n_{Steps} per threshold suffice in the inner loop of Algorithm 2; often it is only one step per threshold.

Gilli et al. [25] suggest to take a random walk through the data instead, recording the changes in the objective function value at every iteration. The thresholds are then a number of decreasing quantiles of these changes. See Algorithm 4.

Algorithm 4 Computing the Threshold Sequence—Variant 2

set n_{Rounds} (# of thresholds), n_{Deltas} (# of random steps)
randomly generate current solution x^c
for $i = 1 : n_{\text{Deltas}}$ **do**
  generate $x^c = N(x^c)$ and compute $\Delta_i = |\phi(x^n) - \phi(x^c)|$
  $x^c = x^n$
end for
compute empirical distribution (CDF) of Δ_i, $i = 1, \ldots, n_{\text{Deltas}}$
compute threshold sequence $\tau_k = \text{CDF}^{-1}\left(\frac{n_{\text{Rounds}} - k}{n_{\text{Rounds}}}\right), k = 1, \ldots, n_{\text{Rounds}}$

10.4.2.4 Constraints

Several generic approaches for handling constraints were discussed in Sect. 10.3.3 above; we will typically use a mixture of those. The budget constraint for example

[6] Similar techniques are used to obtain settings for Simulated Annealing; see for instance Johnson et al. [33].

is automatically enforced by the specification of the neighbourhood. Gilli et al. [27] provide complete implementations for various other constraints.

10.5 An Example: Portfolio Selection with TA

To give an example[7] about the modelling approach, guided by the two principles mentioned in the introduction, i.e., (1) the application matters, and (2) go experiment, we construct a portfolio with the objective to 'at least' replicate a given hedge fund index. In the following we concentrate only on the modelling process.

10.5.1 Data, Backtesting Scheme and Reporting of Results

The index to replicate is the Credit Suisse/Tremont Hedge Fund Index (CST) available at www.hedgeindex.com. According to the information onsite this index is asset-weighted, includes more than 5000 funds with a minimum of US$ 50 million under management. The observations are monthly and cover the period from January 1999 to October 2009.

The instruments used for the replication comprise equity, commodity and bond indices. In the set of equity indices we have about 54 series including broad market, blue chips, sector as well as size and style indices. There are 12 commodity indices and 12 bond indices from government, corporate and emerging markets. The set of data has been collected from Bloomberg.

To analyze the performance of the suggested portfolios we backtest the strategies over 10 years with rebalancing where we account for 10 basis points of transaction costs. The rolling window has a historical length of H and a holding period of F. In this application H is 1 year and F is 3 months. This leads to trajectories of portfolios values where the portfolios have been rebalanced forty times. The scheme below summarises the technique.

To gain insight into the stochastics of the simulated portfolios we jackknife from the historical observations so as to compute a set of results[8] for which we then consider empirical distributions for different features of the portfolio.

For each objective function we report the characteristics of the backtested portfolio in three figures and one table, i.e.: (1) plot of median path of the portfolio value,[9] (2) plot of the kernel estimation of the density of the empirical distribution of the mean yearly return, (3) plot of the empirical distribution of the correlation of

[7]The example builds on Gilli et al. [26].

[8]In this case we computed 100 trajectories for each specification of the objective function.

[9]The median path is defined with respect to the final wealth of the portfolios generated with the jackknifing.

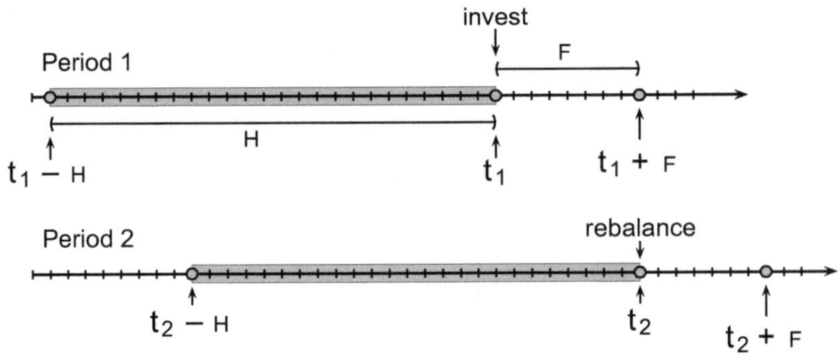

the portfolio with the market and (4) a table showing the tracking error *TE* defined as the standard deviation of the excess return r_E, the Sharpe ratio S, the annualised return and volatility $\bar{r}$, *vol* and the correlation ρ_{r_P,r_M} with the market.

10.5.2 'Genesis' of a Model

Instead of starting with a predefined idea about what the 'best' model should be we chose to begin with the simplest ideas and subsequently add (or remove) elements in the objective function ϕ according to whether or not they have a desirable impact on the results obtained by backtesting.

The general optimisation problem we have to solve for the different objective functions ϕ is then

$$\min_x \ \phi(x) \tag{10.4}$$

$$\sum_{j \in \mathscr{J}} x_j p_{0j} = v_0 \tag{10.4$'$}$$

$$x_j^{\inf} \le x_j \le x_j^{\sup} \qquad j \in \mathscr{J} \tag{10.4$''$}$$

$$K_{\inf} \le \#\{\mathscr{J}\} \le K_{\sup} \tag{10.4$'''$}$$

$$\vdots$$

where x is a vector with x_j being the quantity of asset j in the portfolio. The optimisation is subject to a set of constraints with in particular the budget constraint (10.4′) with v_0 the investable wealth and p_{0j} the price of asset j at the beginning of the investment period. Constraint (10.4″) specifies minimum and maximum holding size for the set of asset ($\mathscr{J}$) in the portfolio. Next we have the cardinality constraint (10.4‴) which allows for tractability of the resulting portfolios. Further constraints might be included such as total transaction cost, sector constraints and other liquidity issues. For all our portfolios the minimum holding and maximum holding for an asset is 1 % respectively 20 % and maximum cardinality is limited to $K_{\text{sup}} = 10$. Solutions will be computed with the Threshold Accepting (TA) heuristic described previously in Sect. 10.4.

10.5.3 Step 1: Optimisation of Tracking Error and Excess Return

A first and straightforward idea is to construct a tracking portfolio, i.e., a portfolio that minimises the distance between historical portfolio returns and the index returns. We denote r_P the historical return vector of the tracking portfolio, r_I the index returns and $r_E = r_P - r_I$ the excess return of the portfolio. In order not to penalise upside deviations for a portfolio we also consider the mean excess return $\overline{r_E} = \frac{1}{n}\sum(r_P - r_I)$ leading to the following objective function

$$\lambda \, \|r_E\|^2 - (1 - \lambda)\, \overline{r_E} \qquad (10.5)$$

where $\lambda \in [0, 1]$ defines a linear combination between tracking error and excess return.[10]

We computed results for the objective (10.5) for varying values of the parameter λ controlling the weighting between tracking error and reward. Figure 10.2 shows the median paths for 100 simulations for $\lambda = 1$ (dark line) which corresponds to minimizing only tracking error. We can trade tracking error against final wealth by decreasing λ. A good compromise is obtained for $\lambda = 0.75$ yielding portfolios close to the index whereas higher weights for the reward (lower values for λ) result in higher final wealth but also higher volatility.

Figure 10.3 illustrates another feature of the simulated portfolios. It shows the plot of the kernel estimation of the density of the empirical distribution of the mean yearly return of the tracking portfolio (left panel). The vertical line indicates the mean yearly return of the index. The right panel in Fig. 10.3 shows the empirical distribution of the correlation of the optimised portfolios with the market. The dotted line corresponds to the correlation of the index with the market. For the tracking portfolios we observe higher values. Table 10.1 summarises the performance for the portfolios obtained with objective function (10.5).

The portfolio for $\lambda = 0.75$ has an average return increased by 50 % compared with $\lambda = 1$ in exchange of insignificant loss in tracking performance and small increase of volatility. Furthermore the correlation of the portfolio with the market decreases.

---

[10]Such an approach has been explored in Gilli and Këllezi [19] using artificial data.

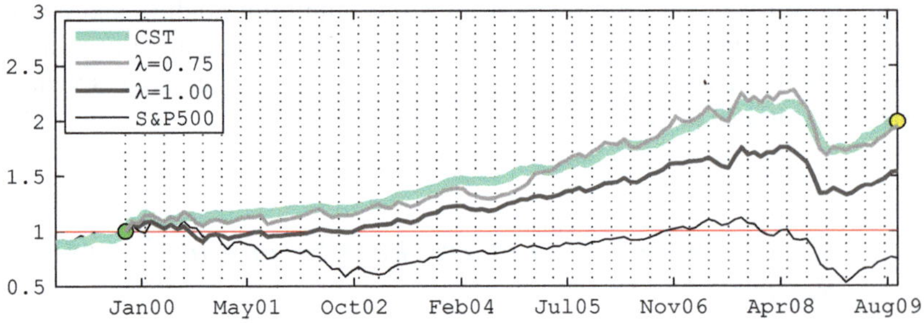

Fig. 10.2 Median paths for portfolios minimizing objective function (10.5) for $\lambda = 0.75$ and $\lambda = 1$. For reference the performance of the CS Tremont (CST) and the S&P500 is also shown. *Dotted vertical lines* indicate rebalancing dates

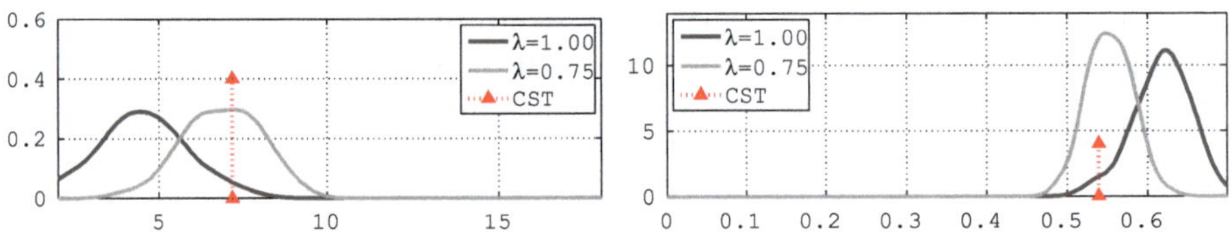

Fig. 10.3 Densities of mean yearly return $\overline{r}$ (*left panel*) and correlation ρ_{r_P,r_M} (*right panel*)

Table 10.1 Results for median path of simulated portfolios for objective function (10.5) (Tracking error *TE*, Sharpe ratio *S*, annualised return and volatility $\overline{r}$, *vol*, correlation with market ρ_{r_P,r_M})

	TE	*S*	$\overline{r}$	*vol*	ρ_{r_P,r_M}
$\lambda = 0.75$	2 %	0.62	6 %	10 %	0.55
$\lambda = 1$	2 %	0.45	4 %	9 %	0.62
CST	–	1.04	7 %	7 %	0.54

10.5.4 Step 2: Optimisation of Tracking Error, Excess Return and ρ_{r_P,r_M}

The goal is to construct a portfolio following the index as close as possible but being little sensitive to adverse market movements. This suggests to include the correlation between tracking portfolio and market into the objective function

$$\lambda \, \|r_E\|^2 - (1-\lambda)\,\overline{r_E} + \rho_{r_P,r_M} \tag{10.6}$$

where ρ_{r_P,r_M} denotes this correlation computed from the historical data. Minimizing the objective function minimises this correlation. The results for the portfolios minimizing the objective function (10.6) are given in Fig. 10.4. For $\lambda = 1$, where no excess return enters the optimisation, we observe a particularly smooth median path almost not affected by the drop in the S&P500 at the end of 2008 (Table 10.2).

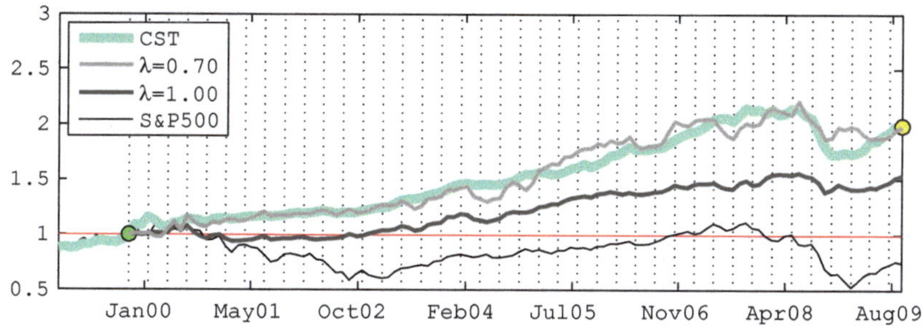

Fig. 10.4 Median paths for portfolios minimizing the objective function (10.6) controlling correlation ρ_{r_P,r_M} with the market

Table 10.2 Results for the median path of the simulated portfolios for objective function (10.6)

	TE	S	$\overline{r}$	vol	ρ_{r_P,r_M}
$\lambda = 0.70$	3 %	0.63	7 %	10 %	0.09
$\lambda = 1$	2 %	0.60	4 %	7 %	0.35
CST	–	1.04	7 %	7 %	0.54

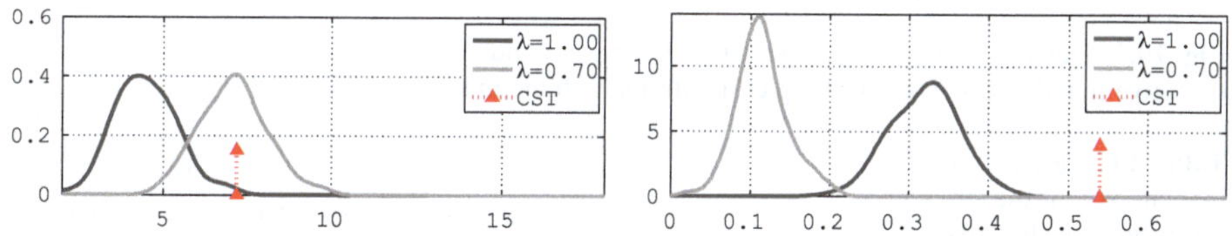

Fig. 10.5 Density of mean yearly return $\overline{r}$ (*left panel*) and correlation ρ_{r_P,r_M} (*right panel*) for the objective function (10.6) controlling correlation with the market

Results in the right panel of Fig. 10.5 are remarkable when compared with the ones in Fig. 10.3 as the distributions indicate that, while maintaining the same level of returns, correlation is now well controlled.

10.5.5 Step 3: Optimisation of Tracking Error, Excess Return, ρ_{r_P,r_M} and ρ_{r_P,r_I}

Model (10.5) focusing on the tracking error, already leads to a portfolio highly correlated with the index. Nevertheless one can think to control this correlation more specifically by introducing it into the objective function. Denoting ρ_{r_P,r_I} this correlation between portfolio and the index, it can be maximised with the new objective function

$$\lambda \, \|r_E\|^2 - (1 - \lambda)\, \overline{r_E} + \rho_{r_P,r_M} - \rho_{r_P,r_I} \, . \tag{10.7}$$

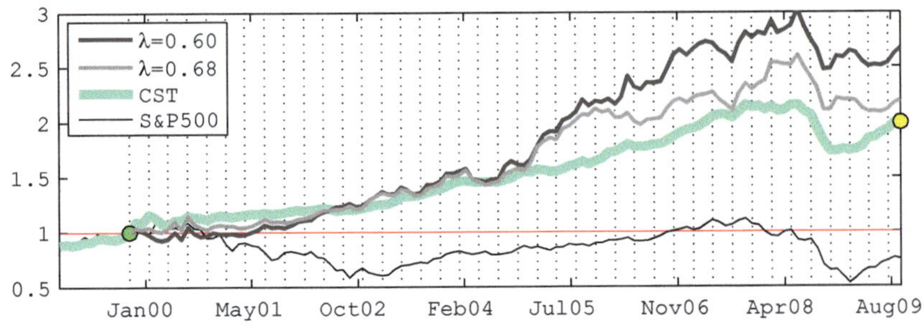

Fig. 10.6 Median paths for portfolios minimizing the objective function (10.7) controlling correlation ρ_{r_P,r_M} with the market and ρ_{r_P,r_I} with the CST

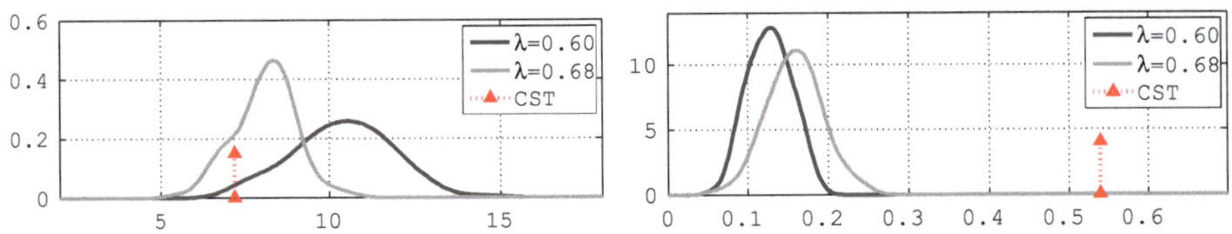

Fig. 10.7 Density of mean yearly return $\bar{r}$ (*left panel*) and correlation ρ_{r_P,r_M} (*right panel*) for the objective function (10.7) controlling correlations with the market and the CST

Table 10.3 Results for the median path of the simulated portfolios for objective function (10.7)

	TE	*S*	$\bar{r}$	*vol*	ρ_{r_P,r_M}
$\lambda = 0.60$	3 %	0.88	10 %	11 %	0.09
$\lambda = 0.68$	3 %	0.74	8 %	10 %	0.17
CST	–	1.04	7 %	7 %	0.54

Notice that the effect of λ is not comparable between the different objective functions due to the impact of the additional terms. The results indicate no significant change in overall performance, we rather observe a shift to improved Sharpe ratios for all values of λ. Again lower values for λ, i.e., higher weighting for excess return leads to portfolios with higher final wealth but of course at the cost of increased volatility (Fig. 10.6).

As visible in Fig. 10.7, from the point of view of returns $\lambda = 0.6$ produces attractive portfolios which moreover show quite low correlation with the market and high Sharpe ratios (Table 10.3).

10.5.6 Step 4: Optimisation of Tracking Error, Excess Return, ρ_{r_P,r_M}, ρ_{r_P,r_I} and $\mathcal{D}_{\max}$

A further desirable feature of a portfolio would be to have low drawdown. For a series of portfolio values v_t, $t = 0, 1, 2 \ldots T$ the drawdown is defined as

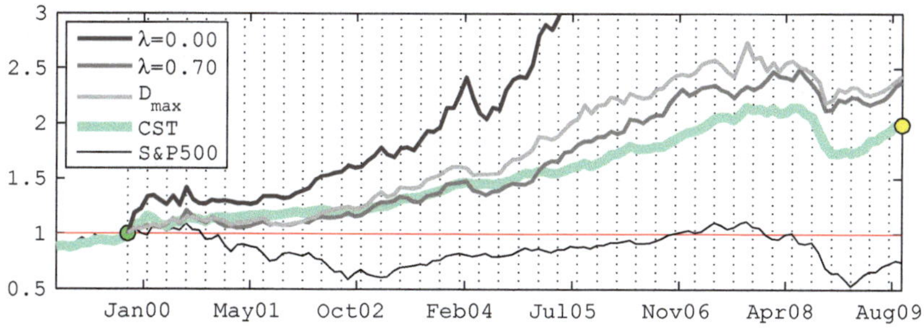

Fig. 10.8 Median paths for portfolios minimizing the objective function (10.8) controlling correlations ρ_{r_P,r_M}, ρ_{r_P,r_I} and the maximum drawdown

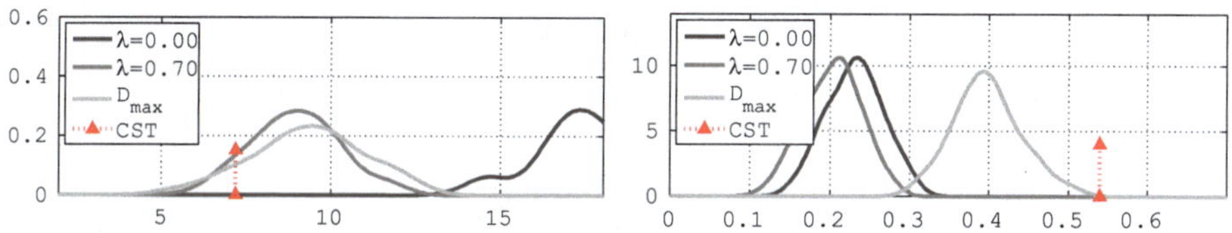

Fig. 10.9 Density of mean yearly return (*left panel*) and density of correlation with the market (*right panel*) for the objective function (10.8) controlling correlations ρ_{r_P,r_M}, ρ_{r_P,r_I} and the maximum drawdown

$$\mathscr{D}_t = v_t^{\max} - v_t$$

where $v_t^{\max}$ is the running maximum, i.e., $v_t^{\max} = \max\{v_s | s \in [0,t]\}$. Following this definition $\mathscr{D}$ is a vector for which we can compute the mean, standard deviation or the maximum element $\mathscr{D}_{\max} = \max(\mathscr{D})$ which is the one we use in our next objective function. In other words $\mathscr{D}_{\max}$ measures the largest drop of the portfolio value over the time horizon.

In a first step we consider only the minimisation of the maximum drawdown. In a second step we combine maximum drawdown minimisation with the objective function defined in (10.7) yielding

$$\lambda \, \|r_E\|^2 - (1-\lambda)\, \overline{r_E} + \rho_{r_P,r_M} - \rho_{r_P,r_I} + \mathscr{D}_{\max} \,. \tag{10.8}$$

Figure 10.8 below reports results for pure maximum drawdown minimisation as well as results for objective function (10.8) with different weighting of tracking error and excess return. As previously, higher weights for excess return (low λ values) produce high final wealth portfolios.

Looking at the final wealth distribution given in Fig. 10.9 we see that for $\lambda = 0$ we get an impressive shift to the right of the distribution (Table 10.4).

In the light of these results portfolios obtained from this last model offer properties suitable to substitute the Credit Suisse/Tremont hedge fund index. In particular for $\lambda = 0.7$ we have comparable Sharpe ratio but higher average return

Table 10.4 Results for the median path of the simulated portfolios for objective function (10.8)

	TE	*S*	$\overline{r}$	*vol*	ρ_{r_P,r_M}
$\lambda = 0$	4 %	0.94	16 %	17 %	0.23
$\lambda = 0.7$	2 %	0.97	9 %	9 %	0.17
$DD_{\max}$	2 %	1.01	9 %	9 %	0.37
CST	–	1.04	7 %	7 %	0.54

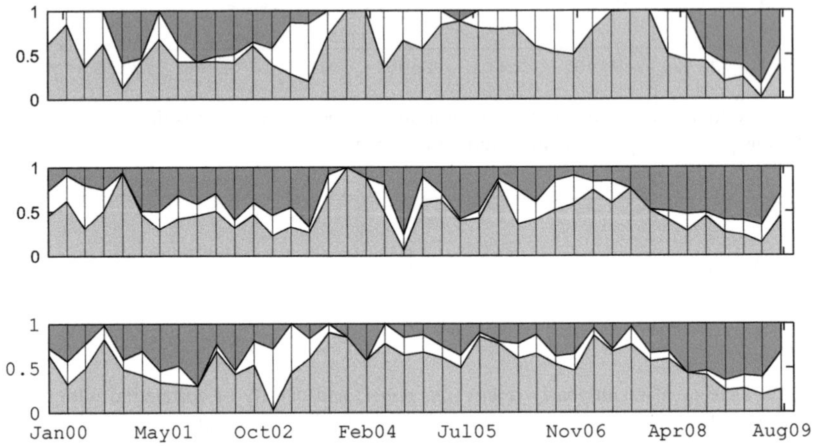

Fig. 10.10 Median portfolio composition in terms of asset classes (equities in *light gray*, commodities in white and bonds in *dark gray*) for the objective function (10.8). From *top* to *bottom* $\lambda = 0$, $\lambda = 1$ and $\mathscr{D}_{\max}$

and lower correlation with the market. However the portfolio can be modulated choosing different values for λ in order to meet different preferences or risk aversion of an investor.

It might be interesting to show how the different asset classes are represented in the median portfolios. This is plotted in Fig. 10.10 and we notice how the model reacts to market conditions. For instance, in periods of distress, the weight of fixed income instruments gains importance at the expense of equities. Also, the portfolio where excess return is favored ($\lambda = 0$) has relatively higher weighted commodities and equities.

10.6 Conclusion

John Tukey once said that an analyst of data needs both tools and understanding. In this chapter, we have given a brief, selective introduction to heuristics in portfolio selection. Heuristics are tools. Powerful tools; but even powerful tools cannot replace understanding.

As we said in the Introduction, the tasks that the analyst faces can be classified into three broad topics: modelling, forecasting and optimisation. In our view, optimisation is the least problematic of these tasks. Of course, that has not always been the case, but on today's computers with today's software, we can handle models and amounts of data that people could not imagine 20 years ago. (Actually, the amounts of data are so large that even today it is difficult to *imagine* them.)

In our view, less progress has been made when it comes to modelling and forecasting. Students are still taught financial theories 'as is', with often only parenthetical reference to practical problems, most notably when it comes to forecasting.

Useful research in portfolio selection should put more emphasis on data handling and the empirical testing of models, thus better re-aligning financial theory with the nature of financial data. That means, in particular, that less emphasis should be put on obtaining numerically-precise solutions. As we said above, it is not reasonable to think that 'only the optimum' will work well: slightly changing the solution should not really change the results. (If it does, we should better not trust the results, anyway.) This is in line with the empirical evidence (e.g., Gilli and Schumann [22]), which suggests that 'the optimum' is not required: good solutions are enough. And those are exactly the solutions that heuristics provide.

References

1. D. Acker, N.W. Duck, Reference-day risk and the use of monthly returns data. J. Account. Audit. Financ. **22**(4), 527–557 (2007)
2. I. Althöfer, K.-U. Koschnick, On the convergence of "Threshold Accepting". Appl. Math. Optim. **24**(1), 183–195 (1991)
3. J.S. Armstrong, Forecasting with econometric methods: Folklore versus fact. J. Bus. **51**(4), 549–564 (1978)
4. R.S. Barr, B.L. Golden, J.P. Kelly, M.G.C. Resende, W.R. Stewart, Designing and reporting on computational experiments with heuristic methods. J. Heuristics **1**(1), 9–32 (1995)
5. M.W. Brandt, Portfolio choice problems, in *Handbook of Financial Econometrics*, vol. 1, ed. by Y. Aït-Sahalia, L.P. Hansen (Elsevier, Amsterdam, 2009)
6. P. Burns, Random portfolios for performance measurement, in *Optimisation, Econometric and Financial Analysis*, ed. by E.J. Kontoghiorghes, C. Gatu (Springer, Berlin, 2010)
7. J. Cohen, The earth is round ($p < .05$). Am. Psychol. **49**(12), 997–1003 (1994)
8. R.M. Dawes, The robust beauty of improper linear models in decision making. Am. Psychol. **34**(7), 571–582 (1979)
9. R.M. Dawes, *House of Cards – Psychology and Psychotherapy Built on Myth* (Free Press, New York, 1994)
10. R.S. Dembo, Scenario optimization. Ann. Oper. Res. **30**(1), 63–80 (1991)
11. G. Dueck, T. Scheuer, Threshold accepting. A general purpose optimization algorithm superior to simulated annealing. J. Comput. Phys. **90**(1), 161–175 (1990)
12. G. Dueck, P. Winker, New concepts and algorithms for portfolio choice. Appl. Stoch. Models Data Anal. **8**(3), 159–178 (1992)
13. E.J. Elton, M.J. Gruber, Estimating the dependence structure of share prices – Implications for portfolio selection. J. Financ. **28**(5), 1203–1232 (1973)
14. E.J. Elton, M.J. Gruber, T.J. Urich, Are betas best? J. Financ. **33**(5), 1375–1384 (1978)

15. P.A. Frost, J.E. Savarino, For better performance: constrain portfolio weights. J. Portf. Manag. **15**(1), 29–34 (1988)
16. S.B. Gelfand, S.K. Mitter, Analysis of simulated annealing for optimization. Technical Report LIDS-P-1494, MIT (1985)
17. G. Gigerenzer, Fast and frugal heuristics: the tools of bounded rationality, in *Blackwell Handbook of Judgment and Decision Making*, chap. 4, ed. by D.J. Koehler, N. Harvey (Blackwell Publishing, Oxford, 2004), pp. 62–88
18. G. Gigerenzer, Why heuristics work. Perspect. Psychol. Sci. **3**(1), 20–29 (2008)
19. M. Gilli, E. Këllezi, The threshold accepting heuristic for index tracking, in *Financial Engineering, E-Commerce and Supply Chain*, ed. by P. Pardalos, V.K. Tsitsiringos. Applied Optimization Series (Kluwer Academic Publishers, Boston, 2002), pp. 1–18
20. M. Gilli, E. Schumann, An empirical analysis of alternative portfolio selection criteria. Swiss Finance Institute Research Paper No. 09-06 (2009)
21. M. Gilli, E. Schumann, Optimization in financial engineering – an essay on 'good' solutions and misplaced exactitude. J. Financ. Transformation **28**, 117–122 (2010)
22. M. Gilli, E. Schumann, Optimal enough? J. Heuristics **17**(4), 373–387 (2011). Available from http://dx.doi.org/10.1007/s10732-010-9138-y
23. M. Gilli, E. Schumann, Risk–reward optimisation for long-run investors: an empirical analysis. Eur. Actuar. J. **1**(1), 303–327, Supplement 2 (2011).
24. P.E. Gill, W. Murray, M.H. Wright, *Practical Optimization* (Elsevier, Amsterdam, 1986)
25. M. Gilli, E. Këllezi, H. Hysi, A data-driven optimization heuristic for downside risk minimization. J. Risk **8**(3), 1–18 (2006)
26. M. Gilli, E. Schumann, G. Cabej, J. Lula, Replicating hedge fund indices with optimization heuristics. Swiss Finance Institute Research Paper No. 10–22 (2010)
27. M. Gilli, D. Maringer, E. Schumann, *Numerical Methods and Optimization in Finance* (Academic, New York, 2011)
28. M. Gilli, E. Schumann, G. di Tollo, G. Cabej, Constructing 130/30-portfolios with the omega ratio. J. Asset Manag. **12**(2), 94–108 (2011). http://dx.doi.org/10.1057/jam.2010.25
29. D.G. Goldstein, G. Gigerenzer, Fast and frugal forecasting. Int. J. Forecast. **25**(4), 760–772 (2009)
30. H. Grootveld, W. Hallerbach, Variance vs downside risk: is there really that much difference? Eur. J. Oper. Res. **114**(2), 304–319 (1999)
31. W.J. Gutjahr, A graph-based Ant System and its convergence. Futur. Gener. Comput. Syst. **16**(9), 873–888 (2000)
32. R. Jagannathan, T. Ma, Risk reduction in large portfolios: why imposing the wrong constraints helps. J. Financ. **58**(4), 1651–1683 (2003)
33. D.S. Johnson, C.R. Aragon, L.A. McGeoch, C. Schevon, Optimization by simulated annealing: an experimental evaluation; part I, graph partitioning. Oper. Res. **37**(6), 865–892 (1989)
34. E. Jondeau, S.-H. Poon, M. Rockinger, *Financial Modeling Under Non-Gaussian Distributions* (Springer, Berlin, 2007)
35. S. Kirkpatrick, C.D. Gelatt, M.P. Vecchi, Optimization by simulated annealing. Science **220**(4598), 671–680 (1983)
36. B. LeBaron, A.S. Weigend, A bootstrap evaluation of the effect of data splitting on financial time series. IEEE Trans. Neural Netw. **9**(1), 213–220 (1998). Available from citeseer.ist.psu.edu/lebaron98bootstrap.html
37. A.D. Lovie, P. Lovie, The flat maximum effect and linear scoring models for prediction. J. Forecasting **5**(3), 159–168 (1986)
38. S. Makridakis, M. Hibon, The M3-competition: results, conclusions and implications. Int. J. Forecasting **16**(4), 451–476 (2000)
39. S. Makridakis, M. Hibon, C. Moser, Accuracy of forecasting: an empirical investigation. J. R. Stat. Soc. Ser. A (General) **142**(2), 97–145 (1979)
40. D. Maringer, *Portfolio Management with Heuristic Optimization* (Springer, Berlin, 2005)
41. H.M. Markowitz, Portfolio selection. J. Financ. **7**(1), 77–91 (1952)
42. H.M. Markowitz, *Portfolio Selection* (Wiley, New York, 1959)

43. O. Morgenstern, *On the Accuracy of Economic Observations*, 2nd edn. (Princeton University Press, Princeton, 1963)
44. P. Moscato, J.F. Fontanari, Stochastic versus deterministic update in simulated annealing. Phys. Lett. A **146**(4), 204–208 (1990)
45. J. Pearl, *Heuristics* (Addison-Wesley, Reading, 1984)
46. G. Pólya, *How to Solve it*, 2nd edn. (Princeton University Press, Princeton, 1957)
47. M.J.D. Powell, Problems related to unconstrained optimization, in *Numerical Methods for Unconstrained Optimization*, ed. by W. Murray (Academic, New York, 1972)
48. G. Rudolph, Convergence analysis of canonical genetic algorithms. IEEE Trans. Neural Netw. **5**(1), 96–101 (1994)
49. E. Schumann, Take-the-best in portfolio selection. Available from http://enricoschumann.net (2013)
50. A. Scozzari, F. Tardella, S. Paterlini, T. Krink, Exact and heuristic approaches for the index tracking problem with UCITS constraints. Ann. Oper. Res. **205**(1), 235–250 (2013)
51. T. Stützle, M. Dorigo, A short convergence proof for a class of Ant Colony Optimization algorithms. IEEE Trans. Evol. Comput. **6**(4), 358–365 (2002)
52. L.N. Trefethen, Numerical analysis, in *Princeton Companion to Mathematics*, ed. by T. Gowers, J. Barrow-Green, I. Leader (Princeton University Press, Princeton, 2008)
53. A. Tversky, D. Kahneman, Judgment under uncertainty: heuristics and biases. Science **185**(4157), 1124–1131 (1974)
54. F. van den Bergh, A.P. Engelbrecht, A study of particle swarm optimization particle trajectories. Inf. Sci. **176**(8), 937–971 (2006)
55. J. von Neumann, H.H. Goldstine, Numerical inverting of matrices of high order. Bull. Am. Math. Soc. **53**(11), 1021–1099 (1947)
56. P. Winker, K.-T. Fang, Application of Threshold-Accepting to the evaluation of the discrepancy of a set of points. SIAM J. Numer. Anal. **34**(5), 2028–2042 (1997)
57. P. Winker, D. Maringer, The Threshold Accepting optimisation algorithm in economics and statistics, in *Optimisation, Econometric and Financial Analysis*, vol. 9, ed. by E.J. Kontoghiorghes, C. Gatu. *Advances in Computational Management Science* (Springer, Berlin, 2007), pp. 107–125
58. S.H. Zanakis, J.R. Evans, Heuristic "optimization": Why, when, and how to use it. Interfaces **11**(5), 84–91 (1981)

Chapter 11
Optimal Financial Decision Making Under Uncertainty

Giorgio Consigli, Daniel Kuhn, and Paolo Brandimarte

Abstract We use a fairly general framework to analyze a rich variety of financial optimization models presented in the literature, with emphasis on contributions included in this volume and a related special issue of OR Spectrum. We do not aim at providing readers with an exhaustive survey, rather we focus on a limited but significant set of modeling and methodological issues. The framework is based on a benchmark discrete-time stochastic control optimization framework, and a benchmark financial problem, asset-liability management, whose generality is considered in this chapter. A wide set of financial problems, ranging from asset allocation to financial engineering problems, is outlined, in terms of objectives, risk models, solution methods, and model users. We pay special attention to the interplay between alternative uncertainty representations and solution methods, which have an impact on the kind of solution which is obtained. Finally, we outline relevant directions for further research and optimization paradigms integration.

Keywords Stochastic control • Dynamic programming • Multistage stochastic programming • Robust optimization • Distributionally robust optimization • Decision rules • Asset-liability management • Pension fund management

G. Consigli (✉)
Department of Management, Economics and Quantitative Methods, University of Bergamo, Via dei Caniana 2, 24127 Bergamo, Italy
e-mail: giorgio.consigli@unibg.it

D. Kuhn
Risk Analytics and Optimization Chair, École Polytechnique Fédérale de Lausanne, CH-1015 Lausanne, Switzerland
e-mail: daniel.kuhn@epfl.ch

P. Brandimarte
Dipartimento di Scienze Matematiche, Politecnico di Torino, Corso Duca degli Abruzzi 24, 10129 Torino, Italy
e-mail: paolo.brandimarte@polito.it

© Springer International Publishing Switzerland 2017
G. Consigli et al. (eds.), *Optimal Financial Decision Making under Uncertainty*, International Series in Operations Research & Management Science 245, DOI 10.1007/978-3-319-41613-7_11

11.1 The Domain of Financial Optimization

Since the seminal work of Markowitz [71], the literature on the application of optimization models to financial decision problems has witnessed an astonishing growth. The contributions presented in this volume and the companion special issue (SI) of OR Spectrum [38] offer a broad picture illustrating a variety of problems and solution approaches that have been the subject of recent research, from both a theoretical and an applied perspective. The main purpose of this chapter is to review the building blocks of recent research on optimization models in finance. We do not aim at giving an exhaustive literature survey, though, and due emphasis is given to contributions to this volume and the SI. The (less ambitious) aim is to reconsider the contributions within a common framework in order to spot research directions and integration opportunities. This should be especially valuable to practitioners and newcomers, possibly Ph.D. students, who may find the heterogeneity of literature somewhat confusing.

Papers dealing with optimization in finance may be characterized according to different features:

- The kind of financial problem that is addressed, such as portfolio selection, asset pricing, hedging, or asset-liability management. These are sometimes considered as different problems but, actually, there is an interplay among them and we see in what follows that indeed *asset-liability management* (ALM) models provide a rather general modeling framework. The specific practical problem tackled restricts the choice of decisions to be made, the constraints that decisions must satisfy, and the criteria to evaluate solution quality. For a given financial problem, alternative modeling and solution approaches may be available.
- We have to define a risk model, which consists of a set of relevant risk factors and a representation of their uncertainty. Classical models rely on a probabilistic characterization of uncertainty, but there is a growing stream of contributions dealing with robustness and ambiguity, as well as model-free, data-driven approaches. Some of these paradigms are indeed non-stochastic, and the choice of a risk model must be compatible with the available information used for estimation and fitting purposes.
- The approach taken to manage a risk-reward tradeoff. Modern Portfolio Theory (MPT) relies on standard deviation of return as a risk measure, which is traded off against expected return within a static model. This is an example of the mean-risk approach, which may be generalized by using other risk measures. However, alternative approaches may be taken, relying on classical utility functions, or on stochastic dominance concepts. MPT revolves around a static portfolio choice, and the extension of risk measures to multiple periods turns out to be rather tricky. Mostly due to evolving regulatory frameworks and increasing tail risk, the set of possible risk-reward trade-offs has increased dramatically over the last two decades.
- All of the previous choices yield an optimization model formulation, which, among other things, may be static or dynamic, discrete-time or continuous-time.

Model selection and problem formulation, however, cannot be made without paying due attention to computational viability. The essential trade-off is in this context between finding the exact solution of a maybe wrong model, or an approximate solution of the right model.
- The set of available solution approaches has widened considerably over the years, due to increasing hardware power and the development of convex optimization techniques. The Markowitz model for instance boils down to a simple and deterministic convex optimization problem and, from a strictly algorithmic viewpoint, is not quite challenging. Sophisticated methods for multistage stochastic programming with recourse, dynamic programming, and robust optimization are called for when tackling challenging problems, and they determine the exact kind of solution that we find and the way in which it can be used.
- Last, but not least, the solution must somehow be validated, and the overall modeling approach must be questioned and critically assessed.

By combining the above features, a wide variety of optimization models can be developed, which should be viewed within a common framework, as a far as possible. Finding a generic formulation that can be instantiated to yield any conceivable model is a hopeless endeavour, but we believe that it is useful to set a benchmark model as a reference, in order to compare different model instances and methodological challenges. From a formal viewpoint, a discrete-time stochastic control model is arguably a good reference framework, whereas, from a financial viewpoint, an asset-liability management model is a suitably general problem.

No optimization model can clearly be considered without reference to a real-world financial context that practitioners have to face in their day-by-day activity. Hence, we start by summarizing a few relevant trends that affect financial decision making in Sect. 11.2. Then, in Sect. 11.3 we describe a benchmark modeling framework and the essential elements in building a financial decision model under uncertainty. There, we emphasize the interplay between the model building and the model solving approaches, which is particularly critical in the context of multistage decision problems. Then, in Sect. 11.4 we take a financial view, and consider how several classes of financial optimization problems may be considered as specific ALM cases. We introduce a pension fund ALM problem as representative of a variety of financial management problems. In Sect. 11.5 we follow up discussing a limited set of solution approaches suitable for dynamic problems: stochastic programming with recourse, distributionally robust optimization and learning decision rules. Finally, we outline possible research directions in Sect. 11.6, and we draw some conclusions in Sect. 11.7.

11.2 A Changing Financial Landscape

The contributions in this volume and the SI, to a certain extent stimulated by ongoing and continuously refined risk management regulatory frameworks in the banking and insurance industry, reflect a structural transformation of agents' management options and decision paradigms within a new financial environment [38]. Such transition may be analyzed from different viewpoints: here we consider primarily practitioners' and modelers' perspectives, the latter with associated computational and numerical implications. When considered in terms of *supply* and *demand* sides, as is common in financial economics, we see that market evolution can be conveniently linked to agents' decision processes: long positions in the market are associated with the demand side and lead to an optimal asset management problem, in which agents' risk attitude need to be considered and a given representation of risk is required. On the contrary short positions are associated with the supply side and lead to an optimal liability management problem and a related pricing issue. The two indeed embed two ALM problems, as we shall see here below. A distinctive element of the financial markets' growth is associated with increasing product diversification, which calls for modeling approaches able to accommodate non-trivial decision spaces, see [1, 46, 54, 66]. The dimension of the investment universe in real applications has an impact on both the adopted uncertainty model and the available optimization options, particularly in dynamic problems [42, 78, 81].

The recent literature on financial decision making reflects the emergence of new market features that call for a revision of traditional assumptions:

- The persistence of an unprecedented period of interest rate curves flattening at almost zero level, which leads to a quest for increased sources of return, and the consequent emergence of alternative investment opportunities, possibly involving illiquid assets.
- A revision of the set of relevant risk factors (risk as exposure to uncertain outcomes that can be assigned probabilities) like sovereign risk, even in OECD countries, longevity risk, systemic risk, as well as model risk, and ambiguity (interpreted as exposure to uncertain outcomes that admit no probabilistic description) [25, 46].
- The heterogeneity of agents' planning horizons, as different agents may be concerned with long vs. short time horizons, and the need to balance long-term objectives (as typical of pension funds), with short-term performance.

The above features have clear implications on the formulation of the associated optimization problem:

- Financial markets need not be consistent with canonical information efficiency assumptions and financial optimization approaches may very well adapt to alternative assumptions. The evolution of risk premia in equity, bond and alternative markets carrying different liquidity is complex to model and forecast but can hardly be avoided [39, 70, 78, 93].
- Asset pricing models need not be based on the assumption of market completeness [1, 46, 54].
- Decision horizons and rebalancing frequency matter: when moving from a single-period, myopic setting to a dynamic one, the risk modelling effort increases substantially and the issue of the optimal policies' and risk measures'

dynamic time consistency becomes central [34, 49]. Further the issue of an increasing dependence of the optimal policy on functions of *risk-adjusted performance* measures [21, 44, 78] becomes relevant over increasing planning horizons. On the other hand extending the horizon requires short-term risk management and profitability not to be jeopardized [1].

Modeling implications are relevant. Indeed, the contributions in this volume and the SI [38] may be distinguished as carrying scenario-based [42] or parametric [21, 58] or set-based [81] representations of uncertainty; model-driven [39, 42, 58, 70, 93] or data-driven approaches [25, 32, 55, 81]; reflecting equilibrium conditions—*arbitrage free* [1, 54, 70] or just based on statistical criteria [25, 31, 32].

From a computational viewpoint, the central approximation issue when dealing with dynamics typically carrying a continuous probability space must be addressed. A trade-off problem lies at the heart of such approximation effort, where an increasingly accurate characterization of the uncertainty may not be consistent with a realistic problem description, particularly in real-world applications [10]. A *sufficient* approximation, achieved through a computationally efficient approach, is regarded as a necessary condition to determine accurate risk control strategies. A substantial literature in stochastic programming addresses scenario generation issues [37, 48, 82]. The following evidences emerge in this respect in this volume:

- robust approaches [25, 81] rely on an approximation scheme which reduces the computational burden in terms of scenario management at the cost of increasing computational cost in the optimization phase [81],
- stochastic programming approaches rely very much on scenario *reduction* methods and the trade-off between approximation quality and stability of the optimal solution drives the computational effort [42].

To a certain extent independently from the adopted problem formulation and specific features, the definition of an effective decision process in contemporary markets relies very much on a sequence of steps: from the analysis of the decision universe, the definition of decision criteria driving the optimization process, the introduction of a statistical model and the derivation of internal (to the agents' economy) and external constraints.

11.3 The Elements of a Decision Model

The original mean-variance model is static and aims at managing an asset portfolio. As a result, decision variables are just portfolio weights and the distributional inputs are (seemingly) modest: a vector of expected returns and a covariance matrix. As it turns out, even this distributional information is not trivial to give in a robust way, but the matter gets much more complicated in a more general dynamic framework. In this section we describe a reference optimization model, namely, a generic discrete-time stochastic control problem. This cannot be regarded as an abstract model from which any relevant financial optimization model can be instantiated, but it allows us to spot the key elements of an optimization model, which have been specified in a wide variety of ways in the literature. From an applied financial perspective, a fairly general model framework can be identified in the *asset-liability* management (ALM) domain.

11.3.1 Discrete-Time Stochastic Control

We consider the formulation of an optimal financial planning problem as the optimal control of a dynamic stochastic system whose essential elements in discrete-time are:

- A discrete and finite sequence of time instants, $\mathscr{T} := \{0, \dots, T\}$. The discretization may arise from a suitable partition of a continuous time interval, or as a consequence of a decision process where choices are only made at specific time instants.
- A sequence of control/decision variables denoted by $\mathbf{x}_t$, $t = 0, \dots, T-1$. In general no decision is allowed at the end of the time horizon T, where we only check the last outcome.
- A sequence of state variables $\mathbf{s}_t$, $t = 0, \dots, T$; $\mathbf{s}_T$ is the terminal state.
- A sequence of random variables $\boldsymbol{\xi}_t(\omega)$, $t = 1, \dots, T$. This is the stochastic process followed by risk factors, where $\omega \in \Omega$ corresponds to a sample path. The stochastic process may be discrete-state, continuous-state, or a hybrid. In most financial optimization problems it is assumed that uncertainty is purely exogenous, i.e., risk factors are not influenced by control actions.

Controlling the system means making a decision, i.e., choosing a control $\mathbf{x}_t$ at each time t, $t = 0, \dots, T-1$, after observing the state $\mathbf{s}_t$. At $t = 0$ only $\mathbf{s}_0$ is known and the first control $\mathbf{x}_0$ is taken under complete uncertainty. The next state $\mathbf{s}_{t+1}$ may depend in a possibly complex way on the state and control trajectory up to t, as well as on the realization $\boldsymbol{\xi}_{t+1}(\omega)$ of the risk factors. A relatively simple case applies when the Markov property holds, which implies that we may introduce a state transition function Φ_t, at time t, such that

$$\mathbf{s}_{t+1} = \Phi_t(\mathbf{s}_t, \mathbf{x}_t, \boldsymbol{\xi}_{t+1}(\omega)). \tag{11.1}$$

The last control action $\mathbf{x}_{T-1}$ will result in the last state $\mathbf{s}_T$. The first requirement of a control action is feasibility, and we may denote the feasible set at time t as $A_t(\mathbf{s}_t)$, emphasizing dependence on the current state. Note that the feasible set is random, as it depends on risk factors through the state variable. The sequence of controls must be a *good* one, where quality may be measured by introducing a sequence of functions $f_t(\mathbf{s}_t, \mathbf{x}_t)$, $t = 1, \dots, T-1$, and a function $F_T(\mathbf{s}_T)$ to evaluate the terminal state. These functions may be maximized or minimized depending on their nature, and we end up with a somewhat loose and abstract formulation:

$$\min_{\mathbf{x}_t} \quad \mathrm{E}\left[\sum_{t=0}^{T-1} f_t(\mathbf{s}_t, \mathbf{x}_t) + F_T(\mathbf{s}_T)\right] \tag{11.2}$$

$$\text{s.t.} \quad \mathbf{x}_t \in A_t(\mathbf{s}_t) \tag{11.3}$$

$$\mathbf{s}_{t+1} = \Phi_t(\mathbf{s}_t, \mathbf{x}_t, \boldsymbol{\xi}_{t+1}(\omega)). \tag{11.4}$$

This formulation is quite intuitive, but loose, as it does not clarify a few important features (see also [34]).

- The nature of the solution. In a deterministic multiperiod problem, we fix the whole set of control actions $\mathbf{x}_t$ at time $t = 0$. Thus, decisions are a sequence of vectors. On the contrary, in a stochastic *multistage* problem the sequence of decisions is a random process, as it depends on the unfolding of uncertainty. They may be thought of as a function of the current state, which gives a solution in feedback form, or as a function of the underlying stochastic process. In the last case, care must be taken to enforce a sensible non-anticipativity condition. We further discuss the nature of the solution in Sect. 11.3.2.
- The objective function (11.2) is additive with respect to time, and might be specified in a way in which only the terminal state matters, or the whole trajectory. The involved functions could be used to capture the risk-reward tradeoff, which may be expressed by risk measures or utility functions. However, we cannot take for granted that a simple additive structure will be able to capture complex trade-offs often driving decision processes.
- The satisfaction of constraints (11.3) must be further qualified. If the representation of uncertainty relies on a discrete tree process constraints must be always satisfied, but when dealing with continuous factors, *almost sure* (a.s., with probability 1) feasibility is required. This condition may be relaxed by requiring that the constraint is satisfied with a suitable probability, leading to chance-constrained problems.
- The state transition function (11.4) captures the uncertain evolution of the state variable, but we leave the evolution of the risk factors implicit. A critical modeling choice is the selection of a suitable set of driving risk factors that affect, e.g., prices (interest rate and credit spreads affect bond prices). The underlying stochastic process ranges from a simple sequence of i.i.d. variables to a complex process exhibiting path dependencies, going through the intermediate case of a Markov process. The expectation in the objective function is taken with respect to a probability measure associated with this model. However, an increasingly relevant amount of research questions our ability to pinpoint a probability measure reliably. Distributional ambiguity and non-stochastic characterizations of uncertainty have been proposed.

In the following we will delve more deeply into these issues, but before doing so, a simple but relevant example is necessary.

11.3.2 *The Interplay Between Model Building and Solution Method*

When looking at the model formulation (11.2), it is not quite apparent which kind of solution we may get and how it can be used. This depends on how uncertainty is expressed and often *approximated* to make the problem computationally viable. The dynamics of $\mathbf{s}_t$ and $\boldsymbol{\xi}(\boldsymbol{\omega})$ will depend on such assumptions. Alternative approaches to express uncertainty include:

- Scenario trees, which we may think as discrete characterization of a continuous-time and/or continuous-state model as well as generated by discrete processes [37];
- Model-free, data-driven approaches [25, 32, 46];
- Uncertainty sets maybe associated with probability distribution supports.

These alternatives, discussed below in more detail, may lead to different solution strategies in the stochastic control problem:

- Multistage stochastic programming. Emphasis is given to the first stage or *implementable* decision $\mathbf{x}_0$.
- Dynamic programming, where we typically, recover by backward recursion an optimal policy, whose time and state evolution will in general depend on the statistical assumptions of the underlying random process,
- Robust optimization has been extended recently to multistage problems [81], called adjustable robust optimization. Robust approaches yield optimal strategies with respect to uncertainty sets. Distributional robustness instead originate from uncertainty over probability domains.
- The adoption of decision rules as shown below has received increasing interest to address dynamic optimization problems [90].

It is useful to clarify first the key distinctive elements of those modeling and solution options with respect to the adopted representation of uncertainty and the assumptions on the underlying data process.

11.3.2.1 Scenario Tree, Non-anticipativity and Information

The discrete evolution of a dynamic stochastic system is often described by a *tree* process, which captures the random dynamics of $\boldsymbol{\xi}_t(\omega)$. Scenario trees reflect the dynamic interaction between control actions—the first taken under full uncertainty, the others always under residual uncertainty—and revelation of uncertainty over a maybe very long time horizon [1, 27, 42, 66, 81]. The tree clearly shows the information to which the decision process $\mathbf{x}_t(\omega)$ must be adapted. A filtered probability space is associated with the tree. Consider the simple example in Fig. 11.1. At time $t = 1$ we essentially see a σ-algebra generated by the sets $\{\omega_1, \omega_2, \omega_3, \omega_4\}$ and $\{\omega_5, \omega_6, \omega_7, \omega_8\}$, whereas at time $t = 2$ random variables

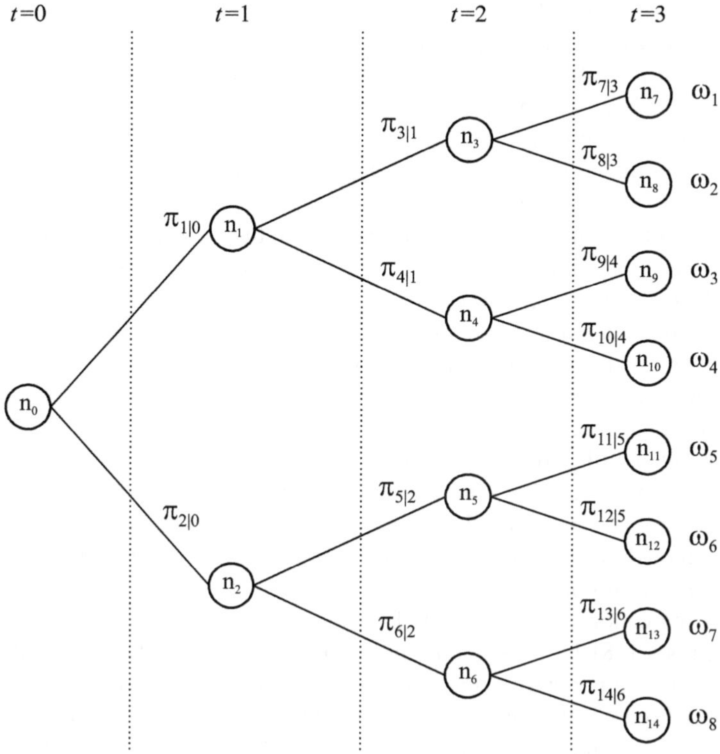

Fig. 11.1 A typical scenario tree

must be measurable with respect to the σ-algebra generated by $\{\omega_1, \omega_2\}$, $\{\omega_3, \omega_4\}$, $\{\omega_5, \omega_6\}$, and $\{\omega_7, \omega_8\}$. As such, the sequence of σ-algebras provides an information model and for consistency, we require decisions to be adapted, measurable, with respect to those information flows.

The decision process is non-anticipative, which means that a decision at any node in the tree must be the same for any scenario that is indistinguishable up to that time instant. As we shall see in Sect. 11.4.2, this requirement may be built in the formulation of the model by an appropriate choice of variables, which are associated with nodes. A noticeable feature of multistage stochastic programming is that the solution is a stochastic decision process, namely a tree, where the decision at the root node, the implementable decision, is what matters [10]. On the one hand, this gives stochastic programming an operational flavor. On the other hand, this may also be a disadvantage with respect to other approaches, due to decreasing computational efficiency. The conditional behavior of a (non-recombining) scenario tree process is reflected in a specific labeling convention which is useful to summarize. We adopt a particularly simple convention and denote with $\mathscr{N}$ the set of nodes in the tree; n_0 is the root node. The (unique) predecessor of node $n \in \mathscr{N} \setminus \{n_0\}$ is denoted by $a(n)$, while the set of terminal nodes is denoted by $\mathscr{S}$. Each node belongs to a scenario, which is the sequence of event nodes along the unique paths leading from

n_0 to $s \in \mathscr{S}$, with probability π^s. $\mathscr{T} = \mathscr{N} \setminus (\{n_0\} \cup \mathscr{S})$ is the set of intermediate trading nodes. We refer to [54] for a more structured labeling scheme. The product of conditional probabilities in Fig. 11.1 will determine the scenario probabilities. In presence of vector tree processes, as common in financial applications, path probabilities are assumed to be equal with conditional probabilities at each stage equal to 1 over the number of branches departing from that node.

The above stochastic structure is common for multistage stochastic programs [10] and provides an intuitive model to understand the interplay between underlying random process evolution, the associated information process and the resulting optimal decision process, defined by a root decision and a sequence of *recourse* decisions. An extended set of financial models carrying specific properties has been adopted over a wide range of financial applications [1, 38, 39, 42, 70]. Validation by stability analysis is necessary and an additional complication in finance: scenarios must be arbitrage-free.

Estimating a full-fledged uncertainty model is certainly not trivial, and one should question its validity. One possible alternative, is to resort to a data-driven approach. Then, a more natural stochastic dynamics can be built according to the following model without any conditional structure, as such referred to as *linear* scenarios or scenario *fan*.

This is the canonical output of risk measurement applications based on Monte Carlo methods [18] in which, for given initial input portfolio, the evaluation of relevant statistical percentiles may be performed. Here we are primarily interested in this model representation in connection with the solution of a stochastic control problem. In a dynamic setting, the model is indeed also associated with a stochastic program where a condition of *perfect foresight* is assumed, corresponding to the so-called wait-and-see problem [15]. The solution of problem (11.2) under the wait-and-see assumption provides a lower bound to the stochastic solution: such difference reflects the expected value of perfect information [15]. As the set of $\mathscr{S}$ increases, the associated solution $\mathbf{x}_{n_0}$ is expected to converge to a stable solution. In financial optimization problems it is common to enforce a *non-arbitrage* condition [42, 54] to require a branching degree at each stage bounded below by the cardinality of the investment universe.

Alternatively assume a simple one-period tree following the dynamics in Fig. 11.2. This would be a possible representation of a discrete probability space with finite and countable support and typically a probability measure giving to each path the same probability. Such model would be stage-wise consistent with a (stochastic) dynamic programming solution approach [34] within a backward recursion algorithm [18].

Robust models and distributionally robust models do not require a specification of an underlying process' sample paths and neither of them will depend on such discrete approximation: instead they will focus either on an uncertainty domain associated with ω or, more precisely $\boldsymbol{\xi}(\boldsymbol{\omega})$, or on uncertainty affecting the probability measure to be adopted to describe the problem stochasticity. When, still under the (collapsed 1-period) process characterization in Fig. 11.2, a data-driven approach is adopted then as common in risk management applications relying on *historical*

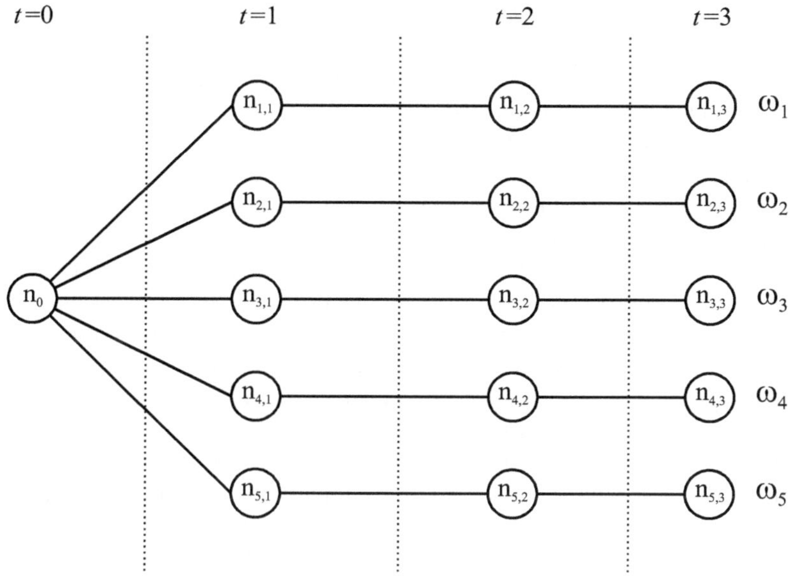

Fig. 11.2 Linear scenarios

simulation, past data realizations are used, typically within a bootstrapping approach [18, 25, 32], to populate the sample space in the problem definition. The associated probability distribution is in this case also referred to as the *empirical* distribution. Empirical distributions are generated by mapping into future random events a given data history, relying on a discrete probability distribution.

Summarizing, either probabilistic models are adopted in financial optimization problems, but the model based on what we introduced as linear scenarios, may be considered either in one period control problems or in dynamic problems but in such case relying on backward recursion approaches. Accordingly, not to violate the required non-anticipativity condition or measurability condition of the optimal strategy (with respect to the current information structure), any decision must be taken in face of residual uncertainty. Here next we consider how different assumptions on the probability space translate into two very popular optimization paradigms, eventually leading, jointly, to an increasingly popular problem formulation.

11.3.2.2 Stochastic, Robust and Distributionally Robust Optimization

A compact representation of problem (11.2)—where for sake of simplicity all constraints are embedded in the decision space $\mathscr{X}$ definition, and a random process ξ, defined in an appropriate probability space, is assumed to characterize the problem's overall uncertainty—is the following stochastic program:

$$\min_{\mathbf{x}\in\mathscr{X}} \mathbb{E}_{\mathbb{P}}\left[f(\mathbf{x},\boldsymbol{\xi})\right]. \qquad (11.5)$$

Expectation is taken under the probability measure $\mathbb{P}$ and f is a functional, or risk function, mapping the interaction between decisions and random events into a given payoff or risk estimate, we wish to control. We refer to [34] in this volume for a thorough analysis of dynamic risk measures. The formulation (11.5) is indeed consistent with a dynamic stochastic program where the decision space is described as a product space and non-anticipativity is enforced implicitly. The probability space is endowed with a filtration, that we may assume generated by the process ξ, and expectation is taken at the end of the planning horizon [49]. For a planning horizon $T = 1$, the above problem formulation collapses to the one-period static case [21, 25, 32, 55]. We consider here general properties of those formulations.

It is widely recognized that multistage stochastic programs have represented now for quite sometime an effective mathematical framework for decision making under uncertainty in several application domains with noticeable examples specifically in finance and energy [10]. A distinctive element of this approach may be found in its ability through a scenario tree representation to combine an optimal decision tree process with specific assumptions on the underlying random process evolution. A discrete framework allows an accurate and rich mathematical representation of the decision problem with a sufficient description of the underlying sources of uncertainty [37]. In presence of a risk exposure generated by continuously evolving market conditions (e.g. commodity prices for energy problems, financial returns in portfolio management and so forth), randomness evolves continuously in time and an approximation issue arises, as well known. The adoption of sampling methods and in general scenario reduction and generation methods aimed at minimizing the *cost* associated with such discrete approximation has represented a relevant research focus in this context [10]. The trade-off between computational viability and scenario tree expansion is central to the modeling effort.

Statistical modeling, particularly in the case of financial applications, has been typically welcome by practitioners as a longly established way to incorporate economic and financial stylised evidences and solid quantitative analysis in day-by-day decision making and within optimization models. Applied research in risk management, based on advanced statistical models had been indeed extremely successful to capture financial portfolios' risk exposure for years [91] and it was stimulated by regulatory institutions.

A fundamental challenge in stochastic programming formulations, however, is that the distribution of ξ is in general not directly observable but must be estimated from historical time series data. In fact, not even the simplest moments such as the means and covariances can be estimated to within an acceptable precision [69, 72]. Estimation errors are problematic because financial optimization problems tend to be highly sensitive to the distributional input parameters. Consequently, estimation errors in input parameters are amplified in the optimization (e.g., assets with overestimated mean returns are given too much weight and assets with underestimated means are given too little weight in the optimal portfolio), which results in unstable portfolios that perform poorly in out-of-sample experiments [20, 35, 73]. This phenomenon is akin to overfitting in statistics: a model that is perfectly optimized for in-sample data has little explanatory power and displays poor generalizability to out-of-sample data.

A popular approach to combat estimation errors in input parameters of financial optimization models is to adopt a robust approach. Here, the uncertain input parameters are assumed to reside within an uncertainty set $\mathcal{U}$ that captures the agent's prior knowledge of the uncertainty. The objective of robust optimization models is to identify decisions that are optimal under the worst possible parameter realizations within the uncertainty set [9, 12]. This gives rise to worst-case optimization problems of the form

$$\min_{\mathbf{x}\in\mathcal{X}} \max_{\boldsymbol{\xi}\in\mathcal{U}} f(\mathbf{x}, \boldsymbol{\xi}). \tag{11.6}$$

The resulting decisions display an attractive non-inferiority property, that is, their out-of-sample performance is necessarily better (lower) than the optimal value of the robust optimization problem provided that the uncertain parameters materialize within the prescribed uncertainty set. The shape of the uncertainty set in problems of the type (11.6) should be chosen judiciously as it can heavily impact the quality of the resulting optimal decisions and the tractability of the robust optimization problem. Popular uncertainty sets that have been studied in a financial context are the box uncertainty set [92], the ellipsoidal uncertainty set [50, 52], the budget uncertainty set [4, 5], the factor model uncertainty set [57] as well as a class of data-driven uncertainty sets that are constructed using statistical hypothesis tests [14]. In robust portfolio optimization, for instance, there is strong evidence that robust portfolios are less susceptible to overfitting effects than classical Markowitz portfolios and therefore display an improved out-of-sample performance [30]. Generalized robust portfolio optimization models that include both stocks and European-style options have been discussed in [98]. This model offers two layers of robustness guarantees: a weak guarantee that holds whenever the asset returns materialize within a prescribed confidence set reflecting normal market conditions and a strong guarantee that becomes effective when the asset returns materialize outside of the confidence set. The model is therefore akin to the comprehensive robust counterpart model [8], which allows for a controlled deterioration in performance when the data falls outside of the uncertainty set. Robust models with objective functions that are linear in the decisions but convex nonlinear in the uncertain parameters have been proposed in [62]. This versatile model can capture nonlinear dependencies between prices and returns as they are common in classical stochastic stock price models.

Despite the ostensible simplicity of modeling uncertainty through sets, robust optimization has been exceptionally successful in providing high-quality and efficiently computable solutions for a broad spectrum of decision problems ranging from engineering design, finance, and machine learning to policy making and business analytics [9]. Nevertheless, it has been observed that robust optimization models can lead to an under-specification of uncertainty as they fail to exploit prior distributional information that may be available. In these situations, robust optimization models may lead to over-conservative decisions. By also exploiting properties of stochastic programming models, distributionally robust models address this issue.

Since the pioneering work of Keynes [63] and Knight [65], it is common in decision theory to distinguish the concepts of risk and ambiguity. Classical stochastic programming can only be used in a risky environment. Indeed, the objective of stochastic programming is to minimize the expectation or some risk measure of the cost $f(\mathbf{x}, \boldsymbol{\xi})$, where the expectation is taken with respect to the distribution $\mathbb{P}$ of $\boldsymbol{\xi}$, which is assumed to be known. If we identify the uncertainty set $\mathscr{U}$ in the robust optimization problem (11.6) with the support of the probability distribution $\mathbb{P}$, then it becomes apparent that stochastic programming and robust optimization offer complementary models for the decision maker's risk attitude: however, when viewed through a decision theory lens, neither stochastic programming nor robust optimization address ambiguity.

Distributionally robust optimization is a natural generalization of both stochastic programming and robust optimization, which accounts both for the decision maker's attitude towards risk and ambiguity. In particular, a generic distributionally robust optimization problem con be formulated as

$$\min_{\mathbf{x} \in \mathscr{X}} \max_{\mathbb{P} \in \mathscr{P}} \mathbb{E}_{\mathbb{P}} \left[f(\mathbf{x}, \boldsymbol{\xi}) \right] , \tag{11.7}$$

where the probability distribution $\mathbb{P}$ of $\boldsymbol{\xi}$ is only known to belong to a prescribed ambiguity set $\mathscr{P}$, that is, a family of (possibly infinitely many) probability distributions consistent with the available raw data or prior structural information. The distributionally robust optimization model can be interpreted as a game against 'nature'. In this game, the agent first selects a decision with the goal to minimize expected costs, in response to which 'nature' selects a distribution from within the ambiguity set with the goal to inflict maximum harm to the agent. This setup prompts the agent to select worst-case optimal decisions that offer performance guarantees valid for all distributions in the ambiguity set.

Notice that the introduced min-max problem (11.7) belongs indeed also to the tradition of stochastic programming problems from the seminal work in [47]. Distributionally robust optimization leads to less conservative decisions than classical robust optimization, and it enables modelers to incorporate information about estimation errors into optimization problems. Therefore, it results in a more realistic account of uncertainty. Moreover, maybe surprisingly, distributionally robust optimization problems can often be solved exactly and in polynomial time—very much like the simpler robust optimization models. For a general introduction to distributionally robust optimization we refer to [40, 56, 95].

Many innovations in distributionally robust optimization have originated from the study of financial decision problem. For instance, one of the earliest modern distributionally robust optimization models in the literature studies the construction of portfolios that are optimal in terms of worst-case Value-at-Risk, where the worst case is taken over all asset return distributions with given means and covariances [50]. Worst-case expected utility maximization models are addressed in [80], while worst-case Value-at-Risk minimization models for nonlinear portfolios containing stocks and options are described in [97]. Moreover, distributionally robust portfolio optimization models using an ambiguity set in which some marginal distributions are known, while the global dependency structure or copula is uncertain, are studied in [45]. While many papers focus on ambiguity sets described by first- and second-order moment, possibly complemented by support information, asymmetric distributional information in the form of forward- and backward-deviation measures are described in [79]. Distributionally robust growth-optimal portfolios that offer attractive performance guarantees across several investment periods are described in [86], where the asset returns are assumed to follow a weak sense white noise process, which means that the ambiguity set contains all distributions under which the asset returns are serially uncorrelated and have period-wise identical first and second-order moments. Besides moment-based ambiguity sets, which abound in the current literature and offer attractive tractability properties, a different stream of research has focused on the construction of ambiguity sets using probability metrics. Here, the idea is to construct ambiguity sets that can be viewed as balls in the space of probability distributions with respect to a probability distance function such as the Prohorov metric [51], the Kullback-Leibler divergence [60], or the Wasserstein metric [75, 83] etc. Such metric- based ambiguity sets contain all distributions that are sufficiently close to a prescribed nominal distribution with respect to the prescribed probability metric. This setup allows the modeler to control the degree of conservatism of the underlying optimization problem by tuning the radius of the ambiguity set. In particular, if the radius is set to zero, the ambiguity set collapses to a singleton that contains only the nominal distribution. Then, the distributionally robust optimization problem reduces to the classical stochastic program (11.5).

In the final section the above general concepts are considered in the domain of a financial application. We complete this short methodological summary by considering data-driven approaches as in [25, 32, 81] and their compatibility with robust and stochastic programming formulations.

11.3.2.3 Data-Driven Optimization

Most classical stochastic programs as well as the vast majority of robust and distributionally robust optimization models are constructed with a parametric uncertainty model in mind. Thus, the distribution of $\boldsymbol{\xi}$ is estimated statistically, or it is constructed on the basis of expert information or known structural properties. The estimated distribution $\hat{\mathbb{P}}$ may then be directly used in the stochastic program (11.5). Alternatively, the support or an appropriate confidence set of $\hat{\mathbb{P}}$ can be viewed as an uncertainty set $\mathscr{U}$ for the robust optimization problem (11.6). Yet another possibility is to use $\hat{\mathbb{P}}$ to construct an ambiguity set $\mathscr{P}$ for the distributionally robust optimization problem (11.7) (e.g., by defining a moment ambiguity set via some standard or generalized moments of $\hat{\mathbb{P}}$ or by designating $\hat{\mathbb{P}}$ as the center of a spherical ambiguity set with respect to a probability metric). In contrast, data-driven optimization uses any available data directly in the optimization model—without the detour of calibrating a statistical model.

Assume, for instance, that we have access to a set of independent samples $\boldsymbol{\xi}^{(1)}, \dots, \boldsymbol{\xi}^{(N)}$ drawn uniformly from the (unknown) distribution $\mathbb{P}$. We now outline how one can use such a time series to construct data-driven variants of the stochastic, robust, and distributionally robust optimization problems (11.5)–(11.7), respectively. First, the celebrated sample average approximation [89, Chap. 5]

$$\min_{\mathbf{x} \in \mathscr{X}} \frac{1}{N} \sum_{i=1}^{N} f(\mathbf{x}, \boldsymbol{\xi}^{(i)}) \tag{11.8}$$

provides a data-driven approximation of the stochastic program (11.5). Here, the unknown true distribution $\mathbb{P}$ is effectively replaced by the uniform distribution on the data points $\boldsymbol{\xi}^{(1)}, \dots, \boldsymbol{\xi}^{(N)}$, which is the empirical distribution. The sample average approximation has been used with great success in financial engineering, risk management and economics [96]. For a modern textbook treatment of the broader area of Monte Carlo simulation we refer to [18].

A data-driven counterpart of the robust optimization problem (11.6) is given by the scenario program

$$\min_{\mathbf{x} \in \mathscr{X}} \max_{i=1,\dots,N} f(\mathbf{x}, \boldsymbol{\xi}^{(i)}), \tag{11.9}$$

which is in fact a variant of (11.6) where the uncertainty set $\mathscr{U}$ has been replaced by the support of the empirical distribution. Even though it is conceptually simple, the scenario program has many desirable theoretical properties. On the one hand, the scenario program (11.9) is efficiently solvable even in situations where the robust problem (11.6) with a polyhedral or ellipsoidal uncertainty set $\mathscr{U}$ is intractable. Due to the inner approximation entailed by the sampling, the solution of (11.9) underestimates the optimal value of (11.6), which typically involves an infinite number of scenarios. However, it has been shown that the cost of any optimal solution of (11.9) under a new data point $\boldsymbol{\xi}^{(N+1)}$ is bounded by the optimal value of (11.9) with high probability provided the number of samples N is sufficiently large [26, 28]. This result is remarkable as it holds independently of the distribution $\mathbb{P}$ of the samples and therefore is applicable even in situations where $\mathbb{P}$ is unknown—as is typically the case in financial applications. Thus, we can interpret (11.9) as an optimization problem that minimizes the cost threshold that can be exceeded only with a certain prescribed probability, which implies that (11.9) is closely related to a value-at-risk minimization problem. For a more detailed discussion of this connection we refer to [24]; applications in portfolio analysis and design are described in [25]. Another modern approach to data-driven robust optimization seeks decisions that are robust with respect to the set of all parameters that pass a prescribed statistical hypothesis test [14].

There are different approaches to deriving data-driven counterparts of the distributionally robust optimization problem (11.7). In [40] it is shown how time series data can be used to construct confidence sets for the first two moments of $\boldsymbol{\xi}$. The ambiguity set of all distributions whose first two moments reside within these confidence sets is guaranteed to contain the unknown true data-generating distribution $\mathbb{P}$ at the prescribed confidence level. The time series data can also be used in more direct ways to construct ambiguity sets. Specifically, several authors have proposed to study the following data-driven distributionally robust optimization problem

$$\min_{\mathbf{x}\in\mathscr{X}} \max_{\mathbb{P}\in\mathbb{B}_r(\hat{\mathbb{P}})} \mathbb{E}_{\mathbb{P}}\left[f(\mathbf{x},\boldsymbol{\xi})\right], \tag{11.10}$$

where the ambiguity set $\mathbb{B}_r(\hat{\mathbb{P}})$ is defined as the ball of radius r around the empirical distribution $\hat{\mathbb{P}}$ with respect to some probability metric. Note that problem (11.10) reduces to the sample average approximation (11.8) when the radius r drops to zero. In practice, the crux is to select the radius r judiciously such that the unknown data-generating distribution $\mathbb{P}$ belongs to the ambiguity set $\mathbb{B}_r(\hat{\mathbb{P}})$ with high confidence. The optimal solution of (11.10) thus displays attractive out-of-sample guarantees, but it may be difficult to compute. In order to ease the computational burden, it has been shown that a robust problem (11.6) whose uncertainty set $\mathscr{U}$ is a union of norm balls centered at the N sample points may provide a close and computationally tractable approximation for (11.10). In [75] it has been shown that the data-driven distributionally robust optimization problem (11.10) is in fact tractable when the Wasserstein metric is used in the definition of $\mathbb{B}_r(\hat{\mathbb{P}})$. In this case no approximation is required.

The use of the Wasserstein metric is also attractive conceptually. Indeed, if (11.10) is a distributionally robust portfolio selection problem where $\mathbb{B}_r(\hat{\mathbb{P}})$ represents the Wasserstein ball of radius r around some nominal distribution $\hat{\mathbb{P}}$, then the optimal solution of (11.10) converges to the equally-weighted portfolio as r tends to infinity (i.e., in the case of extreme ambiguity) [84]. This result, within a financial context, is appealing because it has been observed in empirical studies that the equally-weighted portfolio consistently outperforms many classical Markowitz-type portfolios in terms of Sharpe ratio, certainty-equivalent return or turnover [41]. The fact that the equally-weighed portfolio is optimal under extreme ambiguity therefore provides a solid theoretical justification for a surprising empirical observation.

The above concepts are now considered in the specific domain of a simple, though rather general specification of an asset-liability management problem.

11.4 Asset-Liability Management

In this section we describe a simple *asset-liability management* (ALM) problem, as an example of a stochastic control problem. This class of models subsumes asset management models where, for given market prices, we want to select the best investment policy under budget, turnover, inventory balance constraints, according to some utility function or a risk-reward tradeoff, over a given time horizon.

As well as liability management problems or liability-driven investment models [1] and genuine asset-liability models [10, 42, 81]. Asset management problems are relevant in fund management applications, active and passive portfolio management with or without index tracking [21, 58], allocation among liquid or illiquid asset classes including [78], and so forth. In full generality and in consideration of the contributions in this volume and the special issue, we formulate a generic pension fund problem. Here the pension fund manager seeks an optimal portfolio allocation in such a way as to meet a stream of (uncertain) liabilities. Alternatively for given payoff promised upon retirement by a certain retirement product, she/he might just search an optimal replication strategy [1, 54]. Such replication problem is equivalent to a hedging problem in financial engineering, and its solution depends on the assumptions made on market completeness. This class of problems can thus also be formulated as ALM problems. We provide next a set of examples canonical in finance to convey such generality.

11.4.1 An Overview of Financial Planning Problems

Different assumptions on the specification of problem (11.2) will result in different types of financial optimization problems and qualify most of the collected contributions.

Example 1 (Portfolio Management). The definition of an optimal portfolio strategy under a set of constraints and within different methodological assumptions is considered by Calafiore [25], Gyorfi and Ottucsak [58], Gilli and Schumann [55], Mulvey et al. [78] in this volume and by Pagliarani and Vargiolu [93], Desmettre et al. [44], Kopa and Post [67], Bruni et al. [21] and Cetinkaya and Thiele [32] in the special issue.

Here, as time $t \in \mathcal{T}$ evolves, the state of the system $\mathbf{s}_t$ is captured by the dynamics of portfolio losses or returns, the control $\mathbf{x_t}$ reflects the dynamic portfolio rebalancing decisions within the feasibility region determined by $A_t(\mathbf{s}_t)$. Under parametric assumptions on the sample space or model driven approaches, the characterization of the probability measure $\mathbb{P} \in \mathcal{P}$ is given and no min-max approach is needed. In the more general case of data-driven, model-free approaches or within an uncertainty set characterization, typical of robust formulations, we allow the minimum loss to be associated with a distributionally worst case characterization of the underlying stochastics. The problem collapses to one-period static optimization if $T = 1$: under which case, the optimality of a myopic policy will depend primarily on the statistical properties of the underlying return process. It is sufficient to allow for short cash positions or explicitly for borrowing choices that this turns into an ALM problem. Increasingly in the fund management industry the cost function includes an indexation scheme [21] resulting into an optimal tracking problem.

Example 2 (Asset Pricing). We reconsider here pricing problems, in the sense of asset pricing and derivative contracts valuation as well as the structuring of new financial payoffs, specifically from the market makers, protection seller, viewpoints. MacLean and Zhao [70], Aro and Pennanen [1], Giandomenico and Pinar [54], Mulvey et al. [78] in this volume, Konicz et al. [66], Trigeorgis et al. [46] in the SI, provide different approaches to address such issue. The *fair* valuation of a contract can be conveniently regarded as the solution of an optimization problem leading today to a minimal investment cost of a portfolio that at the maturity of the contract will deliver the contract payoff, independently from the market condition at expiry. A characterizing condition of this problem is that the portfolio strategy leading to such payoff is actually self-financed.

For $t \in \mathscr{T}$, the state of the system $\mathbf{s}_t$ will in this case be represented by say the derivatives' *moneyness*, the control $\mathbf{x_t}$ will determine the dynamic portfolio hedging strategy with $A_t(\mathbf{s}_t)$ to denote the self-financing condition and other technical constraints. The state transition operator Φ_t will reflect the contract complexity and f_t the portfolio replication cost function to be minimized. The time space of the problem here may typically be described by a sufficiently thinned discretization. In general such hedging problem admits a unique solution under Black and Scholes assumptions, whereas in real markets and operational contexts hardly payoffs can be perfectly replicated giving raise to what is referred to as the cost induced by market incompleteness [1, 54]. As a by-product of such optimization approach, the probability measure governing the stochastic dynamics of the problem needs not be an input, rather an output of the dual problem solution. A rather interesting application is considered in Aro and Pennanen [1] and Pachamanova et al. [81] where retirement products are considered. Here assets are used to hedge a given liability whose payoff needs to be replicated. The stochastic program solution leads to both a minimal hedging cost trajectory and a fair value of the contract. The characterization of the underlying stochasticity is typically attained through some analytical description of the underlying risk process but ambiguity over models selection is accommodated in Trigeorgis et al. [46]. In rather general terms the numerical approximation of time and states needs to be carefully determined and this is an issue.

Example 3 (Canonical Asset-Liability Management). A decision maker is here confronted with an optimization problem involving both assets and liabilities: the universe of possible decisions including their dynamics over time is of primary concern to the decision maker. Historically ALM problems have gained popularity as natural modeling frameworks for the description of enterprise-wide long-term management problems, in which the dynamic interplay between asset returns and liability costs plays a central role in the definition of a strategic policy. Increasingly however and from what was said above, from a mathematical standpoint this is possibly the most general modeling framework available today. Individuals' consumption-investment problems [66] and pension fund ALM problems [1, 81] provide different perspectives falling in this application domain. Contributions by Pachamanova et al [81], Aro and Pennanen [1], Dempster et al. [42] in this volume, Dupacova and Kozmik [49], Konicz et al. [66], Davis and Lleo [39] fall in this area.

Here $t \in \mathcal{T}$, is typically discrete and the state of the system $\mathbf{s}_t$ will be represented by an A-L ratio, such as a funding or solvency ratio for an insurance company, the fund return for a Pension fund and alike, the control $\mathbf{x_t}$ will include in the most general case asset and liability decisions to be taken over time. $A_t(\mathbf{s}_t)$ a maybe extended set of conditions on the optimal policy, with regulatory constrains and limitations on the investment and borrowing policies. The chapter by Pachamanova et al. [81] as well as several contributions in [10] provide good examples of such set-up. In [36] an ALM problem for a large property and casualty (P/C) division is considered based on an extended set of both liquid as well as illiquid assets and liability streams associated with the P/C activity. The model provides a good example of the impact on classical ALM applications of the current integration of a capital adequacy regulatory framework.

11.4.2 A Simple ALM Model

Examples 1–3 in Sect. 11.4.1 may be given, with possibly a few additional constraints, a mathematical formulation similar to the one proposed in (11.11)–(11.16) and be interpreted as specific ALM instances. The area of pension fund management provides an interesting case where most of the above theoretical and modeling issues can be traced within a unified framework. Indeed, depending on the pension scheme, we have here relevant regulatory constraints, long-term asset and liabilities, short-term solvency and liquidity constraints, as well as financial engineering applications due to the growing role of *defined contribution* (DC) and mixed DC-*defined benefit* (DB) or *hybrid* schemes. From the perspective of the stochastic system evolution in this segment we also see how indeed policy makers, individuals and pension providers (public and private) interact to determine jointly the system dynamics and influencing its evolution. Konicz et al. [66] consider the individual perspective, Aro and Pennanen [1] as well as Pachamanova et al. [81] the Pension fund perspective.

The model is based on a scenario tree, the standard representation of uncertainty used in stochastic programming, as illustrated in Fig. 11.1. This should be contrasted against the linear scenario arrangement in Fig. 11.2. Consider the following variables specification under the introduced nodal labeling convention.

- L^n are pension payments in node $n \in \mathcal{N}$,
- Λ^n is the pension fund liability in node $n \in \mathcal{N}$: this is the discounted value of all future pension payments,
- c is a (symmetric) percentage transaction cost,
- $\overline{h}_i^{n0}$ is the initial holding for asset $i = 1, \ldots, I$ at the root node.
- r_i^n is the (price) return of asset i in node n.
- z_i^n is the amount of asset i purchased in n.
- y_i^n is the amount of asset i sold in n.
- x_i^n is the amount of asset i we hold at node n, after rebalancing. Accordingly $X^n = \sum_{i=1}^{I} x_i^n$

- ϕ^n is the pension fund's funding ratio in node n: $\phi^n = \frac{X^n}{\Lambda^n}$
- W^s is the pension fund terminal deficit $W^s = \Lambda^s - X^s \; s \in \mathcal{S}$.
- $\rho(w)$ is a terminal functional on w. It is a risk measure that in (11.2) is considered in additive form while here is just applied to the terminal funding gap.

The pension fund manager seeks the minimization of a function of the fund's deficit:

$$\min_{x_i^n, y_i^n, z_i^n} \quad \sum_{s \in S} \pi^s \rho(W^s) \tag{11.11}$$

$$\text{s.t.} \quad x_i^{n_0} = \overline{h}_i^{n_0} + z_i^{n_0} - y_i^{n_0}, \qquad \forall i \tag{11.12}$$

$$x_i^n = x_i^{a(n)}(1 + r_i^n) + z_i^n - y_i^n, \qquad \forall i, \forall n \in \mathcal{T} \tag{11.13}$$

$$(1-c)\sum_{i=1}^{I} y_i^n - (1+c)\sum_{i=1}^{I} z_i^n = L^n, \qquad \forall n \in \mathcal{T} \tag{11.14}$$

$$X^n \geq \phi \Lambda^n \qquad \forall n \in \mathcal{T} \tag{11.15}$$

$$W^s = \sum_{i=1}^{I} x_i^{a(s)}(1 + r_i^s) - L^s, \qquad \forall s \in \mathcal{S} \tag{11.16}$$

$$x_i^n, z_i^n, y_i^n, W^s \geq 0, y_i^s = z_i^s = 0 \tag{11.17}$$

Equation (11.12) expresses the initial asset balance, taking the current holdings into account; the asset balance at intermediate trading dates is taken into account by Eq. (11.13). Equation (11.14) ensures that enough cash is generated by selling assets in order to meet current liabilities. Reinvestment at each stage is allowed until the start of the last stage. No buying or selling decisions are possible at the horizon where the pension fund surplus is computed. Upon selling or buying the pension fund manager faces transaction costs as indicated in the cash balance constraint (11.14). Equation (11.16) is used to evaluate terminal surplus at leaf nodes. Pension payments in this simple model are net of received contributions, which for this reason do not appear in the problem formulation.

In this rather simple ALM problem formulation, no cash is actually generated by current holdings and at each decision stage liabilities are funded through portfolio re-balancing decisions. Nor is it possible to borrow. This is typically the case when one faces an investment universe based on total return indices with no cash flows over the planning horizon. A minimal funding ratio ϕ is allowed over the horizon. Such a constraint may reflect pending regulatory conditions in the market. With no borrowing and no cash account it is possible to satisfy all liabilities only by selling assets, and thus an issue of the fund's solvency actually arises. The problem resembles very much a fund manager's problem when issuing an annuity with given random cash flows (that may depend on exogenous elements) and resulting in an optimal self-financing portfolio strategy over a given horizon. From a financial standpoint this problem falls in the class of *liability-driven* investment (LDI) problems [1].

Formulation (11.11)–(11.16) is in general consistent with a random model for asset returns and pension payments. Liabilities are stochastic, and they depend on both future pension dynamics as well as on interest rate dynamics. Pension payments on the other hand will typically depend on future inflation and employees' careers. All these random elements must be considered when generating a scenario tree instance, and they will determine the computational effort in the problem solution [1, 42, 81]. The constraints are linear and assumed to be satisfied *almost surely*, as is common in stochastic programming models.

From a modeling viewpoint, a key distinction is between a (DB) pension scheme as the one above and a (DC) scheme: in the latter, pension payments and the accumulation of pension rights by an active member of the fund will depend on the fund's return year-by-year. Under such assumption, payments L^n will depend on the strategy to date inducing a non-linearity in the problem and a non-convex funding ratio control problem, with both assets and liabilities which are now decision-dependent. This is one of the challenges when modeling a *pure* DC pension problem. Under such scheme, furthermore, the pension fund manager does not carry any market risk unless a minimum guarantee is attached to the problem [81], in which case the problem would be associated with a DC *protected* scheme.

A continuously evolving market of retirement products includes hybrid pension schemes and life insurance contracts combining optionalities with classical annuities. The problem instance (11.11)–(11.16) may be adapted to the associated ALM problems. Let in particular $\rho(W^s)$ represent the payoff at the terminal horizon of an annuity. Then the life insurer will seek the least expensive *replicating* portfolio to attain such payoff. The present cost of such portfolio, under self-financing conditions, will also provide the current fair value of the contract. As in [1, 54] this is the case of a pricing problem in incomplete markets. Market incompleteness implies the absence of a sufficient set of market instruments to perfectly replicate the contract payoff. A financial cost will arise whose minimization may be included in the objective function formulation.

We take the above pension fund management problem as sufficiently general specification to focus on modeling and solution issues specifically associated with the problem's dynamic formulation. We are primarily interested in the way in which the different concepts of risk are considered in this case, limiting ourselves to how the PF ALM problem is handled from a computational viewpoint. It will also provide a way to identify a set of open numerical and modeling issues. A discrete, multi-stage problem is considered and as specified from the beginning, we maintain a dynamic system risk control perspective. The following risk sources need to be taken into account:

- On the asset side, market risk factors affecting the price evolution of assets' total returns and their probability space characterization including the case of associated distributional uncertainty. Here we go from factors such as interest rates and credit spreads for fixed-income assets to risk premia for equity or auto-regressive models for market indices or factor models in robust approaches and so forth [37, 39, 81].

- Liability risk also depends on factors such as interest rates and inflation, but more specifically on the future behavior of pension members survival intensities and resulting longevity risk [1] as well as on the contributors to pensioners ratios, whose dynamics may be difficult to capture.
- Joint A-L risks focus on the correlation between the above risk factors: a sudden divergence of asset values from liability values resulting into an unexpected fall of the fund's solvency ratio may compromise from 1 day to the next a market's systemic equilibrium and for such reason is constantly monitored by regulatory institutions. It is very much in the actuarial tradition to focus on asset-liability duration mismatching, and this is easily accommodated within a dynamic control problem such as (11.11)–(11.16).
- Increasingly however all the above elements are captured within a critical modeling approach in which model risk and thus ambiguity issues are properly considered [46, 68]. Such effort has been motivated by both the serious deficiencies reported by canonical modeling approaches in the recent 2007–2010 global financial crisis and by the emerging possibility to achieve a more effective overall risk control with a far less expensive computational effort.

In Sect. 11.5 we relate the above points to those methodological approaches currently attracting most interest also in the fund management industry and the only ones able to handle effectively a dynamic management problem: dynamic stochastic programming (DSP) and distributionally robust optimization (DRO) based on decision rules. The reference ALM problem is the one introduced above in this Sect. 11.4.2.

In the final Sect. 11.6 we summarize a set of open modeling issues as well as desirable developments aimed at facilitating the practical adoption of dynamic approaches now-a-days still limited to few, even if rather relevant industry cases.

11.5 Solution Methods and Decision Support

The ALM problem formulation (11.11)–(11.16) is preliminary to the adoption of one or another solution approach. In an operational context, by solving the problem, a pension fund manager aims at identifying an effective strategic asset allocation able to preserve the fund's solvency, payout all liabilities and hedge appropriately all asset and liability risks. In a liability-driven approach she/he would be primarily concerned with a cost-effective liability replication and the generation of an excess return through active asset management. From a methodological viewpoint, depending on the adopted optimization approach, Eqs. (11.11)–(11.16) will be subject to further refinements. Increasingly in market practice the solution of such an optimization problem is embedded in the definition of a decision support system [36] whose key modules are represented by a user interface, a problem instance generator, a solution algorithm and a set of output analyses aimed at supporting the decision process. Hardly any such solution will translate straight-away into an actual decision: it will rather provide one of the few fundamental informations for that. Here below, without going into the complexities associated with the definition of a decision support tool, we focus specifically on key methodological issues. The four basic *building blocks* of a quantitative ALM approach—problem formulation, stochastics, solution and decisions—must be kept in mind [42].

A DSP approach preserves all four elements, puts a clear emphasis on each such block and however, specifically with respect to *stochastics* remains subject to model risk. Increasingly, validation through appropriate statistical measures and out-of-sample backtesting of stochastic models and decisions is part of the process. A robust approach, instead aims at avoiding model risk by incorporating the stochastics within a reformulation of the original optimization problem. Either approach is aimed at generating an effective decision process and, in a dynamic set-up key to a consistent problem solution, it focuses on the interaction between decisions and sequential revelation of uncertainty as expressed by the formulation (11.2)–(11.4).

11.5.1 Stochastic Programming

Consider the optimization problem introduced in Sect. 11.4.2. Let $\mathbf{x} \in \mathscr{X}$ be the control vector or portfolio allocation process and $[\rho_T(.)|\Sigma]$ be a terminal risk measure evaluated at the end of the planning horizon, conditional on the information Σ. The focus of the pension fund manager is on the risk associated with possibly deteriorating funding conditions as captured by the *funding ratio* $\frac{X_n}{\Lambda_n}$ along the scenario tree. All random elements of the problem enter the definition of the coefficient tree process $\boldsymbol{\xi}(\boldsymbol{\omega})$ defined on a probability space $(\Omega, \Sigma, \mathbb{P})$. The specification of a dynamic risk measures $\rho : \mathscr{X} \times \Omega \mapsto \mathbb{R}$ as well as the specification of conditional expectations $\mathbb{E}[.|\Sigma_t]$ for increasing t implies the nesting of the associated functionals [34]. Any decision is taken relying on current information and facing a residual uncertainty, as captured by the subtree originating from that node. Under Markovian assumptions on the decision problem and convexity of the risk measure ρ with respect to the decisions, it is possible to solve the problem recursively relying on a sequence of nested dynamic programs or Benders decomposition [42, 49]. In the tree representation there is always a one-to-one relationship between each state of the process—a node in the tree—and its history: the path from the root to the current node.

The decision space $\mathscr{X}$ may accommodate an extended set of asset and liability classes, accordingly it will increase the dimension of $\boldsymbol{\xi}$. The specification of the dependence of $\boldsymbol{\xi}$ on ω—the stochastics—depends on the adopted statistical assumptions. Specifically in financial management and pricing applications, the tree structure is constrained to carry a sufficient branching structure to accommodate equilibrium conditions such as arbitrage-free conditions [54, 64] leading to the well-known *curse-of-dimensionality* problem. An approximation issue arises also considering the relationship between $\boldsymbol{\xi}$ and ω and their underlying, maybe, continuous time companions.

Under this framework the scenario tree structure [37, 54] will determine both the stochastic *input* information for the optimization problem and the optimal decision process or contingency plan *output* [42]. Along a given scenario, carrying a specific probability of occurrence, the pension fund manager may collect a remarkable amount of information, in terms of:

- asset and liability returns, pension payments and contribution flows, funding ratio and funding gap dynamics,
- optimal investment portfolio rebalancing to replicate a given liability,
- liquidity shortages and need of extraordinary sponsors' contributions,
- evolving risk exposure as determined by inflation or interest rates,
- sensitivity to exogenous market or liability shocks and so forth.

The optimal decision process will provide the best possible strategy in the face of the uncertainty captured by the scenario tree process. The discrete framework enables an extended and thorough analysis of all relationships within the given ALM model specification and those relationships may be analyzed and validated from a quantitative as well as qualitative viewpoint by the decision maker [36].

Once the model risks related to a maybe rough analytical specification of the underlying risk process ω and derivation of $\boldsymbol{\xi}(\boldsymbol{\omega})$ have been identified, several solution algorithms may be adopted on decomposed problems or deterministic equivalent instances:

- Under assumptions of stage-wise independence of the coefficient process, [49] present an application of *stochastic dual dynamic programming* (SDDP) to solve a set of sample average approximations of multiperiod risk measure minimization problems such as a multiperiod CVaR model [34].
- Under sufficiently general assumptions on the underlying process and convex constraints [36] employ CPLEX's quadratically constrained quadratic programming solver, on its own and combined with a conic solver, on the *deterministic equivalent* of an insurance ALM problem.
- If the objective function is separable and displays a nested structure with a Markovian constraint region, nested Benders decomposition may still provide an efficient solution approach [43].
- Within a rather general modeling framework for a pension planning problem with power utility, [66] propose a combined stochastic programming and optimal control approach. As in [36] the authors rely here on GAMS and Matlab as algebraic language and modeling tool and on MOSEK as conic solver to handle the non-linearities.

Under different assumptions on the problem specification, the above may be employed to generate an optimal decision tree process from a given input information. The mixed approach proposed in [66] assumes a partition of the decision horizon between a first horizon for the multistage stochastic program and a second horizon for the optimal control: in this second period an optimal HJB approach is adopted relying on a set of constraint relaxations.

A deterministic equivalent instance of the PF ALM problem is commonly derived for convex programming problems of large dimensions without sacrificing any modeling detail but facing the complexities associated with effective sampling methods [37, 42] and in any case the costs that may be induced by a statistical model misspecification and high sensitivity of the optimal strategy to the input coefficient tree process. Still for an ALM problem, departing from a MSP formulation and related solution, [78] suggest a further direction to minimize the costs associated with the tree approximation, through the introduction of *policy rules*: those are specified in coherence with financial management practice and *evaluated* by simulation over maybe complex investment universes. The adoption of Monte Carlo simulation prior to optimization, for scenario construction, and after for policy rules evaluation provides a way to control the approximation costs associated with a scenario-based optimization.

The estimation of the in-sample and out-of-sample stability of a stochastic program also goes in the same direction [42]. We have in-sample stability when relying on a given sampling method we generate and solve an instance of the optimization problem and then repeating such process several times, for different scenario tree instances we report a relatively stable optimal value of the problem maybe generated by different optimal implementable decisions. Lack of this type of stability (which is easy to detect) would result for a given statistical model and sampling method that different problem instances would result into different objective optimal values and controls. Out-of-sample stability implies that for different scenario trees and associated optimal decisions, given each such decision we employ a procedure to evaluate what would be the solution of the optimization problem when reducing asymptotically the sampling error (increasing the number of scenarios): if different input solution vectors result into similar *true* objective values we will have out-of-sample stability.

Overall the adoption of a combined set of optimization and simulation techniques appears highly desirable in presence of financial planning problems, such as a pension fund ALM problem, specified over long horizons (up to 30 years) and with a long experience on simulation techniques for risk assessment but not yet fully accustomed to dynamic optimization approaches.

11.5.2 Dynamic Optimization Via Decision Rules

The ALM model introduced in Sect. 11.4.2 can be viewed as a scenario-tree approximation of a fundamental control model of the type (11.2)–(11.4). Instead of using a scenario-tree approximation, however, such control models can be rendered tractable by using a decision rule approximation. As decision rules are most frequently used in robust optimization, we will explain the mechanics of the decision rule approximation on the example of the worst-case optimization problem (11.6).

To capture the flow of information mathematically, we assume that the components of the uncertain parameter vector ξ are revealed sequentially as time progresses. As in multistage stochastic programming, future decisions are then modeled as functions of the observable data. For ease of exposition, assume that *all* decisions must be chosen after a linear transformation $P\boldsymbol{\xi}$ of the uncertain parameter $\boldsymbol{\xi}$ (e.g., a projection on some of the components of $\boldsymbol{\xi}$) has been observed. Mathematically, this means that the decision must be modeled as an element of the function space $\mathscr{L}^0(P\Xi;\mathscr{X})$, which contains all measurable functions mapping $P\Xi$ to $\mathscr{X}$. In other words, the decision becomes a function that assigns to each possible data observation $P\boldsymbol{\xi} \in P\Xi$ a feasible action $\mathbf{x}(P\boldsymbol{\xi}) \in \mathscr{X}$. In this situation, the robust program (11.6) becomes an infinite-dimensional functional optimization problem of the form

$$\min_{\mathbf{x}\in\mathscr{L}^0(P\Xi;\mathscr{X})} \max_{\boldsymbol{\xi}\in\mathscr{U}} f(\mathbf{x}(P\boldsymbol{\xi}),\boldsymbol{\xi}). \tag{11.18}$$

The dynamic versions of the stochastic program (11.5) and the distributionally robust model (11.7) are constructed similarly in the obvious way. All of these models easily extend to more general decision-making situations where different transformations $P_1\boldsymbol{\xi},\ldots,P_T\boldsymbol{\xi}$ of the uncertain parameter $\boldsymbol{\xi}$ are observed at times $1,\ldots,T$ and where $\mathbf{x}$ admits a temporal decomposition of the form $(\mathbf{x}_1,\ldots,\mathbf{x}_T)$, with $\mathbf{x}_t$ capturing the subvector of the decisions that must be taken at time t, respectively. This is exactly what happens in the ALM model of Sect. 11.4.2, where $P_t\xi$ coincides with the collection of all market and liability risk factors revealed at time t. To keep the presentation simple, however, we will focus on the generic model (11.6) in the remainder.

Dynamic models are often far beyond the reach of analytical methods or classical numerical techniques plagued by the notorious curse of dimensionality. Rigorous complexity results indicate that, for fundamental reasons, dynamic models need to be approximated in order to become computationally tractable [90]. For instance, one could approximate the adaptive model (11.18) by the corresponding static robust optimization problem (11.6). This approximation is conservative—in the sense that it artificially limits the decision maker's flexibility and therefore leads to an upper bound on the optimal value of (11.18). While this approximation may seem overly crude, it has been shown to yield nearly optimal portfolios in various multi-period asset allocation problems and has distinct computational advantages over scenario-tree-based stochastic programming models [6]. Static robust formulations of the multiperiod portfolio selection problem with transaction costs are also considered in [11], where it is shown empirically that robust polyhedral optimization can enhance the performance of single period and deterministic multiperiod portfolio optimization methods.

Instead of naively approximating adaptive by static models, one can alternatively impose a linear structure on the recourse decisions, that is, one may approximate the measurable function $\mathbf{x}(P\boldsymbol{\xi})$ by a linear function $XP\boldsymbol{\xi}$, which is completely determined by the matrix X. The sensitivity matrix X determines the rate of change of each scalar decision with respect to changes in the uncertain parameters. Under this *linear decision rule approximation*, the adaptive robust optimization problem (11.18) reduces to

$$\begin{aligned} &\min_{X} \max_{\xi \in \mathcal{U}} f(XP\xi, \xi) \\ &\text{s.t.} \quad XP\xi \in \mathcal{X} \;\; \forall \xi \in \mathcal{U}. \end{aligned} \tag{11.19}$$

Linear decision rules have attracted considerable interest in recent years because they inherit the favorable scalability properties of static robust optimization models. Indeed, (11.19) is formally equivalent to a static robust optimization problem where the sensitivity matrix X represents the static (here-and-now) decision.

The linear decision rule approximation has been popularized with the seminal paper [7]. Fuelled by their success in robust optimization, linear decision rules have also gained renewed interest from the stochastic programming community; see, e.g., [90]. To improve the approximation quality, several authors have proposed more flexible *non*-linear decision rules such as deflected and segregated decision rules, which can be interpreted as linear decision rules on an augmented probability space and therefore display favorable scalability properties [33, 56]. An efficient procedure to quantify the degree of suboptimality of linear and non-linear decision rules has been proposed in [68] and [53], respectively.

Linear decision rules have been used with great success in dynamic asset allocation with transaction costs [22, 23]. They have also been used in the context of portfolio execution, where they were shown to achieve near optimal performance [74]. Both linear and non-linear decision rules lend themselves ideally to integration with constraint sampling techniques from data-driven optimization [94]. Interestingly, there is ample evidence suggesting that simple decision rules with few degrees of freedom may systematically outperform more versatile decision rules which are prone to overfitting and error-maximization phenomena. For instance, in [19] the portfolio weight of each stock is modeled as a function of the firm's market capitalization, book-to-market ratio, or lagged return. As in data-driven optimization, the coefficients of this simple portfolio rule are found by maximizing the average utility an investor would have obtained by implementing the policy over a historical sample period. This approach is simple to implement and produces outstanding results in and out-of-sample experiments. Similarly, attractive out-of-sample results for an index tracking application were obtained with the robust data-driven dynamic programming approach proposed in [59].

11.6 Open Issues

We have considered a rich set of modeling and theoretical issues from an extended set of contributions to clarify key diverging points that may emerge when similar financial optimisation problems are addressed relying on different methodological assumptions. Without going into a rigorous treatment, the generality of ALM approaches to accommodate a wide range of financial management problems, from portfolio selection to derivatives pricing in incomplete markets, has been emphasised. Specifically in relation with a pension fund ALM problem, we analysed the decision processes as a sequence of steps, each one carrying its own modeling and methodological complexity, but jointly leading to the definition of quantitatively-based optimal decisions. Indeed we have also remarked that efficient decision making does not necessarily require excessive modeling complexity and a *friendly* approach to computational management, based on simple probabilistic assumptions, may be sufficient. By proposing a unified modeling framework, our interest was primarily on the potentials offered by the integration within a given method of results and advances may be achieved under other modeling and optimization paradigms [38].

Without ambition to be exhaustive in this respect, in what follows we would like to indicate a set of relevant open issues which may emerge from the introduced financial domains but also carrying implications to other application areas. We focus on the two fundamental elements of decision making under uncertainty: stochastics as featured by $\boldsymbol{\xi}(\omega)$ and solutions $\mathbf{x} \in \mathscr{X}$ in (11.2)–(11.4).

11.6.1 Probability Distributions and Optimization

A compact formulation of a stochastic program under distributional uncertainty has been introduced as

$$\min_{\mathbf{x}\in\mathscr{X}} \max_{\mathbb{P}\in\mathscr{P}} \mathbb{E}_{\mathbb{P}}[f(\mathbf{x}, \boldsymbol{\xi})]. \tag{11.20}$$

If the ambiguity set $\mathscr{P}$ is given by a singleton $\{\mathbb{P}\}$, we remain within a canonical stochastic programming framework: here under maybe an extremely challenging computational environment, relying on the discrete scenario tree representation key issues such as *arbitrage free* market conditions, market *incompleteness*, accurate stress-testing and sensitivity analysis (along a set of scenarios), as well as a rich sequence of I/O analyses can be conducted in a sufficiently straight-forward way. The optimal risk control x takes the form of an optimal tree decision process and contingency plan. This is done at an approximation cost and somehow aware of the potential damages that might be induced by estimation errors in parametric models or the adoption of generating processes for $\boldsymbol{\xi}$ that may be contradicted by market dynamics. Such risks would persist under in-sample and out-of-sample stable programs.

For general ambiguity sets that may contain more than one distribution, the worst-case expectation with respect to $\mathscr{P}$ corresponds to a coherent risk measure, that is, $\max_{\mathbb{P}\in\mathscr{P}} \mathbb{E}_{\mathbb{P}}[f(\mathbf{x}, \boldsymbol{\xi})]$ can be viewed as the *risk* of the random loss $f(\mathbf{x}, \boldsymbol{\xi})$ under the coherent risk measure induced by $\mathscr{P}$ [2]. In a distributionally robust setting, arbitrage-freeness of the market is usually no serious concern. In fact, the existence of arbitrage portfolios is ruled out if the underlying financial market is arbitrage-free under at least one distribution $\mathbb{P} \in \mathscr{P}$; and this is generically the case if $\mathscr{P}$ is defined through moment constraints or as a ball in the space of probability distributions. To our best knowledge, market incompleteness has not yet been studied systematically in a distributionally robust framework. However, problem (11.20) lends itself to various stress-tests and sensitivity experiments. For example, it is well-known that the extremal distribution achieving the worst-case expectation in (11.20) for a fixed decision **x** can be computed systematically and efficiently for most commonly used ambiguity sets; see e.g. [86] for an explicit example. This worst-case distribution can be used to conduct contamination or stress-test experiments. Specifically, the performance of a fixed decision **x** can be evaluated under a nominal distribution contaminated with an extremal distribution. The sensitivity of the resulting out-of-sample performance with respect to the contamination parameter provides valuable information about the robustness of the decision **x**. Another striking benefit of the distributionally robust model (11.20) is that it can sometimes be solved more efficiently than the corresponding stochastic program with a nominal distribution [40].

In principle data-driven approaches may be used to construct $\mathscr{P}$ [14, 40]. It is also possible to use the *empirical* distribution directly in the construction of the ambiguity set; see e.g. [75]. Serial independence of the training data may be needed in this case, however, to establish rigorous out-of-sample performance guarantees. But such an assumption may not be justified in reality. An interesting avenue for future research is to develop out-of-sample guarantees for serially dependent training data; e.g., for data generated by a Markov or an autoregressive process.

11.6.2 Dynamic Time Consistency

The concept of time-consistent dynamic risk measures [34, 49] is key to the definition of a dynamic risk control problem and may be analysed with respect to the evolution of the process $\boldsymbol{\xi}(\boldsymbol{\omega})$ as well as to the optimal control **x** [34]: in this latter case loosely speaking time consistency implies that as time evolves $t = 0, 1, 2, \ldots, T$ by solving a sequence of stochastic control problems leading to optimal controls $\{x\}_{s=t}^{T}$ optimality would be preserved throughout for each t. In a discrete setting for scenario tree processes such condition should be considered conditionally with respect to actual realizations. As for time-consistency of a risk measure with respect to $\boldsymbol{\xi}(\boldsymbol{\omega})$ this would imply the persistence as time evolves of previously assessed risk rankings conditionally on the sequence of associated σ-fields. In this case we have a form of monotonicity since we require that as information accumulates, successive evaluations of a random process by the risk measure remain homogenous.

In either cases, the evolution of information plays a central role: this is captured for risk measures by the evolution of the process filtration, while in optimal control problems as in (11.2)–(11.4) the key information would be associated with the evolution of the state variable $\mathbf{s}_t$. In [34] we discuss the conditions to connect the two concepts within a recursive approach based on dynamic programming principles. Recursivity of risk measures is not always attainable [34].

In [49] conditions for time-consistency of an optimization problem based on nested CVaR measures is studied, leading to a solution approach based on SDDP.

In this context, the concept of *cost* of time-inconsistency appears key to the discussion as introduced in [34] and may lead to an effective evaluation of its implications on the optimization problem. In a discrete scenario tree setting the discussion should be specialised with respect to conditional risk measures. In this direction, across the different dynamic optimization paradigms, research is continuing.

11.6.3 Practical Financial Optimization

The sheer amount of literature on financial optimization shows a huge academic interest. From the practical point of view, there are in fact some interesting real-life applications documented [3, 13, 29, 36, 77], but it may be argued that the full potential has not been achieved, yet, at least as far as dynamic portfolio optimization is concerned. What are the main issues potentially hindering the application of financial optimization models in practice? Computational efficiency is certainly a concern in multistage stochastic programming models, but it is arguably not the main issue in other cases. One question we should ask is what would be the main sources of concern for a practitioner considering the use of an optimization model as a decision support tool:

- How will I be held accountable for the decision I am going to make?
- Are there any misalignments between the model's objective and how my performance is assessed?
- How can I trust the model and understand the reasons behind its recommendations?
- How can I compare the model output against alternatives?
- How can I test the robustness of the model's recommendation?
- How can I incorporate my own views?

Some of these concerns, being actually related to a *principal-agent* type of problem, may look nontechnical, but they should be addressed somehow in practical optimization. There is little point in considering model objectives that not aligned with the actual incentives of the decision maker. For instance, it is a nontrivial task to align long-term objectives with potential short-term performance evaluation affecting a portfolio manager. Furthermore, in statistical learning there is a well-known trade-off between model's sophistication, which may contribute to its predictive power, and its interpretability [61]. Relying on the output of a sophisticated model may be hard for a manager, if she gets a supposed optimal solution, but it is difficult to understand why that is the optimal solution. Furthermore, sophisticated models may be fragile. They are subject to overfitting issues in statistical learning, and regularized approaches (like lasso regression) may be helpful. We note that the contribution of [42] has a similar flavor. By a similar token, as suggested in [16], under significant uncertainty simpler rules may be preferable [17]. This does not imply that sophisticated models must be given up, but assessing their performance under stress is certainly no easy task. As suggested in [76], when model risk is an issue, even stress-testing per se may be hard to grasp under certain asset pricing models.

Recent trends in decision rules for dynamic optimization offer solutions that may be easily tested under different scenarios, without requiring extensive rolling horizon simulations, which are computationally demanding. Approximate dynamic programming offers similar benefits for a skeptical manager. When choosing a modeling framework, these considerations may be relevant, and in this way trust may be built, possibly leading to an easier adoption and, maybe, the development of more refined approaches over time.

Personal views are relevant to a portfolio manager, and the increasing role of Bayesian-learning approaches in finance and risk management is well documented [85]. The Black–Litterman model is a well-known approach that incorporates Bayesian principles [87]. They are also relevant for risk management [88], as they provide one possible answer to model risk issues. Model robustness and ambiguity are the subject of active research, and an integration of available approaches is called for from a practitioner's point of view, with an emphasis of proof-of-concept methods to evaluate actual out-of-sample performance under stressed market conditions.

11.7 Conclusions

The variety of contributions in this volume and the companion SI shows that, after several years, financial optimization is still a field in flux and a very active research field. New challenges have emerged from the financial side, and new algorithmic frameworks have been developed to address issues in previous models and solution methods. Integration opportunities between solution methods, like stochastic and robust optimization, or dynamic programming and stochastic programming with recourse, are rather evident, as well as the interplay of different financial problems like pricing, hedging, and asset management. We did not cover every possible application of optimization modeling to finance; for instance, we did not consider pricing model calibration and optimal order execution. These and other applications testify the interest and the practical relevance of this research field.

References

1. H. Aro, T. Pennanen, Liability driven investment in longevity risk management. This volume
2. P. Artzner, F. Delbaen, J.-M. Eber, D. Heath, Coherent measures of risk. Math. Financ. **9**, 203–228 (1999)
3. M. Avriel, H. Pri-Zan, R. Meiri, A. Peretz, Opti-money at Bank Hapoalim: a model-based investmentdecision-supportsystem for individual customers Interfaces **34**(1), 39–50 (2004)
4. C. Bandi, D. Bertsimas, Tractable stochastic analysis in high dimensions via robust optimization. Math. Program. **134**(1), 23–70 (2012)
5. C. Bandi, D. Bertsimas, Robust option pricing. Eur. J. Oper. Res. **239**(3), 842–853 (2014)
6. A. Ben-Tal, T. Margalit, A. Nemirovski, Robust modeling of multi-stage portfolio problems, in *High Performance Optimization*, ed. by H. Frenk, K. Roos, T. Terlaky, S. Zhang (Springer US, New York, 2000), pp. 303–328
7. A. Ben-Tal, A. Goryashko, E. Guslitzer, A. Nemirovski, Adjustable robust solutions of uncertain linear programs. Math. Program. **99**(2), 351–376 (2004)
8. A. Ben-Tal, S. Boyd, A. Nemirovski, Extending scope of robust optimization: comprehensive robust counterparts of uncertain problems. Math. Program. **107**(2), 63–89 (2006)
9. A. Ben-Tal, L. El Ghaoui, A. Nemirovski, *Robust Optimization* (Princeton University Press, Princeton, 2009)
10. M. Bertocchi, G. Consigli, M.A.H. Dempster (eds.), *Stochastic Optimization Methods in Finance and Energy. New Financial Products and Energy Market Strategies*. Fred Hillier International Series in Operations Research and Management Science (Springer US, New York, 2011)
11. D. Bertsimas, D. Pachamanova, Robust multiperiod portfolio management in the presence of transaction costs. Comput. Oper. Res. **35**(1), 3–17 (2008)
12. D. Bertsimas, M. Sim, The price of robustness. Oper. Res. **52**(1), 35–53 (2004)
13. D. Bertsimas, C. Darnell, R. Soucy, Portfolio construction through mixed-integer programming at Grantham, Mayo, Van Otterloo and company. Interfaces **29**(1), 49–66 (1999)
14. D. Bertsimas, V. Gupta, N. Kallus, Data-driven robust optimization (2013). Available from arXiv:1401.0212
15. J.R. Birge, F. Louveaux, *Introduction to Stochastic Programming*, 2nd edn. Springer Series in Operations Research and Financial Engineering (Springer, New York, 2011)
16. R. Bookstaber, *A Demon of Our Own Design* (Wiley, New York, 2007)
17. R. Bookstaber, J. Langsam, On the optimality of coarse behavior rules. J. Theor. Biol. **116**(2), 161–193 (1985)
18. P. Brandimarte, *Handbook in Monte Carlo Simulation: Applications in Financial Engineering, Risk Management, and Economics* (Wiley, New York, 2014)
19. M.W. Brandt, P. Santa-Clara, R. Valkanov, Parametric portfolio policies: exploiting characteristics in the cross-section of equity returns. Rev. Financ. Stud. **22**(9), 3411–3447 (2009)
20. M. Broadie, Computing efficient frontiers using estimated parameters. Ann. Oper. Res. **45**(1), 21–58 (1993)
21. R. Bruni, F. Cesarone, A. Scozzari, F. Tardella, A linear risk-return model for enhanced indexation in portfolio optimization. OR Spectr. **37**(3), 735–759 (2015)
22. G.C. Calafiore, Multi-period portfolio optimization with linear control policies. Automatica **44**(10), 2463–2473 (2008)
23. G.C. Calafiore, An affine control method for optimal dynamic asset allocation with transaction costs. SIAM J. Control. Optim. **48**(4), 2254–2274 (2009)
24. G.C. Calafiore, Random convex programs. SIAM J. Optim. **20**(6), 3427–3464 (2010)
25. G.C. Calafiore, Scenario optimization methods in portfolio analysis and design. This volume
26. G.C. Calafiore, M.C. Campi, The scenario approach to robust control design. IEEE Trans. Autom. Control **51**(5), 742–753 (2006)
27. J.Y. Campbell, L.M. Viceira, *Strategic Asset Allocation* (Oxford University Press, Oxford, 2002)

28. M.C. Campi, S. Garatti, The exact feasibility of randomized solutions of uncertain convex programs. SIAM J. Optim. **19**(3), 1211–1230 (2008)
29. D.R. Cariño, T. Kent, D.H. Myers, C. Stacy, M. Sylvanus, A.L. Turner, K. Watanabe, W.T. Ziemba, The Russell-Yasuda Kasai Model: an asset/liability model for a Japanese insurance company using multistage stochastic programming. Interfaces **24**(1), 29–49 (1994)
30. S. Ceria, R. Stubbs, Incorporating estimation errors into portfolio selection: Robust portfolio construction. J. Asset Manag. **7**(2), 109–127 (2006)
31. R. Cerqueti, P. Falbo, G. Guastaroba, C. Pelizzari, Approximating multivariate Markov chains for bootstrapping through contiguous partitions. OR Spectr. **37**(3), 803–841 (2015)
32. E. Çetinkaya, A. Thiele, Data-driven portfolio management with quantile constraints. OR Spectr. **37**(3), 761–786 (2015)
33. X. Chen, M. Sim, P. Sun, A robust optimization perspective on stochastic programming. Oper. Res. **55**(6), 1058–1071 (2007)
34. Z. Chen, G. Consigli, J. Liu, G. Li, T. Fu, Q. Hu, Multi-period risk measures and dynamic risk control. This volume
35. V.K. Chopra, W.T. Ziemba, The effect of errors in means, variances and covariances on optimal portfolio choice. J. Portf. Manag. **19**(2), 6–11 (1993)
36. G. Consigli, V. Moriggia, Applying stochastic programming to insurance portfolios stress-testing. Quant. Financ. Lett. **2**(1), 7–13 (2014)
37. G. Consigli, G. Iaquinta, V. Moriggia, Path-dependent scenario trees for multistage stochastic programs in finance. Quant. Financ. **12**(8), 1265–1281 (2012)
38. G. Consigli, P. Brandimarte, D. Kuhn, Editorial: financial optimization: optimization paradigms and financial planning under uncertainty. OR Spectr. **37**(3), 553–557 (2015)
39. M.H.A. Davis, S. Lleo, Jump-diffusion asset-liability management via risk-sensitive control. OR Spectr. **37**(3), 655–675 (2015)
40. E. Delage, Y. Ye, Distributionally robust optimization under moment uncertainty with application to data-driven problems. Oper. Res. **58**(3), 595–612 (2010)
41. V. DeMiguel, L. Garlappi, R. Uppal, Optimal versus naive diversification: how inefficient is the 1/N portfolio strategy? Rev. Financ. Stud. **22**(5), 1915–1953 (2009)
42. M.A.H. Dempster, E.A. Medova, Y.S. Yong, Stabilizing implementable decisions in dynamic stochastic programming. This volume
43. M.A.H. Dempster, M. Germano, E.A. Medova, M. Villaverde, Global asset liability management. Br. Actuar. J. **9**(1), 137–195 (2003)
44. S. Desmettre, R. Korn, P. Ruckdeschel, F.T. Seifried, Robust worst-case optimal investment. OR Spectr. **37**(3), 677–701 (2015)
45. X. Doan, X. Li, K. Natarajan, Robustness to dependency in portfolio optimization using overlapping marginals. Available on Optimization Online (2013)
46. T. Driouchi, L. Trigeorgis, Y. Gao, Choquet-based European option pricing with stochastic (and fixed) strikes. OR Spectr. **37**(3), 787–802 (2015)
47. J. Dupačová, The minimax approach to stochastic programming and an illustrative application. Stochastics **20**, 73–88 (1987)
48. J. Dupacova, G. Consigli, S.W. Wallace, Scenarios for multistage stochastic programmes. Ann. Oper. Res. **100**, 25–53 (2001)
49. J. Dupačová, V. Kozmík, Structure of risk-averse multistage stochastic programs. OR Spectr. **37**(3), 559–582 (2015)
50. L. El Ghaoui, M. Oks, F. Outstry, Worst-case value-at-risk and robust portfolio optimization: a conic programming approach. Oper. Res. **51**(4), 543–556 (2003)
51. E. Erdoğan, G. Iyengar, Ambiguous chance constrained problems and robust optimization. Math. Program. B **107**(1–2), 37–61 (2006)
52. D. Fabozzi, P. Kolm, D. Pachamanova, *Robust Portfolio Optimization and Management* (Wiley, New York, 2007)
53. A. Georghiou, W. Wiesemann, D. Kuhn, Generalized decision rule approximations for stochastic programming via liftings. Math. Program. **152**(1–2), 301–338 (2015)

54. M. Giandomenico, M. Pinar, Pricing American options: binomial and trinomial non-recombinant trees and two exercise rights. This volume
55. M. Gilli, E. Schumann, Constructing portfolios with heuristics. This volume
56. J. Goh, M. Sim, Distributionally robust optimization and its tractable approximations. Oper. Res. **58**(4), 902–917 (2010)
57. D. Goldfarb, G. Iyengar, Robust portfolio selection problems. Math. Oper. Res. **28**(1), 1–38 (2003)
58. L. Györfi, G. Ottucsak, H. Walk, The growth optimal strategy is secure too. This volume
59. G.A. Hanasusanto, D. Kuhn, Robust data-driven dynamic programming, in *Advances in Neural Information Processing Systems*, pp. 827–835 (2013)
60. Z. Hu, L.J. Hong, Kullback-Leibler divergence constrained distributionally robust optimization. Available from Optimization Online (2013)
61. G. James, D. Witten, T. Hastie, R. Tibshirani, *An Introduction to Statistical Learning: with Applications in R* (Springer, Berlin, 2013)
62. B. Kawas, A. Thiele, A log-robust optimization approach to portfolio management. OR Spectr. **33**(1), 207–233 (2011)
63. J.M. Keynes, *A Treatise on Probability* (MacMillan, London, 1921)
64. A.J. King, Duality and martingales: a stochastic programming perspective on contingent claims. Math. Program. B **91**, 543–562 (2002)
65. F.H. Knight, *Risk, Uncertainty and Profit* (Hart, Schaffner and Marx, Boston, 1921)
66. A.K. Konicz, D. Pisinger, K.M. Rasmussen, A combined stochastic programming and optimal control approach to personal finance and pensions. OR Spectr. **37**(3), 583–616 (2015)
67. M. Kopa, T. Post, A general test for SSD portfolio efficiency. OR Spectr. **37**(3), 703–734 (2015)
68. D. Kuhn, W. Wiesemann, A. Georghiou, Primal and dual linear decision rules in stochastic and robust optimization. Math. Program. **130**(1), 177–209 (2011)
69. D.G. Luenberger, *Investment Science* (Oxford University Press, New York, 1998)
70. L.C. MacLean, Y. Zhao, Asset price dynamics: shocks and regimes. This volume
71. H.M. Markowitz, *Portfolio Selection: Efficient Diversification of Investments*, 2nd edn. (Wiley, New York, 1991)
72. R.C. Merton, On estimating the expected return on the market: an exploratory investigation. J. Financ. Econ. **8**(4), 323–361 (1980)
73. R.O. Michaud, *Efficient Asset Management: A Practical Guide to Stock Portfolio Management and Asset Allocation* (Oxford University Press, Oxford, 2001)
74. C.C. Moallemi, M. Saglam, Dynamic portfolio choice with linear rebalancing rules (2015). Available at SSRN: http://ssrn.com/abstract=2011605
75. P. Mohajerin Esfahani, D. Kuhn, Data-driven distributionally robust optimization using the Wasserstein metric: performance guarantees and tractable reformulations. Available from Optimization Online (2015)
76. M. Morini, *Understanding and Manageing Model RIsk: A Practical Guide for Quants, Traders and Validators* (Wiley, New York, 2011)
77. J.M. Mulvey, G. Gould, C. Morgan, An asset and liability management system for towers Perrin-Tillinghast. Interfaces **30**(1), 96–114 (2000)
78. J.M. Mulvey, W.C. Kim, C. Lin, Optimizing a portfolio of liquid and illiquid assets. This volume
79. K. Natarajan, D. Pachamanova, M. Sim, Incorporating asymmetric distributional information in robust value-at-risk optimization. Manag. Sci. **54**(3), 573–585 (2008)
80. K. Natarajan, M. Sim, J. Uichanco, Tractable robust expected utility and risk models for portfolio optimisation. Math. Financ. **20**(4), 695–731 (2010)
81. D. Pachamanova, N. Gulpinar, E. Canakoglu, Robust data-driven approaches to pension fund asset-liability management under uncertainty. This volume
82. G.C. Pflug, W. Römisch, *Modeling, Measuring and Managing Risk* (World Scientific Publishing, Singapore, 2007)
83. G.C. Pflug, D. Wozabal, Ambiguity in portfolio selection. Quant. Finan. **7**(4), 435–442 (2007)

84. G.C. Pflug, A. Pichler, D. Wozabal, The 1/n investment strategy is optimal under high model ambiguity. J. Bank. Financ. **36**(2), 410–417 (2012)
85. S.T. Rachev, J.S.J. Hsu, B.S. Bagasheva, F.J. Fabozzi, *Bayesian Methods in Finance* (Wiley, New York, 2008)
86. N. Rujeerapaiboon, D. Kuhn, W. Wiesemann, Robust growth-optimal portfolios. Manag. Sci. **62**(7), 2090–2109 (2016)
87. S. Satchell, A. Scowcroft, A demystification of the Black–Litterman model: managing quantitative and traditional portfolio construction. J. Asset Manag. **1**, 138–150 (2000)
88. M. Sekerke, *Bayesian Risk Management: A Guide to Model Risk and Sequential Learning in Financial Markets* (Wiley, New York, 2015)
89. A. Shapiro, D. Dentcheva, A. Ruszczyński, *Lectures on Stochastic Programming: Modeling and Theory* (SIAM, Philadelphia, 2009)
90. A. Shapiro, A. Nemirovski, On complexity of stochastic programming problems, in *Continuous Optimization: Current Trends and Applications*, ed. by V. Jeyakumar, A.M. Rubinov (Springer, Berlin, 2005), pp. 111–144
91. G. Szego (ed.), *Risk Measures for the 21st Century* (Wiley, New York, 2004)
92. R.H. Tütüncü, M. Koenig, Robust asset allocation. Ann. Oper. Res. **132**(1–4), 157–187 (2004)
93. S. Pagliarani, T. Vargiolu, Portfolio optimization in a defaultable Lévy-driven market model. OR Spectr. **37**(3), 617–654 (2015)
94. P. Vayanos, D. Kuhn, B. Rustem, A constraint sampling approach for multi-stage robust optimization. Automatica **48**(3), 459–471 (2012)
95. W. Wiesemann, D. Kuhn, M. Sim, Distributionally robust convex optimization. Oper. Res. **62**(6), 1358–1376 (2014)
96. H. Xu, D. Zhang, Monte Carlo methods for mean-risk optimization and portfolio selection. Comput. Manag. Sci. **9**(1), 3–29 (2012)
97. S. Zymler, D. Kuhn, B. Rustem, Worst-case value at risk of nonlinear portfolios. Manag. Sci. **59**(1), 172–188 (2013)
98. S. Zymler, B. Rustem, D. Kuhn, Robust portfolio optimization with derivative insurance guarantees. Eur. J. Oper. Res. **210**(2), 410–424 (2011)

Index

© Springer International Publishing Switzerland 2017

G. Consigli et al. (eds.), *Optimal Financial Decision Making under Uncertainty*,
International Series in Operations Research & Management Science 245,
DOI 10.1007/978-3-319-41613-7

Q

R

S

MIX
Papier aus verantwortungsvollen Quellen
Paper from responsible sources
FSC® C105338

If you have any concerns about our products,
you can contact us on
ProductSafety@springernature.com

In case Publisher is established outside the EU,
the EU authorized representative is:
Springer Nature Customer Service Center GmbH
Europaplatz 3, 69115 Heidelberg, Germany

Printed by Libri Plureos GmbH
in Hamburg, Germany